LABOR AND RAINFED AGRICULTURE
IN WEST ASIA AND NORTH AFRICA

Labor and Rainfed Agriculture in West Asia and North Africa

Edited by

DENNIS TULLY

The International Center for Agricultural Research in the Dry Areas (ICARDA), Aleppo, Syria

Kluwer Academic Publishers

Dordrecht / Boston / London

Library of Congress Cataloging-in-Publication Data


```
Labor and rainfed agriculture in West Asia and North Africa / edited
  by Dennis Tully.
      p.   cm.

   1. Agricultural laborers--Middle East--Effect of technological
innovations on.   2. Agricultural innovations--Middle East.
3. Agricultural laborers--Africa, West--Effect of technological
innovations on.   4. Agricultural innovations--Africa, West.
I. Tully, Dennis.
HD6331.18.A292M6285   1990
331.25--dc20                                                90-4104
```


ISBN-13:978-94-010-6740-9   e-ISBN-13:978-94-009-0561-0
DOI:10.1007/978-94-009-0561-0

Published by Kluwer Academic Publishers,
P.O. Box 17, 3300 AA Dordrecht, The Netherlands.

Kluwer Academic Publishers incorporates
the publishing programmes of Martinus Nijhoff,
Dr W. Junk, D. Reidel, and MTP Press.

Sold and distributed in the U.S.A. and Canada
by Kluwer Academic Publishers,
101 Philip Drive, Norwell, MA 02061, U.S.A.

In all other countries, sold and distributed
by Kluwer Academic Publishers Group,
P.O. Box 322, 3300 AH Dordrecht, The Netherlands

printed on acid-free paper

Table of Contents

Foreword

The basic objective of agricultural research at ICARDA is to enhance producer and consumer welfare through increasing the productivity, stability, and profitability of agriculture. Improved practices must be technically, economically, and socially suitable to farmer conditions.

The rainfed areas of West Asia and North Africa have highly variable environmental conditions as well as complex social and economic structures. In recent years, the region has been experiencing major changes in the relative availabilities and costs of the classical factors of production: land, labor, and capital. These changes have important implications for the design of new agricultural technology.

On the one hand, the availability of labor may be an important factor determining the acceptability of new technology. On the other, it is important to consider the impact that technology can have on rural employment. To develop a better awareness of these issues and their relevance to technology development, ICARDA initiated a project on Agricultural Labor and Technological Change. The first stage of the project is published here; it is a review of available literature on selected issues of regional importance, combined with more detailed analyses of the situations of eight countries with important rainfed agricultural sectors.

ICARDA greatly appreciates the financial assistance of the Ford Foundation, which allowed us to support the execution of the study and publication of its findings. We also appreciate the great efforts of the authors in the face of often limited data and facilities.

Nasrat Fadda
Director General
ICARDA

Acknowledgements

It has been a privilege to work with the contributors to this publication and I salute them for their willingness to address a very difficult topic. The Ford Foundation made the preparation and publication of these papers possible with generous assistance. I am grateful to Fiona Thomson for her talented editing and for keeping the preparation of the manuscript rolling in a difficult period. Credit for the attractive graphics and cover art goes to La Neu.

The Editor

PART ONE

Regional Overview Papers

1. Labor and the Adoption and Impact of Farm Technology

DENNIS TULLY

The International Center for Agricultural Research in the Dry Areas (ICARDA), P.O. Box 5466, Aleppo, Syria

Rainfed Farming Systems of West Asia and North Africa

West Asia and North Africa are ancient agricultural regions, and the site of domestications of many major food crop and animal species. Sedentary agriculture has historically been the mainstay of life for the vast majority of the region. Although agriculture has become a secondary sector by some measures, contributing less than 25% of GDP even in countries with substantial land resources (Table 1), these figures probably underestimate its importance. Approximately half of the population lives in rural areas, and in most countries one third to one half of the workforce is primarily engaged in agriculture. But the value placed on family labor in GDP calculations is usually low or zero, and family farms also produce a significant part of their own food requirements, an undervalued contribution to the economy. In addition, a substantial proportion of the industrial sector is related to agriculture, either in manufacturing inputs or processing outputs. Fertilizer plants, tractor factories, textile mills, and food processors all depend upon crop and livestock production. Similarly, a large part of the service sector is involved in bulking, breaking, and transporting agricultural inputs and outputs.

Except for Egypt and Iraq, agriculture in this region is principally rainfed (Table 2). In contrast with irrigated cultivation, rainfed agriculture usually has higher risk, marked seasonality, lower productivity per hectare, and, frequently, lower returns to labor and investment. New technology for rainfed agriculture must be designed with awareness of its special social and economic context, particularly as related to labor.

In the Mediterranean area, rainfall comes in winter and early spring, while in the Arabian peninsula, sub-Saharan Africa, and southwest Asia, most rainfed areas have spring and summer rainfall. Rainfed farming systems generally occur in areas with mean annual rainfall of 200–600 mm in winter rainfall zones, and 300–800 mm in summer rainfall zones. Relatively arid zones cover the largest areas and experience the greatest rainfall variability, which is the most important factor associated with the uncertainty of crop production, farm incomes, and agricultural employment.

The dominant crops of the region are cereals, primarily wheat and barley under winter and spring rainfall, and sorghum and millet in areas of summer

Dennis Tully (ed.), Labor and Rainfed Agriculture in West Asia and North Africa, 3–23.
© 1990 *ICARDA*.

Table 1. Agriculture in the national economies.

	Agricultural % of GDP		Agricultural % of labor force		Urban population as % of total	
	1960	1986	1960	1980	1960	1985
Morocco	23	21	62	46	29	44
Algeria	16	12	67	31	30	43
Tunisia	24	16	56	35	36	60
Libya	–	–	53	18	23	56
Egypt	30	20	58	46	38	46
Sudan	–	35	86	71	10	21
Ethiopia	65	48	88	80	6	15
Lebanon	11	–	38	–	40	–
Jordan	–	8	44	10	43	69
Syria	–	22	54	32	37	49
Iraq	17	–	53	30	43	70
Iran	29	–	54	36	34	54
Afghanistan	–	–	85	–	8	–
Pakistan	46	24	61	55	22	29
Turkey	41	18	79	58	30	46
Saudi Arabia	–	4	71	48	30	72
Yemen AR	–	34	83	69	3	19
PDR Yemen	–	–	70	41	28	37

Source: IBRD (1984; 1986).

rainfall. Areas occupied by pulses, vegetables, fruits, and tree crops are small except in the wettest zones (FAO 1984a). Livestock are an integral part of most rainfed systems, transforming crop residues, failed crops, pastures, agro-industrial by-products, and grain into meat and dairy products which humans can use. Part of the needs of livestock are frequently satisfied outside the rainfed cropping systems, in arid or mountainous grazing areas, or in irrigated areas with complementary seasons of feed availability.

Cereal consumption in this region is approximately 200 kg/person/year, in most countries providing over half of the calories and protein consumed. Milk products are the other main source of protein, while meat consumption is limited to 15–20 kg/person/year, except in the oil-rich countries (FAO 1984b). Rural people often consume less than these aggregate figures indicate, and it appears that nutritional deficiencies are endemic in many areas (El-Sherbini 1977; Fikry 1983; Miladi 1983; Mokbel 1985).

With rapidly growing populations, food self-sufficiency has been declining in many countries (AOAD 1983), although Turkey is an exception (Hanson et al. 1982). In response, policies to stimulate the agricultural sector are becoming more common. In general, governments in this region are very involved in agriculture. Prices of major crops are often set by government bodies, and marketing is commonly controlled. The prices, distribution, and

Table 2. Agricultural lands in West Asia and North Africa (× 1,000 ha).

	Total crops	Irrigated	Rainfed	Rainfed area as % of total
Morocco	7,699	426	7,273	94
Algeria	7,469	310	7,159	96
Tunisia	4,860	123	4,737	97
Libya	2,053	200	1,853	90
Egypt	2,800	2,799	1	<1
Sudan	12,160	1,548	10,612	87
Ethiopia	13,730	2,799	10,931	80
Lebanon	348	85	263	76
Jordan	391	36	355	91
Syria	5,725	547	5,178	90
Iraq	5,287	1,572	3,715	70
Iran	16,450	5,913	10,537	64
Afghanistan	8,048	2,480	5,568	69
Pakistan	19,715	13,601	6,114	31
Cyprus	432	94	338	78
Turkey	27,764	1,983	25,781	93
Saudi Arabia	1,179	382	797	68
Yemen AR	3,187	230	2,957	93
PDR Yemen	195	53	142	73

Source: FAO (1984a).

availability of credit for agricultural inputs are policy matters, while official extension services deliver and implement new technology.

Thus most countries have a structure in place which can and does support agriculture. The level of support and the direction it takes are, however, related to national priorities. Agriculture is often perceived as giving a low return on the investment of scarce government resources, including budget allocations, credit, and foreign exchange, compared to other sectors. Even policies which favor agriculture may discourage development of parts of the sector. Policies have often been officially or practically structured to favor large farms over small, irrigated agriculture over rainfed, and high rainfall areas over low (Campbell et al. 1977; El-Ghonemy 1979; Hogan et al. 1984).

These policies are based on perceptions of where agriculture's greatest potential lies. It is widely believed that rainfed areas, particularly the driest ones, are unlikely to make a significant contribution to the national economy. This has led to low investment in research and extension, low returns, and confirmation of low expectations. However, evidence is increasing for an unrealized potential in rainfed areas, and there is growing awareness of the danger of environmental degradation if they are neglected. In addition, the need to raise farm incomes has been recognized as a way to stem rural-urban migration in the face of growing urban unemployment. As a result, interest in improving rainfed farming systems is growing among policy makers, as is the demand for information about these areas.

Development of the Current Agrarian Structure

Much of West Asia and North Africa experienced a similar history under Ottoman and European administrations. In the first half of the 20th century, urban landlords, governors, or tribal leaders often controlled land or land revenues. Although smallholders existed, most of the rural population paid shares of their crops as rent, taxes, or tribute. With the available farm techniques, both production and consumption must have been low; the limiting factor in most cases was the area which could be tilled with available labor and draft power. However, a low population density allowed cultivation to be restricted to the best soils and allowed extensive fallowing to restore soil fertility. Apparently, farmers could support their families from their share of what they cultivated with draft animals, supplemented by other subsistence-oriented activities.

In the past, labor requirements were relatively continuous. Hand harvesting of successively ripening crops followed by transport and threshing occupied several months. This overlapped with preparation of the next year's seedbed, which occupied up to 3 months before planting. The growing season was slack, as crop maintenance tasks such as weed control required little labor, but some livestock activities, crafts, and crop processing were scheduled in this period. Cereal cultivation was and still is dominant, but in the past, there was more cultivation of legumes and tree crops with complementary seasons. It would appear that landlords made decisions about land use which maximized their utilization of a permanent labor force with a fixed cost.

During the Second World War and its aftermath, many profound changes happened simultaneously; it is difficult to know which were causes and which were effects. Wartime demand stimulated agriculture and industry alike. Political independence came to many countries, and new regimes encouraged urbanization to provide the labor force for more industrial development. The introduction of agricultural mechanization enabled, and possibly contributed to, rural-urban migration. But the trend towards large mechanized estates was offset in many cases by land reform policies, in which expropriated estates and colonial farms as well as government and communal land were allocated to smallholders. In some cases landlords sold their land to small farmers to raise capital for urban investment (Aricanli 1986; Dasgupta n.d.; El-Ghonemy 1979; McLachlan 1985; Okazaki 1985; Warriner 1962).

These processes affected only part of the land and population, leaving a composite agrarian structure. Throughout the region, the vast majority of farmers are cultivating holdings of 10 ha or less, and often as little as 1 or 2 ha, although such farms frequently make up less than 25% of the arable land. At the other extreme, farms of 50 ha and over are a small percentage of the total, but may occupy a quarter to a half of the land (AOAD 1984). These large holdings, admittedly small in global terms, are frequently under direct or indirect government administration.

Rural populations have recovered and grown beyond their pre-war levels, but the relationship of the farm labor force to the land has been completely transformed. Rainfed agricultural areas are now populated essentially by landowners, differentiated by holding size. The pattern of sharecropping has been reversed; landlords in share arrangements are now usually smallholders, temporarily or permanently working in cities, who give control of their land to large operators with machinery, or to other smallholders trying to find enough land to make it worth staying on the farm. In other cases the landlords are nomads who have been allocated a small plot of what was formerly grazing land which is now cultivated by machinery owners; the pastoralists take their share and go elsewhere to graze their flocks. Share-cropping of the old pattern is now rarely found except in very productive, usually very rainy or irrigated areas where there is a need to retain a laborer-manager from planting to harvest. Large owners in drier zones rely instead on machinery supplemented by seasonal casual labor.

Farm Size, Labor Availability and Technology Choice

Policy makers, tractor owners, and landlords influence farm decisions, but the last word is with the farmer. Farms are highly variable in their endowments of land, labor, and capital, which leads to different strategies of technology use. In practical terms, only land and labor need to be considered, as large holdings generally have greater financial resources, including better access to credit.

Large farms are usually focused on crop production, while large-scale livestock production is carried out by dairy, fattening, and poultry operators. Management and technology are specialized and there is little opportunity to integrate crop and livestock production. Large farms may be commercial or state farms. Land and other inputs may be available on a concessionary or non-market basis, but in other respects commercial farms act like ordinary firms in selecting a mix of inputs to maximize profits. State farms, on the other hand, may choose to maximize production rather than profit, in pursuit of food self-sufficiency or other policy goals. In both cases manual labor is treated as an external cost, and hired only as needed to meet farm objectives; labor may be replaced if there is a cheaper alternative, or hired in greater quantity if it is more profitable or productive to do so.

On small farms, decisions are more complex. These farmers have less flexibility, and may be expected to maximize income by efficiently using their available resources, rather than hiring inputs to maximize profit from a single enterprise. In contrast to specialized large farms, small farms typically try to

Table 3. Migration and domestic employment in Turkey.

	1962	1975	1977	Mean annual change 1962–75	1975–77
Total employment	12,643	14,668	14,726	156	29
Agricultural employment	9,740	9,463	9,100	−21	−182
Non-agricultural employment	2,903	5,205	5,626	177	211
Workers abroad	13	795	815	60	10

Source: IBRD (1980:131).

increase income and manage risk by diversifying out of crop production into livestock and off-farm employment (Hogan et al. 1984). Livestock contribute as much income as crops, for example, in drier areas of Morocco and Cyprus, and they play important roles in years of crop failure (Campbell et al. 1977:84; Papachristodoulou 1979; Tully 1986). Technological options which increase production of either crops or livestock at the expense of the other conflict with the smallholder's desire for diversification.

Labor is often the greatest resource of the farm family, but it may be best used in some mixture of on-farm and off-farm activities. Off-farm employment is widely available, as a result of well developed industrial and service sectors and opportunities for international migration. In Turkey, for example, agricultural employment has decreased in absolute terms in favor of non-agricultural employment and emigration (Table 3), and throughout the region the proportion of the labor force in agriculture has declined (Table 1). Off-farm income has been estimated at 24–37% of household income in several studies in Syria (Rassam 1984; Tully and Rassam 1986; Somel et al. 1984). In Cyprus the majority of farmers consider agriculture a secondary occupation (Papachristodoulou 1979), while in a Tunisian survey 67% of households with less than 10 ha had off-farm income (ICARDA 1985:58). Off-farm employment adds stability to rural incomes, but it also reduces the family labor which is available for agriculture (Hogan and Hansen 1983). New technologies which interfere with off-farm opportunities will not be widely adopted on small farms.

Farmers with small holdings are sometimes considered an obstacle to change. However, they have shown willingness to adopt new practices when suitable technologies have been developed, and when information and financing have been made available. Turkey's farmers, 73% of whom cultivate 5 ha or less, are among the highest input users of the region. Syria's small-scale wheat farmers also show high rates of use of fertilizers, herbicides, and improved seed (Aricanli and Somel 1979; Tully and Rassam 1986).

One of the key elements underlying technology adoption by small farms is labor. On the surface it would appear that small farms with a high ratio of people to land would have a labor surplus and avoid labor-replacing technology, while accepting technology which requires more labor. However, even if aggregate underemployment is reported in a rural area, the labor which is present is not necessarily the correct type to use the most productive technologies on offer. Within the farm family, labor is a complicated mix of persons, each with his or her own characteristics and opportunities. Farm labor requirements must be considered in terms of the sex and age of workers required, levels of skill needed, and the timing of labor inputs.

The demographic structure of the family can affect the choice of technology. Most societies have at least a partial division of labor by sex and age, of varying rigidity. In northwestern Syria, for example, tillage and seed broadcasting were traditionally male activities, weeding and spreading organic fertilizer were female, and both men and women took part in harvesting and threshing. Both children and adult men herded livestock, while milking was done either by males or by females depending on which ethnic group is considered. New technologies have generally been assigned to men; men still till, with tractors, but they also carry out chemical fertilization and weed control. Meanwhile, hand harvesting is becoming a female activity, as it is generally carried out by hired labor and women represent an increasing proportion of the hired labor force.

The processes by which tasks are culturally assigned to one sex or another are not well understood but in general scientists and policy makers have accepted them as given rather than trying to introduce a new technology requiring a cultural change of unknown difficulty. The fact that new technology is often the provenance of males, however, makes the shortage of adult male labor on farms that much more problematic. In the face of emigration by men, the possibility of family labor replacing them varies with cultural norms.

Skill requirements also differentiate the labor pool. New technologies often require special training, and shortages of associated skilled workers, such as drivers and mechanics, are common. Skilled agricultural work is available on large and small farms, and the pay is usually comparable to urban employment, providing an incentive to young men to learn these jobs and stay in rural areas. However the number of such jobs is limited, and to the extent they do exist, they contribute to the shortage of adult men for manual work.

At a more general level, the abilities to read directions, measure and mix chemicals, and handle machinery are coming to be more commonly needed by all farmers. These are acquired through education and experience, which tend to be most available to males, reinforcing the cultural allocation of new technology to men. Also, technologies usually require a certain amount of local adaptation to be most effectively used, and to do this well farmers need to understand the principles behind them. Thus adoption of a given technology may depend on the presence of someone in the family who can visualize its potential on the family's own farm, and feels comfortable learning how to adapt it. Exactly what personal characteristics are involved in technology adoption, how they are acquired, and how often such persons are found in the right place at the right time are not well understood.

The third labor issue pertaining to the choice of technologies by farm families is timing. Labor requirements in rainfed agriculture are highly seasonal. In the Mediterranean, rainfed agriculture generally demands most labor at the beginning (for tillage and seeding) and at the end (for harvest and threshing) of the cropping cycle, although in wetter and some summer-rainfall areas weed control during the growing season may require most labor. In either case, labor inputs have a seasonal high, generally called the bottleneck period, in which the greatest demand is made on the available labor pool. The bottleneck may result from several activities needing labor at the same time; thus in studying the labor cycle, labor inputs to all crops and livestock must be taken into account, as well as migration schedules. School schedules can also limit children's availability for work.

Elimination of a bottleneck through technological change may alter the structure of the labor cycle, but it will remain seasonal. For example, mechanization of tillage in much of the Mediterranean has led to a new labor bottleneck at harvest. Labor demand for legume harvesting is particularly intense because a mature crop rapidly incurs losses due to pod shattering and leaf drop. In general, a technological introduction which would reduce a bottleneck, for example a non-shattering lentil which permits a longer harvest season without crop loss, will be attractive to medium-sized and large farms because it will allow more of the work to be done by family labor, and decrease the problem of recruiting casual labor. A technology which increases labor needs at a bottleneck season immediately faces potential problems of labor availability and cost.

However, the significance of bottlenecks must always be evaluated for each particular case. Casual labor is not always problematic, and on very small farms the implications of bottlenecks may be completely different. On these farms, mobilizing labor for peak activities is not such a serious problem, because absolute requirements are so small. In fact such farmers might prefer to concentrate farm work in a limited period so they can work on other things later. A technological choice which requires off-season labor can be a greater problem than one which worsens a bottleneck in such cases. For example, in Syria, a substantial portion of males from barley-producing families are absent from their farms through the growing season, seeking seasonal wage labor or tending their livestock, and even a light labor demand mid-season would be unattractive if it required their return to the field. However, with women and children remaining at the site of crop production while men migrate, a technology which can be carried out by female family labor would be practical. Thus if a crop variety requires weed control, for example, this may be most efficiently managed on commercial farms with herbicide, but with manual labor on small farms.

Assuming that the appropriate type of labor exists within a family or can be hired to use a particular agricultural choice, the farm family must decide if it is worth deploying this labor. In feasibility studies for technological changes using partial budgets, this is usually done by assigning a cost to the labor required and then determining if the new choice is more profitable than the current practice, after subtracting costs from revenue. The key issue then is the correct estimate of labor costs, particularly if the labor will come, in whole or in part, from the family without cash outlay. The alternative of using the partial budget to determine returns to labor is less widely used, but may be more relevant to farmer decisions.

The most common procedure is to value family labor at the market rate or assume that all labor is hired at the market rate. If in fact family labor of the appropriate sex and age is expected to be available, this may be overly pessimistic. If labor is costed at market rates, farmers are sometimes observed to be doing things which are unprofitable, such as harvesting by hand when "cheaper" combines are available (Jaubert and Oglah 1985). It is a fairly common phenomenon globally that people are prepared to work for lower pay in order to be their own bosses, and in some rural societies there may also be an element of shame associated with performing manual labor for a wage.

A related approach to evaluating family labor is based on the concept of opportunity cost. A farmer incurs an opportunity cost when he or she has to forego possible income from some other activity to work on the farm. The amount of this income depends on opportunities, which depend on a person's individual qualities. Adult males with skills and education which make them desirable off-farm employees have particularly high opportunity costs, and thus if they work on the farm their labor must be costed at above the market rate.

At the lower limit, family members with no employment opportunities have an opportunity cost of zero. One does sometimes see partial budgets which assume family labor to be freely available without imputed costs, but this is not realistic. Few people will work without the expectation of some benefit, and the characteristic market integration of rural areas in the Middle East region makes people particularly aware of the monetary value of their activities. However, the lack of other opportunities for productive employment by family labor, primarily women and children, may mean that their labor is available at a return substantially below the market wage. Detailed research on family labor availability as a function of economic returns, especially in advance of a technical introduction, is extremely difficult but is the only way to remove a major uncertainty facing many development initiatives.

The segmented nature of the labor market also has an effect on the willingness of farm families to work for low returns. In many cases the amount of off-farm income dilutes the potential contribution of on-farm labor to family welfare, particularly if this income has raised family living standards to a high level by comparison with the past or with neighbors. Family members may expect to be able to reduce their field work as a benefit of relative wealth (Bates and Rassam 1983:151; Myntti 1984), and the market wage may be inadequate to interest members of higher-income families. The fact that some family members obtain high wages may also be discouraging;

a woman may feel that it is a waste of her time to work hard for a week to earn what her husband, father, or son can make in a few hours, or to process farm produce into consumption items such as bread or woven goods when family males can buy cheap substitutes with their cash incomes.

With increasing market integration and monetization of rural life, such negative perceptions of former subsistence activities, based on their low cash value, are becoming more common. Rural youths, in particular, are often reported to consider agriculture a low-status occupation with little attraction. The young are emigrating from rural areas, leaving farming to the older generation (Papachristodoulou 1979; Snobar and Arabiat 1984; USDA 1985:27). One often hears of young men living in rural areas because they cannot find urban work, yet refuse to engage in farm work. It is difficult to put a value on their opportunity cost without detailed information about labor markets. In the short term it would appear to be zero, but they are perhaps thinking of the future, and reacting against gradually committing themselves to farming by taking on contractual obligations, raising family expectations, stopping the search for more desirable work, and being unavailable if an off-farm opportunity should arise.

In practice, these factors vary from region to region and farm to farm, as do family labor endowments, which is why the partial budget continues to be used with all its recognized flaws. However, with information about the farm population and its economic activities, the availability of labor and the predictive power of the budget model can be better judged. In general, labor with a high opportunity cost and family labor in high income households should not be expected to be available, but in large, poor families it is more likely that manual labor will be available, and possible that a return below the market wage would still be interesting. The data needed to evaluate the labor force of a particular farm population in these terms are, however, rarely found.

Farmers as Managers

The discussion so far has assumed that farm labor is either unskilled manual labor or skilled, specialized labor. However, thought must also be given to the role of the farmer as manager. While management is, in some circumstances, the opposite of labor, the activities of a farm family include both labor and management, and both require time.

To get his or her field tilled, the farmer has to make decisions about timing and amount of cultivation, and then negotiate and supervise the hiring of a tractor. Planting involves choices of crops, varieties, seed rates and qualities, decisions about seed treatment and simultaneous fertilizer application, arrangement of the purchase and transport of the inputs, and possibly hiring a person to broadcast seed or to apply it by machine. Every laborer has to be hired, and if a large number of seasonal workers are needed this can be a major task. If inputs or marketing involve government offices, the bureaucracy must be dealt with. The growing crops and livestock must be supervised for signs of rodent, bird, insect, or disease problems requiring urgent treatment, and for responses to new treatments.

All of these jobs require time, effort, and judgment, and usually they cannot be delegated beyond the immediate family. How much time, how to cost it, and how farmers think about it are not well understood. It is clear, however, that management is not scale neutral. A technology such as fertilizer may give the same increase in yield for the same input cost on a per ha basis, irrespective of farm size, and thus would appear to be equally interesting to large and small holdings alike. However, a profit increase worth 10 USD per ha translates into 1,000 USD for the farmer with 100 ha, and 5 USD for the farmer with 0.5 ha. The management time to arrange the purchase, transportation, and application of the input would be approximately the same for the two farm sizes. Clearly the incentive to learn about and apply new technologies will be higher for large farms.

Many technologies require increased amounts of this sort of management activity and this may be true even if labor hours are reduced, as in the case of herbicide application. Management time can be reduced by improving farmer access to inputs and markets through cooperatives, extension agencies, or similar efforts, and these are of greatest importance for reaching small farms. Unless new technology shows very large productivity increases, it must be made very easily available to farmers, or else there is little incentive for smallholders to go to the trouble of experimenting with it.

Another route to farmers' fields leaves the marketing to free enterprise; small operators may create semi-skilled jobs for themselves in the field of technology delivery, as has happened with mechanical tillage and in some areas with herbicide application from knapsack sprayers. This is rarely promoted by policy makers but may be an opportunity to provide at least some rural jobs which are attractive to youth. Larger commercial firms also may bring technology to farmers, but as smallholders are not the most attractive customers they may not benefit as much as large farms.

Employment and Productivity

The preceding sections have focused on the relevance of labor constraints to the adoption of technology by farm families. It is clear that technologies which increase or reschedule labor needs must be developed with due consideration of the availability of family or hired labor and the schedule of competing activities. But the impact of technology on labor is also important. Technologies which reduce labor requirements, such as mechanization or herbicides, may be less desirable if they eliminate important sources of income or increase inequalities in rural income distribution. Most countries do not wish to contribute to the rural exodus by reducing rural employment;

Table 4. Trends in availability of agricultural machinery.

	Tractors			Combine harvesters		
	1974/76	1979/81	1986	1974/76	1979/81	1986
Morocco	19,992	26,000	32,000	2,708	3,137	3,190
Algeria	41,533	46,949	62,000	3,850	4,452	6,500
Tunisia	29,000	25,833	26,100	3,267	2,453	2,580
Libya	6,767	23,500	29,500	–	–	–
Egypt	21,463	36,276	44,000	1,927	2,110	2,250
Sudan	8,767	12,333	19,000	900	1,133	1,190
Ethiopia	3,600	3,933	3,900	130	153	150
Jordan	3,736	4,568	4,840	190	243	340
Syria	15,548	28,090	47,573	1,901	2,449	2,949
Iraq	19,741	23,336	40,915	5,028	3,221	3,032
Iran	43,900	77,981	112,000	2,133	3,056	2,900
Afghanistan	683	770	770	–	–	–
Pakistan	35,905	98,258	164,000	376	500	680
Cyprus	9,767	10,800	13,600	300	397	535
Turkey	241,339	431,548	611,052	11,542	13,117	11,457
Saudi Arabia	800	1,200	1,750	277	410	560
Yemen AR	1,010	2,000	2,170	–	–	–
PDR Yemen	1,190	1,033	1,120	12	14	15

Source: FAO (1984a; 1988a).

on the contrary, there is considerable interest in finding ways to create new jobs in agriculture. This is highly desirable but it will not be easy to use technology to create employment that is attractive and competitive with other opportunities. An examination of the current farm employment situation shows some of the difficulties.

Increasing employment in rainfed agriculture would mean reversing a trend towards labor replacement which has resulted in very low labor inputs. The pace of mechanization has varied from country to country, but mechanized tillage is currently the norm in West Asia and North Africa, with mechanization of other operations proceeding briskly. Custom services are widely available, so mechanization can be efficiently used on medium-sized and small farms as well as large (Arabiat et al. 1983; ICARDA 1985; Johnson 1983; Somel 1979). Machine numbers continue to grow in most countries (Table 4), and tillage with draft animals is mostly limited to drier or mountainous regions and small farms (Aricanli and Somel 1979; Campbell et al. 1977; Parvin and Hic 1984; Snobar and Arabiat 1984).

Seed drills and spinners have been introduced but hand broadcasting of seed and fertilizer persists in most areas, although this requires so little labor that it does not create a problem. Weed control is sometimes manual but herbicide is commonly used on cereals on small farms in Turkey and Syria, and on large farms more generally. In some cases, particularly in North

Africa, weeds are gathered to feed livestock, which requires manual techniques (ICARDA 1985). Legumes are also weeded manually because they are susceptible to most herbicides currently available to farmers. However, suitable herbicides have been identified and may become more available in the future.

Use of combines to harvest and thresh cereals is widespread, but limited in some countries by the lack of suitable machinery for stony or sloping fields, and by the desire of many smallholders with livestock to maximize the harvest of straw for feed. In Iraq 20% of cereal is hand harvested (Hermis and Hussain 1979), while in the barley fields of northern Syria the figure is 31% (Somel et al. 1984). In most cases where cereals are hand harvested, however, mechanical threshing follows. On the other hand, harvesting and processing of legumes, summer crops, and tree crops are largely manual, due to lack of suitable machinery.

Currently, the major demand for hired labor in rainfed agriculture is for harvesting, mostly of non-cereal crops. Other non-mechanized tasks, such as weeding, animal care, dairy activities, and even cereal harvesting have long, low-level labor demands which can be met to a large extent by household labor on small farms. Legumes and tree crops also tend to be localized in environmentally favorable areas, and at harvest large numbers of hired laborers must be brought from outside the area. Harvesting jobs in rainfed areas are available only for a limited period. In the Mediterranean the harvest starts in April or May with lentils, followed by barley, wheat, chickpeas, summer crops, and olives. It is mainly the lentil and olive crops that require extensive hiring and pay high wages because of concentrated harvest seasons. Work in irrigated agriculture provides some additional opportunities, but the absolute amount of work available is still limited.

In addition, there are annual variations in the availability of work. In 1986 Morocco's rainlands had a good cereal harvest, and farmers faced a severe shortage of harvest labor. As this harvest followed 4 years of drought with little employment of labor, it was difficult to mobilize workers when the good year came. This case is extreme, but demand for labor is variable generally, so rural workers are discouraged from keeping themselves available for jobs which may or may not develop. In the absence of any guarantees for workers when employment is unavailable, every downturn will cause more workers to leave rural areas.

Thus labor replacement has already proceeded to such an extent that there is not enough work during the year to support a significant landless labor force and, as a result, those who work as agricultural laborers inevitably do not depend exclusively on such employment. They are generally members of families, possibly owning small farms, with some members employed out of agriculture. This diversification of the rural labor force in turn raises rural wages because of the increasing opportunity costs and the growing capacity of families to survive without income from farm work. Seasonal employment in rainfed agriculture contributes a share, probably small, of family income. It is unlikely that any single technological change would encourage landless laborers to remain in rainfed areas. In terms of generating employment, a more realistic goal is to increase opportunities for poor farm families to use their available labor to make a greater contribution to the family income.

It is sometimes forgotten that it is income that is important, not just keeping idle hands busy. However, farmers can only pay what they can afford and wages are limited by low prices and productivity. In one study in the Aleppo area, the lentil harvest required only 10–12 man-days per ha but cost farmers 46% of the total value of the crop (Tully 1984). Harvest wages were high enough to discourage many farmers from growing the crop, and harvest workers were earning the highest agricultural wages of the year, yet these wages barely reached the lowest urban rate. To improve opportunities for income, significant increases in farm revenues are needed, through increased production, prices, or both.

This section has focused on employment in the sense of the hiring of one person by another, but in the more general sense it includes self-employment on family farms. Both are important in rainfed areas, although family labor is available only if the economic return of the work is attractive. Whether the workers are hired or working on their own farms, the overall capacity of rainfed areas to employ their population in gainful activity depends upon the total value of what can be produced in those areas. Thus, new technology should be judged both on its direct impact on labor and on its effect on system productivity and income.

Technological Improvements for Rainfed Agriculture

The generation of new agricultural technology by research centers almost invariably begins with biological scientists who see possibilities for technical improvements which would increase crop or livestock production or profitability. It is at a somewhat later stage that a technology can be said to be "in the pipeline" and a better idea can be had of its potential effects, and of the socioeconomic conditions under which it will be adopted. At this stage it may be possible to modify the technology to have a better overall effect. Also, it is then time to begin discussing policies needed to improve adoption and impact, and methods for reaching the farmers who can benefit most.

In this respect, labor issues should be considered in detail and one cannot simply assume that labor will be available or that labor should be replaced. Major adoption factors to be considered include seasonal availability, division of labor by sex and age, competing opportunities for labor, and educational level required. In terms of impact, researchers and policy makers should consider overall income and employment effects in addition to productivity per unit area.

New technologies under development in the region include changes in the time and method of seedbed preparation and planting; new types of mechanization; improved techniques of weed control, fertilization, and pest control; alternative crops and rotations; and new crop-livestock interactions. All of these techniques can be assumed to alter labor inputs – sometimes requiring more, sometimes less, and sometimes rescheduling activity. In this section selected technologies will be discussed in terms of related labor issues. More information on the technologies discussed can be found in recent ICARDA reports (ICARDA 1983–1986; FSP 1987; FRMP 1988; PFLP 1987–1988). To avoid any suggestion of external criticism I have not cited much of the very deserving work done by national program scientists in many of the same fields.

Small Changes at Planting

Technologies with the least dependence on labor per se are new cultivars which can be substituted directly for old ones while providing additional disease or drought resistance or small yield increases. Similarly, seed treatments or inputs which can be applied at planting, such as phosphate fertilizer or soil insecticides, should not face major labor problems because they are being applied when farmers are obliged to be present anyway, but are not very busy, except on large farms.

Assuming that price and availability are not problems, the main constraint for this type of technology is likely to be one of farm management time to learn about and acquire the new input. To reach small farms an effective extension or cooperative service will bring the inputs to the villages with sound information and minimal formality. Beyond initial success on the station, researchers should work on the "marketing" of an innovation through on-farm trials to evaluate its success under farmer conditions. The application must be straightforward, easy to learn, and tolerant of mistakes which farmers are likely to make if they are getting the information second or third hand. If a new technology must be used as part of a package, this increases the chance of failure geometrically, as each part of the package has to face the same adaptation process and chance of being used in a way which has not been tested.

Changes in Growing Season Management

Changes in cultivar may require significant changes in crop management in the growing season. For example, some high-yielding varieties require additional weed control, pesticide, and fertilization. In addition, a number of improved growing-season management techniques are being developed, including the determination of optimal nitrogen fertilizer rates according to seasonal rainfall, improved herbicides, and the use, as needed, of insecticides and rodenticides.

In most cases, the increased labor required is small, but there is a greater need for crop observation, awareness of the availability of numerous options, and ability to arrange application of the input. Where farmers or responsible representatives are resident on the farm, improved growing-season management can be achieved with a gradual process of extension and education of farmers, although to reach small farms extension services should be structured around the fact that many farmers will be absent in working hours! In cases where farm families are non-resident, however, the success of such technologies will be less than innovations at planting. Finally, where farm women are resident while men are absent, success will depend on whether they and their community accept the innovation as a women's activity, and whether an effort is made to reach them.

Innovations which significantly increase field labor in the growing season mainly involve manual weeding. A new disease-resistant and cold-tolerant chickpea, for example, can be planted in winter instead of spring and give much better yields, but requires manual weeding. In interviews with farm women in northern Syria, Rassam (1984) found that most were prepared to do the work if promised yield increases were achieved.

Harvesting Changes

In combination, the many small improvements planned for planting and growing seasons with existing crops can be expected to increase overall productivity. Yield increases can present harvest and transport labor problems, although in rainfed systems this is not likely in most years, particularly if the harvest is mechanized, as farmers are accustomed to dealing with highly variable yields. However, improved legume yields could be problematic.

One potential solution is the development of non-shattering lentils and vetches, which can extend the harvest period. Another, well under way for lentils, is legume harvest mechanization using erect varieties, and matching tillage practices with new harvesting machinery to reduce crop losses. This would undoubtedly replace a large number of casual laborers who spend 1–3 weeks per year in this activity. However, in the mechanization of cereal cultivation, labor replacement has been accompanied by production increases due to doubling or tripling of cultivated area, making it possible for large numbers of rural people to stay in their villages. The impact of legume harvest mechanization is not likely to be as great as that, but it is expected that the amount of land on which legumes can profitably be grown instead of fallowing will increase substantially. This could be a boon for the owners of small and medium-sized farms, often poor themselves, who cannot afford to hire labor. However, the significance of this income to the families of current harvest laborers and the impact mechanization will have on them is not known. The provision of alternative sources of income may be considered by policy makers.

Small-scale Mechanization

Some policy makers concerned with income distribution, employment, and productivity effects of mechanization are promoting small-scale machinery, particularly small tractors. The purpose is to enable more farmers to own tractors and till their own land in a more timely fashion, rather than depending upon relatively wealthy owners of large tractors. To introduce small tractors, financial incentives and import restrictions must be substantial to overcome certain inherent disadvantages. Small tractors are rarely as efficient as large ones in terms of initial cost, operating cost, and operator's time. Thus a farmer may find it cheaper to hire tractor services than to buy and operate a small tractor himself. Also, a smallholder who wishes to till his own fields and then provide tractor services for money will face stiff competition from larger, faster machines.

The advantages of independent tractor ownership can be substantial in cases where timeliness of operations is important and where the existing tractor park is concentrated in too few hands to provide a competitive market. In particular, if large landowners also own all the tractors, they can be expected to carry out field operations at the best times on their own fields, while smallholders who hire their services face problems of timeliness. However, small-scale tractors only solve this problem for farms that get a tractor. An alternative strategy is the provision of credit or subsidies to allow smallholders to buy full-size tractors. With enough tractors for hire, particularly in the hands of smallholders, timeliness and quality of work should improve for non-owners.

In terms of employment the two strategies differ in opportunities for skilled labor, namely drivers and mechanics. Small tractors would provide jobs for more drivers but at lower pay because of lower productivity. The cost and benefit of these additional jobs are the significant pieces of information for policy makers.

Fallow Replacement

As a form of intensification of land use, fallow replacement is a more substantial farming system change than the technologies discussed above. Two options are under consideration for rainfed areas: rotation of cereals with annually sown food legumes such as lentils or fodder crops such as vetch and lathyrus, or alternatively with self-seeding pastures. The annually sown crops face problems with expensive harvest labor as mentioned, and also face high input costs; at the levels of productivity currently achieved, they are not very attractive to dryland farmers. However, research is continuing to identify varieties and management practices which will increase yields and allow farmers to pay for harvest labor. If successful, this would increase employment through increasing system productivity.

Medic pastures avoid the high input costs of annually sown crops by reseeding themselves, so once established the main constraint is labor. It is expected that they will be used as grazing for sheep over an extended period from late winter until the autumn rains. In certain seasons, it is very important to control the stocking rate to ensure good seasonal grazing as well as a sufficient deposition of seed in the soil for the pasture to regenerate. In this respect the technology requires certain skills and understanding by the shepherd. More intensive grazing in a restricted area may increase the frequency of sheep diseases and parasites, and this would also require knowledgeable supervision. At present herding of family flocks is usually carried out by children, who are probably unable to appreciate the issues, while herding of village flocks by hired adults involves large numbers of animals which graze on communal pastures.

Several solutions are possible. The pastures might be so successful that they will make it worth the adult male's time to supervise herding in critical seasons, particularly if the flocks and lands are extensive; the job may be defined as an acceptable female activity, at least in some areas, as it is carried out on the family's own farm; measures may be established which allow a farmer to exercise sufficient supervision with daily checking, and leave the actual herding to children; or a new type of shepherd job may be developed for youth. In any event the pastures will generate employment, and in the final analysis it is the returns to that labor which will determine whether it is attractive to large numbers of farm families.

Supplementary Irrigation

The last example to be presented is perhaps the most substantial intervention being planned in arable dryland agriculture. This technology uses applied water as a supplement to rainfall to increase and stabilize crop yields. As the crop depends mostly on rainfall, the area which can be irrigated with a given quantity of water is much greater than under fully irrigated systems, and the problems of drainage and salinity are much less. To be most efficiently applied, supplementary irrigation requires monitoring of soil moisture to determine crop requirements, and thus someone who understands what is needed must be present throughout the growing season. At several critical points in the season equipment must be mobilized to apply the water to the field, and this should be done in a timely manner.

Thus, of the examples given, supplementary irrigation might be seen as the technology facing the greatest labor constraint. However, it also holds very great promise for increased, regular yields and as such is perhaps the best example of a technology which is likely to increase rural employment through increased productivity.

Conclusion

These examples illustrate the differences among technologies in the hopes and concerns they raise with respect to labor. Technology is generally understood to be most applicable and beneficial in the appropriate climatic situation. Similarly, if the definition of farming systems is broadened to include labor variables, some technologies will be more beneficial than others in specific locations. Others will succeed if they are adapted, while some may be totally inappropriate to the labor situation even if climate and soils are ideal.

In many cases these adaptations can be seen as an extension of the job of the biological scientist who, working with farmers as well as colleagues in the social sciences, can try to anticipate problems of adoption and impact and consider any technical fixes to mitigate them, possible social changes which will provide the labor needed, and policy recommendations that can help make a technology attractive. And if examination of the labor situation suggests a bleak future for a technology that works well on station, then it is in the interests of research managers to identify this fact at an early stage.

References

AOAD (Arab Organization for Agricultural Development). 1983–84. Yearbooks of Agricultural Statistics, vols. 3–4. Khartoum: AOAD.

Arabiat, Suleiman, Nygaard, David and Somel, Kutlu. 1983. Factors Affecting Wheat Production in Jordan. Aleppo: ICARDA.

Aricanli, Tosun. 1986. Agrarian Relations in Turkey: A Historical Sketch. Pages 23–67 in Food, States and Peasants: Analyses of the Agrarian Question in the Middle East (Richards, Alan, ed). Boulder, CO: Westview.

Aricanli, Tosun and Somel, Kutlu. 1979. Observations on Developments in Agriculture and Land Distribution in Turkey. (In Turkish.) Studies in Development 6: 91–110.

Bates, Daniel and Rassam, Amal. 1983. Peoples and Cultures of the Middle East. Englewood Cliffs, NJ: Prentice Hall.

Campbell, R.R., Follett, R.H., Howell, H.B., Riddle, R., Stubbendieck, J.T., and Hanway, D.G. 1977. Applied Agronomic Research Program for Dryland Farming in 200–400 mm Rainfall Zone of Morocco. Publication No. PN-AAF-329. Washington, DC: USAID.

Dasgupta, Biplab. n.d. Labour Migration and the Rural Economy. WCARRD Follow-up Programme Report. Rome: FAO.

El-Ghonemy, Mohamed Riad. 1979. Agrarian Reform and Rural Development in the Near East. WCARRD, RNEA Paper No. 5. Rome: FAO.

El-Sherbini, A. A. 1977. Problems of Arid Agriculture in West Asia. World Development 5: 441–446.

FAO (Food and Agriculture Organization). 1984a-88a. 1983–87 FAO Production Yearbooks, vols. 37–41. Rome: FAO.

FAO (Food and Agriculture Organization). 1984b. Food Balance Sheets 1979–81 Average. Rome: FAO.

FRMP (Farm Resource Management Program). 1988. Annual Report for 1987. Publication No. 131 En. Aleppo: ICARDA.

FSP (Farming Systems Program). 1987. Annual Report for 1986. Publication No. 108 En. Aleppo: ICARDA.

Fikry, Mona. 1983. Social Implications of the Wheat Development Program (Project Ble). Pages C1–C47 *in* Tunisia: The Wheat Development Program (Johnson, William Frederick, Ferguson, Carl E., and Fikry, Mona, authors). Publication No. PNAAL329. Washington, DC: USAID.

Hanson, Haldore, Borlaug, Norman E. and Anderson, R. Glenn. 1982. Wheat in the Third World. Boulder, CO: Westview.

Hermis, Yousif A. and Hussain, Sabah A. A. 1979. Constraints to Cereal Production and Possible Solutions in Iraq. Pages 69–78 *in* The Gap Between: Present Farm Yield and the Potential. Volume 1. Proceedings of the Fifth Regional Cereals Workshop, Algiers, 5–9 May 1979. Algiers: Ministry of Agriculture and Agrarian Revolution; Mexico City: CIMMYT, and Aleppo: ICARDA.

Hogan, Edward B., Furtick, William R. and Grayzel, John A. 1984. Morocco Increase in Cereal Production Project. Report to USAID.

Hogan, Edward B., and Hansen, Gary. 1983. Jordan Wheat Research and Production. Report to USAID.

IBRD (International Bank for Reconstruction and Development). 1980. Turkey: Policies and Prospects for Growth. Washington, DC: IBRD.

IBRD (International Bank for Reconstruction and Development). 1984–88. World Development Reports 1984–88. New York: Oxford University Press.

ICARDA (International Center for Agricultural Research in the Dry Areas). 1983–86. Annual Reports 1982–85. Aleppo: ICARDA.

Jaubert, R. and Oglah, M. 1985. Farming Systems Management in the Bueda/Breda Subarea 1983/1984. Research Report. Aleppo: ICARDA.

Johnson, William Frederick. 1983. Institutional, Policy, and Economic Influences of the Wheat Development Program in Tunisia. Pages B1–B40 *in* Tunisia: The Wheat Development Program (Johnson, William Frederick, Ferguson, Carl E. and Fikry, Mona, authors). Publication No. PNAAL329. Washington, DC: USAID.

McLachlan, Keith. 1985. The Agricultural Development of the Middle East: An Overview. Pages 27–50 *in* Agricultural Development in the Middle East (Beaumont, Peter and McLachlan, Keith, eds). Chichester: John Wiley and Sons.

Miladi, Samir. 1983. The State of Food and Nutrition in the Middle East. Pages 35–54 *in* Interfaces Between Agriculture, Food Science and Nutrition (Miladi, Samir, Mahgoub, Salah and Neate, Paul, eds). Aleppo: ICARDA.

Mokbel, Mirella G. 1985. Evaluation of Nutritionally Relevant Indicators in Villages in Aleppo Province, Syria, and Their Relation to Agricultural Development. PhD dissertation, University of Massachusetts.

Myntti, Cynthia. 1984. Yemeni Workers Abroad: The Impact on Women. MERIP Reports No. 14:11–16.

Okazaki, Shoko. 1985. Agricultural Mechanization in Iran. Pages 171–187 *in* Agricultural Development in the Middle East (Beaumont, Peter and McLachlan, Keith, eds). Chichester: John Wiley and Sons.

Parvin, Manoucher and Hic, Mukerrim. 1984. Land Reform Versus Agricultural Reform: Turkish Miracle or Catastrophe Delayed? International Journal of Middle Eastern Studies 16: 207–232.

Papachristodoulou, S. 1979. Socioeconomic Aspects of Rainfed Agriculture in Cyprus. Paper presented at the FAO Regional Seminar on Rainfed Agriculture in the Near East, Amman, 5–10 May 1979.

PFLP (Pasture Forage and Livestock Program). 1987–88. Annual Reports 1986–87. Aleppo: ICARDA.

Rassam, Andree Marie. 1984. Syrian Farm Households: Women's Labor and Impact of Technologies. MA dissertation, University of Western Ontario.

Somel, Kutlu. 1979. Technological Change in Dryland Wheat Production in Turkey. Food Research Studies 17: 51–65.

Somel, Kutlu, Mazid, Ahmed and Hallajian, Maria. 1984. Survey of Barley Producers in Northern Syria 1981/1982. Volume III: Descriptive Statistics. ICARDA Research Report No. 12–III. Aleppo: ICARDA.

Snobar, Bassam A. and Arabiat, Suleiman M. 1984. The Mechanization of Agriculture and Socio-economic Development in Jordan. Dirasat (Agricultural Sciences) 11: 159–200.

Tully, Dennis. 1984. Land Use and Farmer Strategies in Al Bab: The Feasibility of Forage Legumes in Place of Fallow. Research Report No. 13. Aleppo: ICARDA.

Tully, Dennis. 1986. Rainfed Farming Systems of the Near East Region. Discussion Paper No. 17. Aleppo: ICARDA.

Tully, Dennis and Rassam, Andree. 1986. Wheat Production Practices in Northwestern Syria. Pages 43–56 in ICARDA Annual Report 1985. Aleppo: ICARDA.

USDA (United States Department of Agriculture). 1985. Middle East and North Africa Outlook and Situation Report. Washington, DC: Economic Research Service, USDA.

Warriner, Doreen. 1962. Land Reform and Development in the Middle East: A Study of Egypt, Syria and Iraq. 2nd edition. London: Oxford University Press.

2. Labor Markets in North Africa and the Near East: A Survey of Developments Since 1970

SIMON COMMANDER

Economic Development Institute, The World Bank, 1818 H Street, N.W., Washington, DC 20433, U.S.A.

and

SIMON BURGESS

University of Reading, London Road, Reading RG1 51Q, U.K.

Introduction

It is normally the case that the shares of national income and employment accounted for by agriculture are lower in economies with higher aggregate levels of development. Yet, where resources become available to an economy through windfall gains, or where macro-economic policies discriminate against the agricultural sector, certain features of development may occur in transitory or disequilibrated form. Such outcomes have commonly been associated with the discovery and exploitation of natural resources. The subsequent economic booms have not only raised the trend rate of growth but have, where the boom phase was long enough, had an impact on the sectoral distribution of resources. A windfall tends to generate resource shifts through competing uses for factors, associated with resource movement towards non-tradeables and away from tradeables. Where economies have, prior to the boom period, been largely characterized by the dominance of the agricultural sector, this implies a shift of resources away from that sector.

Since 1970, windfall oil gains in some countries have been accompanied by a classical spending effect associated with an appreciating exchange rate, and by an extended period of sectoral resource transfers, with major consequences for labor markets. Both exporters and importers of oil have felt the direct and indirect effects of the oil price boom and subsequent decline. In addition, a number of these economies, such as Sudan and Morocco, have been characterized by trade and macro-economic policies that have been unfavorable for producers of tradeables, particularly in the agricultural sector. Amongst other consequences, this has accelerated migration from rural to urban areas, one result of which has been a growing labor constraint on agricultural production.

It is normally argued that a windfall drives up domestic demand for goods but will, over time, increase prices for home goods relative to tradeables

Dennis Tully (ed.), Labor and Rainfed Agriculture in West Asia and North Africa, 25–47.
© 1990 *ICARDA.*

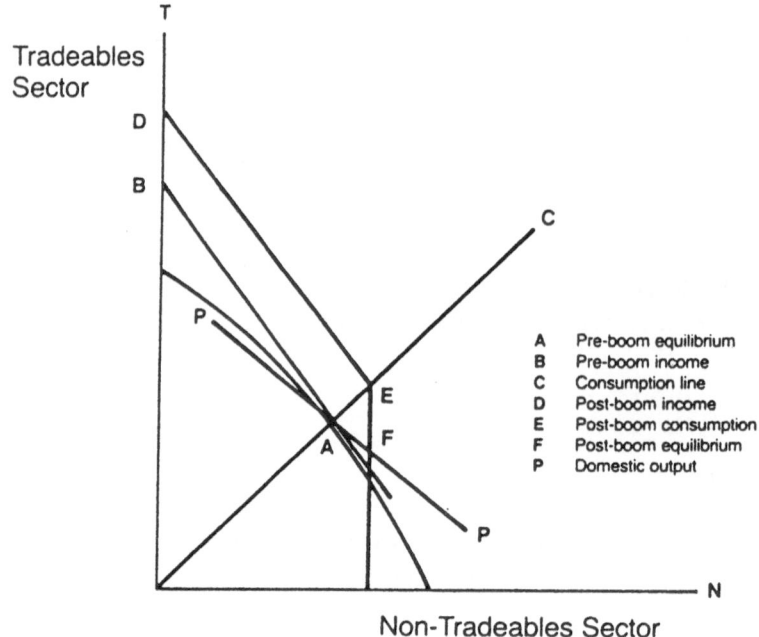

Fig 1. Long run effects of a resource boom.

(Neary and Wijnbergen 1986). This price effect will partly choke off demand while also calling forth a larger output. In the case of tradeables, increased demand will stimulate imports.

Following Bruno and Sachs (1982), Figure 1 indicates the likely consequences of a long run resource boom. A windfall shifts national income by BD with consumption moving along the consumption path to E. Consumption of non-tradeables at this point must be met from domestic output; thus the new equilibrium point is F. Clearly, capital and labor use will have shifted towards the non-tradeables sector and fallen in the tradeables sector.

The implications of a boom for employment and labor markets will largely depend on the production structure of a country prior to the boom and, in particular, the distribution across traded and non-traded goods sectors. Assuming full mobility across sectors, an increase in the relative price of non-tradeables will be associated with a growing demand for labor in that sector, which would tend to force up the real consumption wage over time and cause growing labor shortages. This could be expected where non-tradeables have greater weight in aggregate demand. The reverse would hold where non-tradeables were more important in supply terms. Such a distinction crudely reflects the difference between, say, the Gulf economies or Libya on one hand and Iran or Algeria on the other.

The oil boom directly affected Algeria, Iraq, Iran, and Libya, and also Syria and Tunisia to a far lesser extent. At its peak in 1981, the real price

Table 1. Petroleum exports as a percentage of total exports (value terms).

Year	Algeria	Iran	Iraq	Libya	Syria	Tunisia
1968–70	69.6	89.6	93.3	99.9	13.5[1]	22.0
1971–73	78.3	91.5	95.8	99.6	20.7	27.5
1974–76	91.3	97.2	98.5	99.9	62.6	38.8
1977–79	92.2	97.2	98.6	99.9	63.9	41.2
1980–82	87.7	96.8	99.1	96.0	58.7	48.5
1983–85	73.5[2]	98.1	98.6[3]	99.8	52.2[2]	41.3

Source: IMF, International Financial Statistics.
[1]1969/70 only; [2]1983/84 only; [3]1983 only.

of oil had risen nearly sixfold since 1972; then, between 1985 and early 1987, the real price fell by over 50% (IMF 1987). This price reversal has had damaging consequences for most countries in North Africa and the Near East whose economies have been closely associated with the oil-exporting nations, primarily through falling remittance earnings from migrant workers, reduced capital transfers, and contraction of demand for exports. These wider linkages have necessitated adjustment to external trading conditions, as well as domestic adjustment.

The growth in oil exports can be seen from Table 1. Export revenues derived from the resource boom were, in virtually all cases, fed into the economy largely through the increased resources available to governments. In Algeria, for example, the domestic oil windfall accounted for 27–30% of GDP between 1974 and 1981. There, as for most of the sample countries, the windfall was used to stimulate public investment as well as private consumption.

The expenditure effect associated with the oil boom through the 1970s was manifested in significant change in the domestic composition of national output. The share of agriculture declined significantly in total output in the decades following 1965. However, this decline was by no means unique to the oil-exporting countries. In Turkey, for example, agriculture accounted for under 19% of GDP by 1984 compared with 34% in the mid-1960s. The decline was paralleled by a sharp rise in the share of manufacturing in aggregate output. However, most countries saw a shift away from tradeables toward services and construction. This was obviously reflected in the distribution of labor force (Table 2). In general, for both oil-exporting and importing countries the strongest redistribution was toward the services sector.

These shifts in the structure of production and the composition of labor demand, though strongly associated with a natural resource boom, must be related to additional features including pre-boom factor endowments, the differential impact of government policies, and the availability of technology. In this paper, distinction is made between countries on the basis of access to oil resources, in the periods before and after 1973. Clearly, the distinction between oil-endowed and non-oil economies is simplistic, particularly in terms of labor markets. One of the most striking developments in the 1970s

Table 2. Labor force by sector as percent of total labor force.

	Agriculture		Industry		Services		Labor force growth (%)	
	1965	1980	1965	1980	1965	1980	1973–84	1980–2000
Oil Exporters								
Algeria	57	31	16	27	26	42	3.6	4.1
Libya	40	18	21	30	39	53	4.1	4.1
Iran	49	36	26	33	25	31	3.0	3.6
Iraq	50	31	20	22	30	48	3.1	3.8
Oil Importers								
Syria	52	32	20	32	28	36	3.4	3.9
Tunisia	49	35	21	36	29	29	2.9	2.9
Jordan	36	10	26	26	37	64	1.6	4.7
Yemen AR	79	69	7	9	14	22	2.1	3.2
Morocco	62	46	15	25	24	29	2.6	3.1
Turkey	75	58	11	17	14	25	2.0	2.2
Sudan	82	71	5	7	13	22	2.4	2.8

Source: World Bank (1986b).

was the growth of a strong regional labor market, fueled by the predominantly liberal attitude of governments towards labor migration. Thus the classification is used only to facilitate organization of the available empirical evidence.

Employment in the Oil-Endowed Economies

Pre-1973

Until the late 1960s, oil revenues were modest but rising and the labor demand created by the petroleum sector, largely by foreign oil companies, was satisfied mostly through domestic migration and labor transfers. With the gradual growth in oil revenues and the increasing nationalization of foreign-owned assets, there was a consistent shortage of the skill level required by governments to run the oil industry. In Libya, as well as the Gulf economies, a weak agricultural sector base was characteristic, as were relatively small populations with low participation rates and skill levels. In the early 1970s the total population of Libya fell below 2.25 million with a crude labor force participation rate barely exceeding 20%. Potential oil wealth was thus combined with poverty of human capital.

This was not the case for the other major oil economies of Iran, Iraq, and Algeria, where crude participation rates ranged between 26 and 30%. In addition, the absolute size of the labor force was considerably larger, and agriculture had a more important position in these economies. In the early 1960s agriculture accounted for around 21% of GDP in Algeria, 29% in

Iran, and around 17% in Iraq. Prior to 1973 there was a concerted effort, in Iraq and Libya in particular, to stimulate the agricultural sector through enhanced investment and special programs, although actual allocations to the sector were limited by absorptive constraints. In Iraq actual expenditures fell below 70% of planned allocations during 1968–73, with a similar pattern in Libya and Algeria. While the hydrocarbons sector was largely capital intensive, accounting for no more than 1–2% of the labor force, a reasonably strong drift of labor out of the primary sector had already started prior to the first rise in oil prices (Niblock 1982).

Post-1973

The sharp rise in oil prices led to a rapid influx of resources. In general over 80% of the windfall gain went directly towards support for public finances. This shift in resource availability was accompanied by relaxed restrictions on imports and reduced attention to other sources of government revenue. With the possible exception of Algeria, the fiscal structure outside the oil sector was neglected, impelling a strong dependence on petroleum revenues in public finances. By 1975, oil revenues amounted to 85–90% of total government revenues, except in Algeria where their share just exceeded 60% (Gelb 1986).

Appreciating real exchange rates and expansion of the monetary base, due to an inability to stem the inflow of international reserves, resulted in a sharp rise in annual inflation rates during the 1970s. In Iraq, inflation during 1970-75 was double that of the previous 5 years, while in Algeria average annual inflation rose from under 7% in 1965–70 to over 28% in 1970–75, increasing further in 1975–80 (Bennamane 1980). Trends in the real exchange rate were, however, less pronounced. In Algeria, the exchange rate actually depreciated between the early 1970s and 1978, partly due to the explicit policy of accumulating non-oil balances. It was only in Iran that a marked appreciation occurred, of the order of 18% between 1970 and 1978 (Figure 2). However, in all cases, there was significant appreciation after 1980.

The general growth effect of oil resources and the aggregate level of activity in the world economy meant that national income grew significantly faster through most of the 1970s than in the previous decade. However, this was not matched by performance in the agricultural sector. Between 1970 and 1978 agriculture had a negative growth rate in Iraq of nearly 2% per annum and remained roughly constant in Algeria. In all cases except Libya, agricultural growth rates fell below those for GDP as a whole and considerably below the trend rates registered in countries with broadly comparable levels of development.

Low growth rates in agriculture were dynamically associated with the distribution of public expenditure. At a time when aggregate public revenues were rising sharply, agriculture's share of total public investment tended to fall, and even allocations under recurrent budgets grew at lower than average

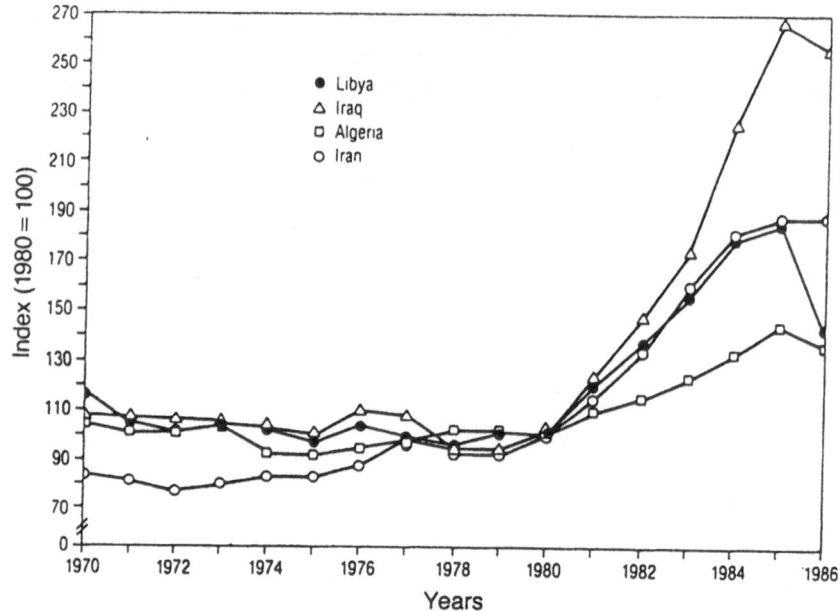

Fig. 2. Real exchange rates of oil-exporting countries, 1970–86. *Source*: IMF data.

levels (Askari and Cummings 1976). This can be seen in Table 3 which maps a variety of effects, the most striking of which relates to the trend in the share of agriculture in non-oil GDP.

In Algeria, the availability of oil-derived resource flows was directly related to a political and economic strategy based on augmenting the share of the public sector in GDP and accelerating the growth of a manufacturing sector along conventional import-substituting lines. There was an explicit neglect of agriculture which had already had a low growth rate prior to 1973 and by the early 1970s, agricultural output per worker was around half that of the previous decade (Edens 1979). This can be explained by biases in public investment and technological constraints in the 80% of the country that fell in the semi-arid and arid tracts of southern Algeria and by the

Table 3. Oil-exporting countries, 1970–80.

Country	Government index[1]	Urbanization index		Labor force in services		Services in non-oil GDP		Agriculture in non-oil GDP	
			Average RC[2]		Average RC		Average RC		Average RC
Libya	70.5	43.0	18.1	47.0	12	21.3	−0.7	5.4	−2.0
Iraq	74.9	64.7	14.7	31.5	1	20.4	1.2	18.5	−2.8
Iran	37.0	45.0	10.3	26.0	2	18.0	6.4	17.4	−8.8
Algeria	32.0	38.0	12.1	43.0	14	11.4	6.3	10.4	−2.7

Source: Stevens (1986).
[1] Share of government expenditure and consumption in non-oil GDP.
[2] RC = Rate of change.

peculiarly disruptive effects of the War for Independence and subsequent measures to bring more land under state control.

Following the sharp rise in oil revenues post-1973, the share of public resources allocated to both the hydrocarbons and industrial sectors rose. These sectors accounted for nearly 57% of total public investment in 1974–78. Agriculture's share fell from just over 8.5% in 1970–73 to around 5.5% in 1974–78. This had inevitable consequences for medium-term productivity trends and for employment. Between 1974 and 1978, agricultural employment grew by only 1% per annum while growth of employment in the hydrocarbons, manufacturing, and construction sectors was 13–23% per annum over the same period. Overall growth rates translated into a large and ultimately debilitating migration out of agriculture. Between 1970 and 1979 it has been estimated that over 1.7 million rural workers migrated to the four largest towns. At the end of the 1970s nearly 44% of the total population was urbanized, and a culture of out-migration had been established in rural areas (World Bank 1982a; Nelson 1979).

In 1977, 35% of the Algerian labor force was employed by the state, rising to 59% by 1984. With public employment growing substantially and increasing wage differentials between sectors, a major flow of labor to the urban sector occurred along standard Harris–Todaro lines. However, there was a relatively low labor absorption potential in the urban sector. The high levels of urban unemployment that resulted were not met with a return flow to rural areas, because of the massive underemployment which continued to exist in the rainfed belts, especially in the south, and the opening up of the French labor market (Trebous 1970). Although the large-scale migration to France and the EEC was relatively short-term, by the mid-1970s it was estimated that as much as 20% of the potential labor force was working abroad. This has caused major manpower constraints due to the loss of skilled labor.

The Algerian example highlights a number of common outcomes of windfall resource gains. First, the boost to public revenues was translated into a far higher level of public expenditure in real terms. Second, this expenditure was directed towards sectors other than agriculture and third, with higher growth rates and labor demand emanating from the non-agricultural sectors, there was a large out-migration from rural areas. This raised the marginal productivity of labor for the remaining workers in the sector but the rate of increment was held back by a combination of other factors, including relatively low producer prices and inadequate investment in infrastructure and institutions in the predominant rainfed regions. Possibly the most visible consequence, itself associated with the usual consumption effects of an upward shift in incomes, was a rapidly widening food deficit. Between 1974 and 1983 per capita food production fell by around 17% (World Bank 1985).

In Iran, the cooption of oil revenues by the state was as marked as in Algeria and Libya, but the revenues were used somewhat differently. Furthermore, the Iranian oil industry was better established and the economy more dependent overall on oil exports. Until 1978, Iran held a market share of around 16% of total crude oil exports, which already comprised over 90% of total exports prior to the 1973 price rise. Between 1974 and 1977 nearly 37% of non-oil GDP was accounted for by fiscal revenues (Gelb 1986) which were actively used to promote very significant investment in both the hydrocarbons sector and infrastructure. The agricultural share of public investment did not exceed 11% during 1972–77. However, unlike in Algeria, the Iranian government did not emphasize the consolidation of a domestic manufacturing sector organized around import-substituting priorities and private consumption was not restrained.

By the second oil price rise, agriculture contributed a mere 9% of GDP compared with 16% in 1972/73 and 27% in 1962/63. In contrast, the oil sector generated over 38% of GDP in 1977/78, having accounted for under 20% at the start of the 1960s. The shift in the composition of aggregate output was mirrored in a very sharp decline in the share of total employment accounted for by the agricultural sector. In 1967 nearly 47% of the labor force worked in agriculture but by 1972 this share had fallen to 37% and by 1982/83 to around 27% (Rachidzadeh 1978; Katouzian 1983; World Bank 1973). This trend can be attributed to several factors. A series of land reform measures had ambiguous consequences and while they reduced the overall skewedness in landholding, a significant number of laborers were left without land. Also, a bias developed toward the relatively small proportion (under 3% of area) of large independent farms and farm corporations established by the government. Thus, of the total agricultural investment budget, nearly 55% was directed to the so-called modern sector, which also received most of the flow of credit and inputs (Katouzian 1983).

Thus, resources were unequally distributed across income groups and regions, with the bulk of the peasantry being excluded from 'official' development. This was emphasized by the parallel interventions that governed the consumption path and availability of goods. As domestic demand for wage goods expanded, primarily in urban areas due to the rapid increase in disposable income, it was increasingly satisfied by recourse to imports of food and other goods. Cereal imports rose substantially between 1974 and 1984 from 2.1 to 5.3 million tons in Iran. This was also the case in Libya, Iraq, and Algeria where imports rose from 0.61 to 1.0 million tons, 0.87 to 4.5 million tons, and 1.8 to 4.2 million tons, respectively (World Bank 1986b). In general, the performance of the Iranian farm corporations and agribusinesses set up by the land reform program remained poor and output of cereals and livestock from the peasant sub-sector was insufficient to meet domestic demand (World Bank 1976). Although Iranian animal products continued to attract a premium price, large-scale rice and wheat imports reduced effective producer prices. During the 1960s rural per capita incomes were around 45% of their urban equivalents and by 1973 had fallen to 40%, then to under 27% at the end of the 1970s. Between 1963 and 1978 the ratio of rural to national consumption on a per capita basis collapsed from 0.7 to 0.4 (Katouzian 1983).

As in Iran, urban incomes in Iraq grew faster than rural incomes: between 1970 and 1983, it has been estimated that rural incomes grew by around 7.2% per annum, with urban incomes growing at over 9% per annum. At the end of this period average rural incomes were around 45% of their urban equivalents (World Bank 1974a; Penrose and Penrose 1978; Niblock 1982).

Iran's explicit focus on urbanization and the direction of public resources to the urban sector, as well as the relatively high unemployment rates that existed in rural areas even in the early 1970s, resulted in a large out-migration from rural areas. A direct consequence was absolute stability in the size of the rural population at a time when the total labor force was growing at around 3% per annum. Urbanization was rapid, accounting for nearly 54% of the population by 1984. The frenetic growth of the urban economy resulting from the resource boom caused short-run labor market maladjustment as chronic labor shortages emerged in the 1970s. But after 1976 the demand for labor fell as major construction and other investment projects were halted in the face of political uncertainties. Urban unemployment rose very sharply, so that by 1982/83 nearly 20% of the total labor force was classified as unemployed. Even this figure is low; there is major over-staffing in the public sector, and employment figures are affected by the high military demand for labor.

Though less extreme, unemployment in the other oil-exporting countries has also grown. In Libya the response has primarily been to eject foreign labor, and many Tunisian and Egyptian laborers have been expelled since 1984/85. In Algeria, around 8% of the labor force lacked work in the mid-1970s and by the early 1980s this figure approached 12%. With public sector employment already accounting for 59% of total employment, the state has had to act, in counter-cyclical fashion, as continuing employer of 'last resort'. But this has not been without its more dynamic, longer term consequences. The public sector remains marked by poor management, over-staffing, and effective underemployment. In response to fiscal constraints, wages have been allowed to stagnate or fall in real terms.

Thus, although very different political and economic strategies were overtly pursued by the governments of the oil-exporting nations, the net result of the post-1973 windfall gains was an accelerated transfer of resources away from the agricultural sector. This appears to have had longer term prejudicial implications, not least of which are severe labor market disequilibria, resulting not only from skill mismatching but also from a premature and substantial transfer of labor into the urban sector.

Oil-Exporting Countries and the Regional Labor Market

With the growth in oil revenues, the Gulf states, Saudi Arabia, Iraq, and Libya became substantial labor importers (ILO 1976b). In Iran, Afghan and other labor was also imported in the 1970s to fill shortages in the agricultural sector. In contrast, Algeria was a net exporter of labor. The most significant labor importer was Libya (El Fathaly and Palmer 1985), where by 1975 there were over 280,000 migrant workers (Serageldin et al. 1983; Birks and Sinclair 1980a), comprising just under 40% of the labor force. By 1980, the number of migrant workers had virtually doubled. For Iraq, the growth in migrant workers occurred mostly after 1980/81. At that time the number of migrant workers was put at around 270,000 but by mid-1986, it was estimated that there were as many as 2 million Egyptian workers in Iraq.

In other countries, the trend since 1980 has been downward with a higher proportion of the migrant labor force being skilled workers (Ferghany 1980; Richards and Martin 1983a). In the initial boom of the 1970s, around 47% of the migrant labor force in several Arab countries was unskilled, and in Libya this share rose to around 55%. This reflected the fact that the principal source of demand for labor was the construction sector, itself primarily driven by the proliferation of public sector projects, in particular large infrastructural projects. Falling public revenues from oil in the 1980s have, however, compromised such strategies and have consequently led to a downturn in the demand for migrant labor as well as increased domestic unemployment. Although recent reliable estimates are not available, evidence on remittances suggests that migration within the Arab region has fallen. This is certainly the case in Egypt, one of the major labor exporters (Commander 1987; Richards and Martin 1983b).

Employment in the Non-Oil Economies

Pre-1973

With the sole exception of Jordan, the non-oil-exporting countries discussed in this paper were overwhelmingly agricultural. In the mid-1960s, considerably over half the labor force worked in agriculture, which generated between a quarter and a third of national income. In Sudan, the economy with the least sectoral diversity, over four fifths of the labor force worked in agriculture, while agriculture in Turkey accounted for 34% of GDP in the mid-1960s (World Bank 1978b; 1979).

Given the diversity of economic and ecological conditions, it is difficult to generalize but there are several common factors. The most significant from the point of view of a labor market study are the land reform measures introduced in several countries, mostly in the 1960s, and the measures taken by governments to diversify by promoting sectors other than agriculture.

The impact of the various land reform programs differed quite significantly. In PDR Yemen, for example, as much as 55% of the total arable area was redistributed by the late 1970s, while in Tunisia and Syria land reform covered only 19 and 13%, respectively, of the cultivated area (World Bank 1981a; 1981b; 1982b; 1986a; Sayigh 1978; Pridham 1982; Grissa 1973).

In Morocco, land reform affected an even smaller area and did little to correct a long-standing bias in policy towards irrigated areas. Yet with substantial public investment in the irrigated sector and a redistribution of land from European control into Moroccan hands, largely in the irrigated zones, output growth in the sector remained high throughout the 1960s and early 1970s. Between 1968 and 1972 agricultural production grew at around 6% per annum, primarily fueled by growth in productivity in the irrigated sector and in the size of the irrigated area (Seddon 1987; Nelson 1978; World Bank 1983a). In the rainfed zones, which are predominant in Morocco, there was little expansion, and this was accentuated by the prior distributional features of the economy as well as regional and infrastructural inequalities. Prior to 1973, dryland agriculture attracted a limited share of public resources, and thus remained characterized by low-productivity cereal farming with limited marketing and considerable seasonal underemployment (World Bank 1974b). Amongst other consequences, this stimulated the outflow of labor from the dryland regions to the towns.

In this period most non-oil countries emphasized the development of a manufacturing sector. In Turkey, such policies dated back to an earlier period and were directly articulated to a policy that emphasized the public sector. By the early 1980s over 35% of the non-agricultural labor force was in the public sector, a figure matched only in Egypt. Nevertheless, despite considerable public investment outlays in industry, by the mid-1960s only 11% of the total labor force had been attracted into the manufacturing sector.

Economic diversification was a similar priority in Syria in the 1960s and early 1970s. By the mid-1960s manufacturing accounted for nearly 17% of the labor force. Much of this employment was in the public sector following a wave of nationalization measures. The structure of the Syrian industrial sector closely paralleled that of other economies striving for import substitution (Sayigh 1978; World Bank 1981b). Apart from agricultural processing enterprises, the main manufacturing sub-sector was textiles. By 1970 food processing and textiles accounted for around 70% of output and value added in manufacturing, and were dominated by public sector enterprises (Meyer 1987; CBS 1986). However, alongside the relatively heavily capitalized public sector units there was a substantial private sector, organized around smaller, more labor-intensive production particularly in textiles, metal working, and wood and furniture manufacture.

Nevertheless, by the early 1970s most of the population was still living and working in rural areas. During 1966–71 agriculture accounted for nearly two thirds of exports and employment in Syria. However, emigration from rural areas began to have an effect in the 1960s due to significant income disparities between urban and rural areas, adverse climatic conditions in the drier regions, stagnant real wage levels in agriculture, and the growing availability of public services and, to a lesser extent, employment in the urban coastal areas.

In other non-oil economies, a similar pattern of continuing agricultural predominance but growing rural-urban migration prevailed. With the exception of the irrigated zones, such as the Gezira scheme in Sudan and the Moroccan and Syrian irrigation schemes, most agricultural regions of the countries surveyed had high output variance, largely due to climatic conditions but also associated with generally low levels of productivity and substantial variation in seasonal employment. High underemployment in rainfed agricultural zones helped to generate a strong drift of labor away from these areas and towards the towns where resource availability and utilization were higher.

External migration had also begun prior to the oil boom. Although the number of Tunisian migrant workers was no more than 30,000 in the mid-1960s, by 1972 there were over 200,000 with more than half this outflow directed toward the French labor market and a further third toward Libya. Most of these workers were aged between 20 and 29 years. In Jordan, sharply rising migration to other Arab states preceded the 1973 oil price rise, creating a tight domestic labor market and a shift of investment towards more capital-intensive activity (Seccombe 1981; Birks and Sinclair 1980b).

Post-1973

The impact of the regional boom that followed the oil price increases of the early 1970s can be traced in a number of ways. In particular, the effects on consumption and employment were profound. This was true in Sudan and Tunisia, even though the direct employment effects were widely different in terms of regional migratory labor flows. At the simplest level, oil revenues were translated into higher levels of direct assistance from the OPEC countries, primarily Arab countries (Table 4) (Underwood 1977). While other Arab countries were the principal beneficiaries, accounting for around 80% of total OPEC aid during 1974–83, the so-called front-line states, Jordan, Egypt, and Syria, took over half the aggregate disbursements. Indeed, prior to the Camp David agreement, Egypt alone accounted for a little under 30% of total OPEC aid.

The overall impact of development assistance, including that from Arab OPEC members, has been varied but for a number of countries the scale of such assistance has been large. In the case of Jordan, official development assistance between 1979 and 1983 amounted to around 38% of GDP with OPEC assistance comprising about 32% of GDP. This is an extreme case, but in Yemen AR and PDR Yemen aid averaged between 12 and 16.5% of GDP, for Syria 10%, and Sudan 9.5%. For all these countries, excluding Sudan, Arab aid accounted for a substantial share of total assistance.

In general, OPEC transfers were overwhelmingly directed toward non-agricultural investments. For the period 1977–80, it appears that of the resources delivered by the three largest OPEC bilateral agencies, between 11 and 19% were directed to agriculture (Shaw 1983). However, multilateral

Table 4. Net bilateral disbursements of OPEC aid, 1974–83.

Recipient country	1974–78		1979–83	
	Million USD	Share (%)	Million USD	Share (%)
Arab countries	13,690	76	17,917	86
Jordan	1,275	7	4,405	21
Syria	2,647	15	6,330	30
Lebanon	322	2	710	3
Yemen AR	786	4	1,099	5
PDR Yemen	304	2	210	1
Morocco	310	2	1,417	7
Tunisia	130	1	186	1
Sudan	623	3	1,131	5
Turkey	30	–	595	3
Total allocated	18,032	100	20,826	100
Total unallocated	5,657	–	11,359	–
Total OPEC aid	23,689	–	32,185	–

Source: OECD, Development Cooperation, Paris 1974–1985.

bodies such as BADEA and IFAD provided assistance to the agricultural sector.

Apart from direct transfers, the principal method for redistributing oil surpluses within the region came through regional migration and remittance flows. While out-migration had commenced in several countries prior to 1973, it was primarily directed towards European labor markets, in particular France and West Germany. Out-migration and its regional displacement accelerated after the first oil price shock. In a number of cases this involved a more complex set of labor flows than merely from oil-poor to oil-rich economies. In Jordan, political, strategic, and economic considerations ensured not only high levels of foreign aid from other Arab countries but also engendered a strong labor demand which, in turn, required inflows of labor from other oil-poor economies in the region (World Bank 1978b). Thus, by the 1980s Jordan was characterized by substantial labor migration to the Gulf, with that labor being replaced mostly by Egyptian labor. Jordanian labor migrants were largely skilled and, including dependents, it has been estimated that by 1984 there were between 800,000 and 1 million Jordanians living and working abroad. In the same year there were 130,000 or more foreign workers employed largely in unskilled work in the primary and construction sectors. Figures on inflows and outflows of earnings suggest that in 1984 net remittances amounted to 338 million JOD, the largest single source of foreign exchange that year.

In Jordan, as elsewhere, the movement of labor across national boundaries

Table 5. Migrant workers in the Arab Middle East, c. 1975.

To \ From	Yemen AR	Jordan	PDR Yemen	Syria	Lebanon	Sudan	Tunisia	Morocco/ Algeria	Turkey
Saudi Arabia	280,400	175,000	55,000	15,000	20,000	35,000	–	–	500
Libya	–	14,150	–	13,000	5,700	7,000	38,500	2,500	9,000
UAE	4,500	4,500	4,500	4,500	1,500	–	14,000	–	–
Kuwait	2,757	47,653	8,658	16,547	7,232	873	49	47	37
Oman	100	1,600	100	400	1,100	500	100	–	–
Iraq	–	5,000	–	–	3,000	200	–	–	–
Qatar	1,250	6,000	1,250	750	500	400	–	–	–
Jordan	–	–	–	20,000	7,500	–	–	–	–
Bahrain	1,121	614	1,122	68	129	400	–	–	–

Source: Birks and Sinclair (1980a).

was determined primarily by the rapid increase in labor demand in the oil-exporting countries. Regionally, both skilled and unskilled labor was involved and, despite the later growth in migration from the Indian sub-continent and southeast Asia, a substantial share of total labor demand was met from within the region. This movement out of the oil-importing economies led to tighter labor markets, particularly for skilled workers, and to skill mismatching and shortages, particularly in Syria, Jordan, and Egypt. However, for much of the 1970s labor market tightening reflected moves towards fuller employment and hence to higher real wages, particularly in urban areas (Sherbiny 1981; Swanson 1979; Paine 1974). In turn, this was dynamically associated with an investment path skewed towards relative capital intensity, a strategy that has had some undesirable long run consequences.

Mapping migration in the region has proved difficult due to poor official statistics but Table 5 provides some estimates for the mid-1970s. The volume of labor exports rose significantly between 1975 and 1980. For Jordan and Yemen AR it appears that over 40 and 24%, respectively, of the potential domestic labor force was working abroad in that period (Serageldin et al. 1983).

For these two countries, remittances were the major source of foreign exchange. Remittances likewise comprised a significant share of such earnings in Sudan and Morocco. In Egypt, remittances grew by over 30% per annum in real terms between 1974 and 1981 and came to be a major factor in the transformation of Egypt's resource base, alongside the hydrocarbons sector, Suez Canal earnings, and foreign capital inflows (Commander 1987). The importance of remittance income can be gauged from Table 6.

However, the migratory flows on which such income transfers were based led to labor market disequilibria in the sending countries. This was particularly pronounced in Morocco, Tunisia, and Syria. Similarly, in Yemen AR and Sudan, despite the fact that most migrants were unskilled, a significant share of the skilled and professional manpower migrated to the Gulf states (Ferghany 1980; Richards and Martin 1983). This represented a heavy social cost to the sending countries in foregone manpower for which they had borne the effective formation costs.

Table 6. Remittances in labor-exporting economies, 1975–77.

Country	Value of remittances 1977 (USD)	Remittances as share of total value of	
		Exports (1977)	Imports (1977)
Jordan	419	62.3	26.0
Morocco	533	29.0	13.0
Syria	92	6.6	3.2
Tunisia	145	10.7	7.3
Turkey	943	40.3	14.9
Yemen AR	914	801.8	103.6
PDR Yemen	179	71.9	43.3

Source: Shaw (1983).

Furthermore, the withdrawal of manpower led to selective shortages in domestic labor markets and consequently had a direct impact on the wage rate. This effect has been most clear in Egypt after the mid-1970s (Commander 1987) but has also occurred in Yemen AR, Sudan, and Jordan. In Yemen AR, for example, by the early 1980s as much as 17% of the rural labor force was working abroad. This accentuated peak season labor scarcity despite substantial substitution by female labor (Morton 1981). Remittance income was estimated to comprise around a third of average rural income (World Bank 1986a; 1986c) and has been associated with upward pressure on wage rates and an apparent raising of the reservation wage. This has been the case in both agriculture and urban areas even though, with current technology and potential labor availability, there is no serious labor shortage. Nevertheless, high male wage rates relative to average and marginal productivities of labor have inhibited the development of commercial agricultural enclaves. Evidence from a number of other economies suggests that labor exporting has a strong upward impact on domestic wage levels, particularly when the outflow of labor has largely been skilled or semi-skilled. When associated with an increase in the monetary base, this has had obvious inflationary implications.

Reliance on exogenous resource transfers had further unintended and undesirable consequences. Perhaps most importantly, fluctuations in the size of such transfers has led to current account problems and undermined public investment programs. This has evidently been the case in Jordan since 1984 following major discrepancies between aid pledges and actual disbursements. Likewise, the trend in remittance earnings has been relatively volatile and subject to decline in real terms since 1983/84. This can be explained by the impact of the oil price fall on the demand for labor in the Gulf states and other labor importers, except Iraq. Official remittances to the Syrian economy have declined consistently, from around 900 million USD in 1979 to 460 million USD in 1983, and 325 million USD in 1984 (CBS 1986). In Yemen AR remittances, which averaged around 1 billion USD in the early 1980s, fell to 867 million USD by 1985 and 600 million USD in 1986.

Table 7. Oil-importing countries: Capital expenditure on agriculture, 1976–81.

Country	Share of capital expenditure to agriculture	Average annual expenditure (USD) per		Ratio N/A
		Agric. worker (A)	Non-agric. worker (N)	
Jordan	15.0	657	1,455	2.2
Morocco	19.4	99	463	4.7
Sudan	31.2	67	553	8.3
Syria	23.3	598	1,885	3.2
Tunisia	11.6	318	1,827	5.7
Yemen AR	15.8	73	1,243	17.0
PDR Yemen	27.7	47	202	4.3

Source: Shaw (1983).

Migration and the Domestic Labor Markets

Urban-rural income disparities remained pronounced and in some cases widened as a direct consequence of skewed investment expenditures. Table 7 provides some information on the capital budgets for the oil-importing countries. Given the average share of the labor force remaining in agriculture, the actual sectoral share of total investment remained small. It appears that neither private income transfers, such as remittances, nor the increased public resources available to governments from aid flows and resource movements were invested in agriculture. Private remittances appear, in general, to have been directed to a very limited extent to productive investment, being biased toward consumption and commercial sector activity. By the mid-1970s rural per capita incomes were less than a quarter of the urban level in Yemen AR, in Tunisia around a third, and in Morocco below 45% (Shaw 1983).

The factors stimulating migration away from rural areas can largely be explained in relation to income, employment, and the availability of public services. In Turkey, urbanization began to increase in the 1950s so that by the mid-1960s around 30% of the population was classified as urban. At the start of the 1980s, this share had risen to around 45%, suggesting that during the 1970s rural areas lost about half their natural rate of increase to migration. This phenomenon cannot be explained by simple income effects. The supply of urban jobs undoubtedly increased, particularly with the growth in the large parastatal sector, while the supply of jobs in rural areas, where seasonal unemployment had always been pronounced, significantly lagged demand. At its peak in the early 1970s, the agricultural labor surplus was around 9–10% of the agricultural labor force but by the early 1980s this had declined to around 7.5% (World Bank 1978a). But these figures say nothing about underemployment: there was very little development of non-agricultural employment in rural areas with barely 12% of rural sector workers being employed in non-farm work. The options for seasonal labor smoothing have thus been limited.

Very similar processes can be detected in the other oil-importing countries, but the level of out-migration has been determined by a variety of factors. Where non-farm employment in rural areas has been enhanced, the migratory flow has been less. Where the labor absorption capacity of both formal and informal sectors in urban areas has been limited as in Sudan, the ability to sustain large-scale migration has also been reduced.

Yet migration needs to be viewed dynamically. With natural rates of population growth of 2–3% annually, and *ex ante* labor surpluses in the agricultural sector, the rate of out-migration would be strongly associated with investment in the sector. This can be narrowly defined in terms of the capital budget for agriculture, but public investment in infrastructure and institutions, such as extension and research systems, should also be considered. This is particularly relevant in rainfed regions where low cropping intensity and productivity limit potential employment. At the same time, institutional factors, such as the degree of inequality in landholding and the average size of farm households, play a major role in labor demand. The land reforms that were passed pre-1973 only partially restructured the framework for labor demand and the outcome of these measures has been strongly colored by the continuing divide between irrigated and rainfed areas. Therefore, with respect to migration and agriculture generally, the main issues remain the relative capabilities of sectors to generate particular welfare levels and the degree to which those capabilities are enhanced or prejudiced by discretionary policy measures.

Exchange Rate and Domestic Pricing Policy

In addition to investment policies, both exchange rate and domestic pricing policies also had major implications for the labor market. Syria, Sudan, and Jordan, like other oil-importing economies, experienced significant exchange rate appreciation post-1973 (Figure 3) which can be attributed to several factors. Firstly, the effect of increased exogenous resources through remittances, aid, and other transfers was comparable to a conventional windfall. Secondly, a number of governments, such as Sudan, maintained fixed exchange rates, which were inadequate to cope with the rapid changes in global and relative prices following the first oil price rise. In Sudan during 1974–78 domestic inflation was over 15% per annum, and with world inflation at only 7%, the purchasing power parity of the currency fell substantially, resulting in an overvaluation of approximately 30% by 1978.

Assuming no shift in the production frontier, overvaluation allows greater

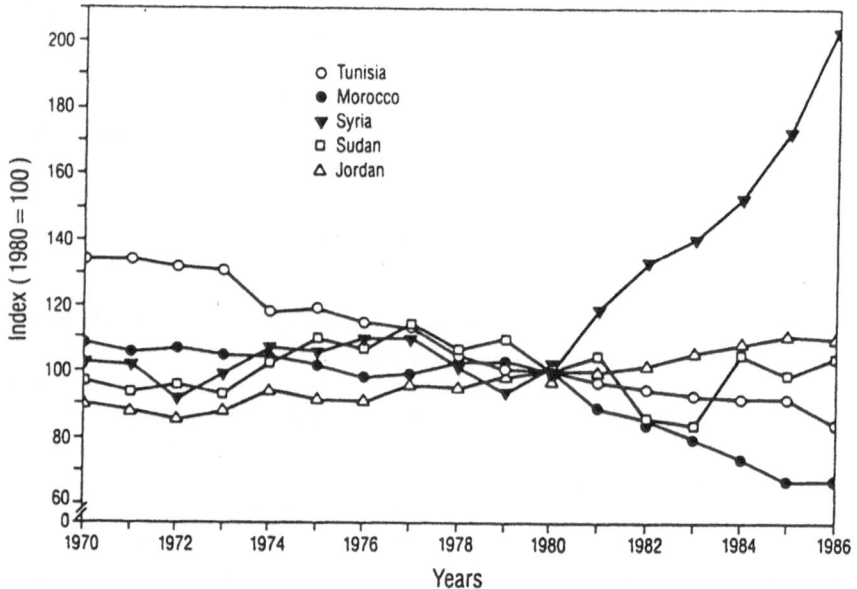

Fig. 3. Real exchange rates of oil-importing countries, 1970–86. *Source*: IMF data.

consumption of imports, which adversely affects domestic output. As agriculture tends to be dominated by tradeables output and an important source of foreign exchange earnings, an appreciating real exchange rate would act prejudicially against the sector. Overvaluation can only be sustained to the extent that current account deficits can be financed. Access to major transfer payments, where the size of those transfers is effectively dissociated from the level of activity in the 'real' economy, can allow for a level of consumption which is not warranted by the underlying capacity of the economy. In several of the oil-importing economies discussed here, this allowed relatively large food imports to be sustained, reducing incentives for domestic producers. At the same time, export competitiveness tended to be compromised.

The response to overvaluation has varied between countries. In Turkey during the 1970s, adjustments to the exchange rate were irregular and generally major. After 1980, the government moved toward a managed float with daily adjustments. This has improved the competitiveness of traditional exports in some cases, but the outcome has been ambiguous given the range of domestic pricing interventions made by the government (Balassa 1982).

Evidence from the Sudan further suggests that abrupt devaluations may not actually engender a sustained real depreciation and, therefore, the spread of gains that should in theory derive from a devaluation (World Bank 1979). The Sudanese case is particularly interesting in that large (15 and 25%) devaluations in 1972 and 1978 sought to increase external competitiveness and boost output in agriculture. By raising the return on investment in agriculture the devaluation measures could have been expected to stimulate employment in the export sub-sectors (Nashashibi 1980). However, with relatively low supply elasticities over the short run (Bond 1983) and inelastic demand functions for imports, the 1978 devaluation in Sudan had only a weak overall impact on external competitiveness. Other constraints, such as labor shortages in peak agricultural seasons, low productivity, infrastructural bottlenecks, and a chronic shortage of foreign exchange, restricted the ability of the economy to respond to the devaluation. Moreover, Hussain and Thirlwall (1984) show that, with competitiveness defined as the ratio of foreign exchange earnings to each unit of domestic resources used, in a regime of low supply elasticities and high elasticity of foreign and domestic prices to devaluation, the latter may lead to actual losses of foreign exchange earnings. Amongst other effects, import prices would rise more rapidly than aggregate and individual export prices.

A further factor inhibiting output and employment growth in the tradeables sector, particularly in Sudan, is the use of administered prices and subsidies (ILO 1976a). Between 1972 and 1981, producer prices for Sudan's principal export crops, cotton, groundnuts, and sesame, ranged between 40 and 86% on average of their border price levels. Producer prices for wheat and sorghum, on the other hand, were 115 and 114%, respectively, of border prices (World Bank 1983b). The net employment implications of this have depended on the extent to which the effective protection given to home producers of wheat, sorghum, and millet has engendered a countervailing response. In the Sudanese case, labor demand in the tradeables sector has exceeded demand for these goods, hence probably causing an overall decline in labor demand within the sector. For both tradeables and non-tradeables producers, this demand has increasingly been satisfied by internal migration. Informational and other market imperfections have led to the incorrect perception that sections of Sudanese agriculture are marked by labor scarcity, resulting in the adoption of labor-substituting policies for the sector, principally through mechanization.

Policies which discriminate against agriculture, primarily through high implicit taxation rates via low administered prices, have reduced real agricultural incomes and consequently stimulated the outflow of labor from agriculture in Turkey, Yemen AR, Morocco, and Tunisia. In contrast, incentive pricing for cotton producers in Syria, combined with continuing substantial public investments in the sector, has stabilized agriculture's share of GDP during 1970–84, although the sector's share of total employment has fallen from 51% of the labor force in 1970 to 39% in 1977 and 25% in 1984 (Meyer 1987). This indicates that the Syrian agricultural sector has responded differentially with most growth being concentrated in the irrigated sector. Low productivity and out-migration have still remained the predominant features of the rainfed zones. Favorable price regimes will not provide sufficient conditions for employment stability, especially in the presence of regional labor markets.

Conclusion

Since the early 1970s, the economies discussed in this paper have, in their various ways, undergone substantive change as a consequence of a massive natural resources boom. The oil producers have experienced the income and consumption implications directly, while non-oil-producing economies have been affected through remittance, income, aid, and other transfers. This has allowed rapid growth in the share of exogenous resources in the economies as well as shifts in resource allocation, and has commonly led to domestic market disequilibria.

Concerning the labor market, the primary characteristic has been the stimulation of domestic migration and the development of a regional labor market. Combined with falling real capital allocations to agriculture, this has tended to be associated with a growth in service sector employment and, almost invariably, in public employment. The ability to sustain these changes has been compromised by the fall in the price of oil. Public investment programs have been cut and the major infrastructural and other projects commenced in the 1970s have been either already completed or cut. Demand for construction labor has been a reasonable barometer of this shift.

A general and alarming consequence of a downturn in the business cycle has been the emergence of higher levels of unemployment, particularly in urban areas. By 1985, over 16% of the labor force in Turkey was unemployed, while in Tunisia estimates for 1983 suggest that a quarter of the labor force was unemployed, rising to nearly 40% for 18–26 year old males. Job creation under the 1982–86 Tunisian Development Plan generated 30% less employment than projected, with current employment growth satisfying just over 60% of the minimum required to stabilize unemployment at 1983 levels (World Bank 1985). In Jordan, as in Tunisia and Morocco, diminishing foreign demand for labor from both European and Arab economies has reduced the flow of remittances. For Jordan, the official unemployment rate doubled between 1983 and 1986 to around 6–7%. This has in turn reduced the non-Jordanian labor force, but the substitution possibilities have been constrained by skill mismatching and the relatively high reservation wage for nationals.

For the oil-exporting countries, unemployment has likewise become a potent force. In Iran, political instability led to rapidly growing levels of unemployment in urban areas in the late 1970s. By 1982/83 official figures indicated that nearly 19% of the workforce lacked work. Moreover, the underlying trend is higher. Major over-staffing exists in the public sector and employment figures are obviously disturbed by the military demand for labor. Cessation of hostilities with Iraq would undoubtedly reveal far higher levels of unemployment than are already acknowledged. Though far less extreme, unemployment in the other oil-exporting countries has also grown. In Libya the response has primarily been to eject foreign labor and substantial expulsions of Tunisian and Egyptian labor have occurred since 1984/85. In Algeria, in the mid-1970s around 8% of the labor force lacked work; by the early 1980s this figure approached 12%. Thus, falling oil revenues tended to leave disequilibrated economic structures where a combination of boom year policies and the maintenance of inappropriate economic policies exaggerated, rather than dampened, the scale of domestic imbalances.

References

Askari, H. and Cummings, J.T. 1976. Middle East Economies in the 1970s. New York: Praeger.
Balassa, B. 1982. Turkey: Industrialisation and Trade Strategy. Washington, D.C.: World Bank.
Bennamane, A. 1980. The Algerian Development Strategy and Employment Policy. Monograph No. 9. Swansea: Centre for Development Studies, University of Swansea.
Birks, J.S. and Sinclair, C.A. 1980a. International Migration and Development in the Arab Region. Geneva: International Labor Office.
Birks, J.S. and Sinclair, C.A. 1980b. Arab Manpower, The Crisis of Development. London: Croom Helm.
Bond, M.E. 1983. Agricultural Responses to Prices in Sub-Saharan African Countries. IMF Staff Papers 30, 4.
Bruno, M. and Sachs, J. 1982. Energy and Resource Allocation: A Dynamic Model of the Dutch Disease. Review of Economic Studies 49: 167–181.
CBS (Central Bureau of Statistics). 1986. Statistical Abstract. Damascus: CBS.
Commander, S.J. 1987. The State and Agricultural Development in Egypt Since 1973. London: Ithaca/ODI.
Edens, D. 1979. Oil and Development. London.
El-Fathaly, O.I. and Palmer, M. 1985. Political Development and Social Change in Libya. Lexington.
Ferghany, N. 1980. The Affluent Years are Over – Emigration and Development in the Yemen Arab Republic. Geneva: ILO. Mimeo.
Gelb, A.H. 1986. Adjustment to Windfall Gains. A Comparative Analysis of Oil-Exporting Countries. In Natural Resources and the Macro Economy (Neary, J.P. and Wijnbergen, J.P., eds). London: Centre for Economic Policy Research.
Grissa, A. 1973. Agricultural Policies and Employment – A Case Study of Tunisia. Development Centre Studies, OECD.
Hussain, M.N. and Thirlwall, A.P. 1984. The IMF Supply-Side Approach to Devaluation: An Assessment with Reference to Sudan. Oxford Bulletin of Economics and Statistics 46 (2): 145–167.
ILO (International Labor Office). 1976a. Growth, Employment and Equity: A Comprehensive Strategy for the Sudan. Geneva: ILO.
ILO (International Labor Office). 1976b. Manpower and Employment in Arab Countries – Some Critical Issues. Geneva: ILO.
IMF (International Monetary Fund). 1987. World Economic Outlook. Washington, D.C.: IMF.
Katouzian, H. 1983. The Agrarian Question in Iraq. Pages 309–357 in Agrarian Reform in Contemporary Developing Countries (Ghose, A.K., ed.). London: Croom Helm.
Meyer, G. 1987. Economic Development in Syria Since 1970. in Politics and the Economy of Syria (Allen, J.A., ed.). London: Centre of Near and Middle Eastern Studies, School of Oriental and African Studies, University of London.
Morton, J. 1981. Emigration in the Yemen Arab Republic. SOAS Development Seminar Working Paper No. 42. University of London.
Nashashibi, K. 1980. A Supply Framework for Exchange Reform in Debt Countries: The Experience of Sudan. IMF Staff Papers 27, 1.
Neary, J.P. and Wijnbergen, J.P. (eds). 1986. Natural Resources and the Macro Economy. Oxford: Basil Blackwell.

46

Nelson, H.D. (ed.). 1978. Morocco: A Country Study. Washington, D.C.

Nelson, H.D. (ed.). 1979. Algeria: A Country Study. Washington, D.C.

Niblock, T. (ed.). 1982. Iraq: The Contemporary State. Exeter: Centre for Arab Gulf Studies, University of Exeter.

Paine, S. 1974. Exporting Workers: The Turkish Case. Cambridge: Cambridge University Press.

Penrose, E. and Penrose, E.F. 1978. Iraq: International Relations and National Development. Boulder: Westview Press.

Pridham, B.R. (ed.). 1982. Economy, Society and Culture in Contemporary Yemen. Exeter: Centre for Arab Gulf Studies, University of Exeter.

Rachidzadeh, E. 1978. Le Secteur Rural et le Developpement Economique: Le Cas de l'Iran. Geneva: Institute Universitaire de Hautes Etudes Internationales, Universite de Geneve.

Richards, A. and Martin, P.L. 1983a. The Laissez Faire Approach to International Labor Migration: The Case of the Arab Middle East. Economic Development and Cultural Change 31(3): 455–474.

Richards, A. and Martin, P.L. (eds). 1983b. Migration, Mechanization and Agricultural Labour Markets in Egypt. Boulder: Westview Press.

Sayigh, Y. 1978. The Determinants of Arab Economic Development. London: RKP.

Seccombe, I. 1981. Manpower and Migration: The Effects of International Labour Migration on Agricultural Development in the East Jordan Valley, 1973–80. Durham: Centre for Middle Eastern and Islamic Studies, University of Durham.

Seddon, D. 1987. Structural Adjustment and Agriculture: Morocco in the 1980s. London: Overseas Development Institute. Mimeo.

Serageldin, I., Socknat, J.A., Birks, S., Li, B. and Sinclair, C.A. 1983. Manpower and International Labour Migration in the Middle East and North Africa. Oxford: Oxford University Press.

Shaw, R.P. 1983. Mobilising Human Resources in the Arab World. London: KPI.

Sherbiny, N.A. 1981. Labour and Capital Flows in the Arab World: A Critical Review. The Industrial Bank of Kuwait. Mimeo.

Stevens, P. 1986. The Impact of Oil on the Role of the State in Economic Development – A Case Study of the Arab World. Surrey: Energy Economics Centre, University of Surrey.

Swanson, J.C. 1979. Emigration and Economic Development: The Case of the Yemen Arab Republic. Boulder: Westview Press.

Trebous, M. 1970. Migration and Development – The Case of Algeria. Development Centre Studies, OECD.

Underwood, A.M. 1977. Inter-Arab Financial Flows. Durham:.Centre for Middle Eastern and Islamic Studies, University of Durham.

World Bank. 1973. Memorandum on Current Economic Development in Iran. WB Report No. 49a-IRN. Washington, D.C.: World Bank.

World Bank. 1974a. Current Economic Position and Prospects of Iraq. WB Report No. 419a-IRQ. Washington, D.C.: World Bank.

World Bank. 1974b. Current Economic Position and Prospects of Morocco. WB Report No. 329-MOR. Washington, D.C.: World Bank.

World Bank. 1976. Non-Farm Activities in Rural Areas and Towns: The Lessons and Experiences of Iran. Studies in Employment and Rural Development No. 31. Washington, D.C.: World Bank.

World Bank. 1978a. Turkey – Agricultural Sector Survey. WB Report No. 1684–TUR. Washington, D.C.: World Bank.

World Bank. 1978b. Country Economic Memorandum on Jordan. WB Report 1738–JO. Washington, D.C.: World Bank.

World Bank. 1979. Sudan – Agricultural Sector Survey. WB Report No. 1836a-SU. Washington, D.C.: World Bank.

World Bank. 1981a. Manpower Development in the YAR. WB Report No. 3181a-YAR. Washington, D.C.: World Bank.

World Bank. 1981b. Syria – Country Economic Memorandum. WB Report No. 3303–SYR. Washington, D.C.: World Bank.

World Bank. 1982a. Algeria – The Five Year Development Plan and the Medium Term Prospects for 1980–84. WB Report No. 3668–AL. Washington, D.C.: World Bank.

World Bank. 1982b. Tunisia – Agricultural Sector Survey. WB Report No. 3876–TUN. Washington, D.C.: World Bank.

World Bank. 1983a. Morocco – Priorities for Public Sector Investment (1981–85). WB Report No. 4156–MOR. Washington, D.C.: World Bank.

World Bank. 1983b. Sudan – Pricing Policies and Structural Balances. WB Report No. 4528a–SU. Washington, D.C.: World Bank.

World Bank. 1985. Tunisia – Country Economic Memorandum: Mid-term Review of the Sixth Development Plan (1982–86). WB Report No. 5328–TUN. Washington, D.C.: World Bank.

World Bank. 1986a. Yemen Arab Republic – Current Position and Positions: Country Economic Memorandum. WB Report No. 5621–YAR. Washington, D.C.: World Bank.

World Bank. 1986b. World Development Report. Washington, D.C.: World Bank.

World Bank. 1986c. Yemen Arab Republic – Agricultural Strategy Paper. WB Report No. 5574–YAR. Washington, D.C.: World Bank.

3. Mechanization, Off-Farm Employment and Agriculture

ALAN RICHARDS
Department of Economics, University of California, Santa Cruz, CA 95064, U.S.A.

and

AHMED RAMEZANI
Department of Agricultural and Resource Economics, 207 Gianni Hall, University of California, Berkely, CA 94720, U.S.A.

Introduction

Mechanization of agriculture has proceeded more rapidly in the Middle East and North Africa than in any of the other major areas of the developing world during the past 15 years (Binswanger 1986). Such technological change has been the response to both market forces and government policies. This paper focuses on the impact of labor market changes in inducing this form of change.

Labor shortages are commonly thought to be widespread in agriculture in the region and the principal cause of farm mechanization. It is true that mechanization is usually capital using and labor saving. It is often argued that the growth of incomes from the oil boom (roughly, 1974–82), which was diffused throughout the region by migration, has made capital more abundant in most national economies, even in rural areas. At the same time, it is commonly argued that the withdrawal of labor from the countryside reduced the supply of agricultural labor. Accordingly, farmers and their governments responded by resorting to more mechanized farming techniques.

In this scenario mechanization is a response to a labor shortage. Farmers mechanize because they cannot find the large numbers of workers they would need if traditional techniques were still used. Since capital is increasingly abundant, this view explains most mechanization as a response to changes in the relative prices of capital and labor. However, the available evidence from both macro data and various micro-level studies suggests that the mechanization phenomenon in the region is considerably more complex than this.

The Pattern of Farm Mechanization in the Region

Mechanization is spreading rapidly in the Middle East and North Africa (Tables 1–3). Much, if not most, of this mechanization has taken the form

Dennis Tully (ed.), Labor and Rainfed Agriculture in West Asia and North Africa, 49–65.
© 1990 *ICARDA.*

Table 1. Ratio of arable land per tractor (ha/tractor).

Country	1960	1965	1970	1975	1980	1985
Algeria	248	226	148	171	171	143
Libya	885	776	694	155	148	76
Morocco	648	633	611	376	327	278
Sudan	3,029	1,765	1,267	1,387	1,129	732
Tunisia	373	349	206	201	176	180
Iran	1,359	984	781	404	236	153
Iraq	1,028	537	366	268	245	179
Jordan	882	635	477	365	305	274
Saudi Arabia	2,854	1,873	1,423	1,387	931	722
Syria	1,064	803	647	367	206	149
Turkey	536	343	260	115	65	49
Yemen AR	–	–	2,980	1,334	675	628
PDR Yemen	272	171	142	150	124	152
Egypt	219	–	156	124	63	54

Source: Calculated from FAO (1961a; 1966a; 1971a; 1976a; 1981a; 1986a).

Table 2. Ratio of arable land per harvester (ha/machine).

Country	1960	1965	1970	1975	1980	1985	Average
Algeria	1,409	1,356	1,163	1,846	1,821	1,503	1,516
Morocco	1,710	2,213	3,001	2,843	2,501	2,636	2,484
Sudan	51,500	27,173	18,800	13,511	10,797	10,639	22,070
Tunisia	1,615	1,518	1,451	2,074	1,869	1,838	1,727
Iran	15,358	11,907	9,547	7,454	5,713	5,113	9,182
Iraq	3,757	1,908	1,287	1,088	1,434	1,981	1,909
Jordan	22,634	13,684	9,460	7,184	5,750	4,660	10,562
Saudi Arabia	6,714	6,082	5,487	4,007	2,792	2,181	4,544
Syria	4,744	4,403	4,190	3,011	2,410	2,014	3,462
Turkey	4,133	3,578	3,244	2,405	2,083	2,030	2,912
PDR Yemen	52,500	30,750	24,833	11,833	11,142	11,133	23,698

Source: As in Table 1. Libya and Yemen AR unavailable.

Table 3. Growth rates of tractor park and combine harvesters in selected countries, 1960–85.

Country	Tractors (%/year)	Combines (%/year)
Algeria	2.5	0.1
Jordan	5.4	7.0
Iran	8.6	4.2
Iraq	7.5	3.1
Morocco	4.0	(<0)
Sudan	8.5	9.1
Syria	7.3	2.9
Tunisia	3.2	(<0)
Turkey	9.8	3.1
Yemen AR	9.81[1]	n.a
PDR Yemen	4.2	8.1

Source: As in Table 1.
[1]1970–85.

of tractorization. Although tractors are highly versatile, they are mainly used for primary tillage, transport, and power for irrigation pumps and portable threshers. The use of combine harvesters has spread in some countries, but they are less common than tractors. More sophisticated machines such as seed drills, fertilizer applicators, and fruit and vegetable harvesters are much less prominent.

This is consistent with the worldwide pattern of farm mechanization. Binswanger (1986) has suggested a very useful typology of farm mechanization by distinguishing between "power-intensive" and "control-intensive" operations. The former include such tasks as water lifting (in irrigated areas), primary tillage, threshing, grinding, milling, crushing, and transport. These tedious tasks require much physical energy, and farmers have used work animals for them for many centuries. In contrast, control-intensive operations "require primarily the control functions of the human mind or judgement" (Binswanger 1986). Such operations include sifting, winnowing, pest control, and the harvesting of cotton, fruits, and vegetables. Grain harvesting and secondary tillage and interculture are intermediate operations. Binswanger argues that the typical pattern of diffusion of mechanization follows the sequence, power-intensive tasks, intermediate jobs, and control-intensive operations. Although the first stage of mechanization may save some human labor, labor scarcity is not the key to the process; the critical economic consideration is the relative cost of machine and animal power. But as mechanization proceeds into the second and third phase, labor saving becomes increasingly important.

In many countries of the Middle East and North Africa, mechanization of power-intensive operations has proceeded quite far, and nearly all arable rainfed land in Mediterranean countries is now plowed by tractors. For example, in the Settat region of Morocco, over 80% of the land is tilled at least once per year by tractors (USAID 1986). On large commercial sorghum farms in the heavy clay belt of central Sudan, also, all plowing is mechanized. In Turkey the ratio of tractors to land area in 1984/85 was 49 ha per tractor, comparable with the 40 ha per tractor in the United States (FAO 1986a). Some experts believe that Turkey has too many tractors per hectare, resulting in under utilization of existing tractors (World Bank 1983).

The use of combine harvesters is spreading more slowly. The rate of growth of tractor numbers has exceeded that of combines except in Sudan, PDR Yemen, and Jordan. In the first two cases, the growth rates are somewhat deceptive because the starting base point is so low. As we would expect, the diffusion of combine harvesters is furthest advanced in countries whose agricultural labor force has actually declined, such as Algeria, Jordan, and Iraq, or is relatively stable, e.g., Turkey and Syria (Tables 2 and 4). However, the prevalence of combine harvesters also seems to reflect the difficulties which large commercial farmers have in supervising large numbers of hand harvesters. For example, combine harvesters have recently spread rapidly on large farms in some regions of Morocco, despite stagnant or falling real

Table 4. Estimates and projections of the agricultural labor force (×1,000) in selected countries of the Near East region, 1960–2000.

Country	1960	1970	1980	1985	1990	2000
Main oil-exporting						
Algeria	1,990	1,394	1,262	1,301	1,342	1,387
Iran	3,220	3,547	4,026	4,082	4,199	4,331
Iraq	972	1,125	1,081	1,043	1,049	1,073
Libya	200	150	137	131	127	118
Oman	97	102	140	163	163	165
Saudi Arabia	817	1,019	1,333	1,490	1,599	1,729
Sub total	7,296	7,337	7,979	8,210	8,479	8,803
Main non-oil-exporting						
Egypt	4,364	4,765	5,158	5,526	5,902	6,786
Jordan	199	162	66	63	59	51
Lebanon	199	130	106	90	86	69
Morocco	2,195	2,333	2,594	2,746	2,860	2,950
Sudan	3,376	3,601	4,331	4,606	4,866	5,348
Syria	685	785	707	713	747	848
Tunisia	663	559	668	648	630	549
Turkey	10,991	11,361	11,146	11,385	11,418	11,335
Yemen AR	973	1,024	1,013	1,103	1,226	1,561
PDR Yemen	201	208	199	203	206	208
Sub total	23,846	24,928	25,988	27,083	28,000	29,705
Total	31,142	32,265	33,967	35,293	36,479	38,508

Source: FAO (1986b).

agricultural wages. Some sources assert that mechanical harvesting techniques offer considerable cost advantages over hand techniques. For example, USAID (1986) reports that combine harvesting in Morocco cost only 25% as much as hand harvesting. However, it is clear that mechanization of harvesting is still far less advanced than tractorization in most countries. Finally, few countries of the region have proceeded very far with the mechanization of control-intensive operations, which continue to be largely performed by hand labor.

This pattern is consistent with the argument that in power-intensive operations, machines are substituting mainly for animal power rather than for human labor (e.g., Egypt, Syria, Turkey). A major reason for this is the very rapid rate of growth in demand for livestock products which in turn is the result of both growth in per capita incomes and changes in local food preferences. From 1973 to 1980 the consumption of meat and milk in the region grew annually at 4.6%, well above the population growth rate, due to the regional preference for animal products, whose income elasticity of demand exceeds one in most countries (Sherbiny 1984). Average incomes have risen rapidly in most countries of the region in the past 15 years, and this has accelerated the growth of demand for animal products. Consequently, the opportunity cost of using animals for work has increased (working animals produce less milk and meat than animals kept in stalls), providing incentives to mechanize power-intensive operations like land preparation.

Mechanization of primary tillage in rainfed agriculture is also driven by agronomic considerations. Firstly, in many areas rainfall is necessary to soften the ground before it can be prepared using animal traction. Given the highly erratic rainfall in many parts of the region, this can delay planting and reduce yields. In Tunisia, for example, a tractor can plow 1 ha of dry upland in 2.5 hours; the same operation using two bullocks and a Mediterranean plow requires 41 hours (USAID 1982). Secondly, only tractors can plow land in the dry season, a practice which reduces weeds and improves moisture retention. Therefore, tractors have complemented the diffusion of higher-yielding varieties of bread wheats in Tunisia and Turkey. Here, too, direct labor costs are at best a secondary consideration in the diffusion of the mechanization of power-intensive operations.

This does not mean that migration and off-farm employment may not have contributed to the pattern of mechanization observed so far. Because many operations require much physical force, social customs have often allocated these tasks mainly to adult men, and it is this type of labor which is primarily affected by migration and off-farm employment. Migration and off-farm employment also contribute to mechanization by providing many rural families with additional income from remittances. When a portion of these higher incomes is spent on meat, this contributes to the incentives for other farmers to mechanize plowing, freeing their animals to produce more milk, cheese, and meat.

Of course, mechanization has also been encouraged by government policy. Most countries of the region directly subsidize tractorization, usually by offering loans at less than the economy-wide opportunity cost of capital and cheap diesel fuel. Imported tractors are either directly subsidized, as in Saudi Arabia and Libya, or are indirectly encouraged by overvalued real exchange rates. For example, in 1982 the Moroccan government began to promote farm mechanization by removing all duties and taxes on agricultural equipment and by expanding credit for such purchases. The real cost of a tractor fell by 30% and not surprisingly, the numbers of tractors rose from 23,000 in 1978 to 40,000 in 1986 (USAID 1986). These direct subsidies are complemented by the paucity of research into improved animal-drawn implements (e.g., in Turkey: World Bank 1983). These policies, combined with growing incomes, more abundant capital, and accelerating demand for animal products, have all promoted the substitution of mechanical for animal power and have contributed to the very rapid mechanization of power-intensive operations throughout the region.

In summary, farm mechanization in the region has so far been mainly of the "power-intensive" tasks. Mechanization of "control-intensive" operations has made relatively little progress, while there are substantial differences across countries in the extent of mechanization of grain harvesting.

Migration, Off-Farm Employment and Labor Shortages

Three types of migration can be distinguished; to non-farm rural markets, to national urban markets, and to foreign markets. In each case, the general cause is the considerable earnings gap between farm and off-farm work. Throughout the region, average urban incomes are at least 1.5 times rural incomes although in countries such as Sudan, the ratio approaches 3.0 (El-Ghonemy 1984). Cereal farmers are often the poorest in these countries. For example, in Tunisia in 1980, the average-sized cereal farm earned about 20% of the national average family expenditure (World Bank 1982). Wage differentials between farm work and unskilled labor in the major oil-exporting countries of the Gulf are still larger. For example, in 1980 construction workers in Saudi Arabia earned wages 13 times greater than those of farm laborers in Egypt.

Such income gaps may arise either because of changes in the farm sector, usually described as a push effect, or changes off the farm, usually called a pull effect. The usual economic explanation of such migration combines these effects into a model which holds that migrants roughly compare the benefits from migration with the costs. The benefits are the sum of expected earnings, i.e., the probability of finding a job times the wage rate, while the costs are the opportunity cost of wages and/or other remuneration in agriculture e.g., family-farm earnings, and the costs of migration itself.

One problem with this framework is that it suggests that over time migration should be associated with a convergence of wage rates in urban areas with those in agriculture. But the available evidence does not support this conjecture very strongly. Many explain the lack of wage convergence by arguing that migration is simply drawing off the additions to the rural labor force, without any net decline in farm labor supply. In addition, open unemployment in urban areas is often quite high and usually considerably greater than in rural areas. For example, unemployment in Morocco is conservatively estimated at 12%; an additional 18% of the workforce is employed for less than two thirds of the year, and the pool of unemployed is estimated to be growing at 18% per year (USAID 1986). In Tunisia an estimated 20–25% of the workforce is either unemployed or employed only part-time (World Bank 1982). Field studies suggest that it takes rural migrants some time to find an industrial job, and that many initially find jobs which are poorly paid in areas where living costs are considerably higher than the countryside.

One recent modification of the basic model has helped to explain the persistence of migration even though there is substantial unemployment in urban areas. Stark (1984) points out that migration is usually part of a family, rather than purely individual, strategy. Economists now recognize that migration and other off-farm employment are parts of a family division of labor: certain members may go to work in local industries, while others are encouraged to seek work in cities or abroad. This implies that the farm family may, in the short run, subsidize the migrant while he looks for a job, which encourages migration.

Although the composition of migrants varies from one country to another, in all countries migrants are predominantly young adult males. For example, an ILO study of Sudanese migrants found that over 90% were men and over 60% were 18–30 years old (Berar-Awad 1984). Similar patterns have been observed in Yemen AR and Jordan (World Bank 1986a; 1986b).

It is widely believed that migration has created pervasive labor shortages which constrain agricultural development and have induced large-scale mechanization in the region. However, the evidence suggests that labor shortages may be temporally and spatially localized. They are mainly shortages of adult male hired labor in seasons of peak demand for certain crops and in specific locations. Family labor is typically not in short supply, although Syria may be an exception (USDA/USAID 1980). Further, these shortages are probably temporary disequilibria rather than long-run structural problems. The very rapid rate of growth of demand for off-farm labor in the 1970s and early 1980s is unlikely to be repeated in the coming decade.

Table 4 shows the actual and projected trends in the absolute size of the agricultural labor force in selected countries of the region. For most countries, the farm workforce grew during the 1970s and is expected to continue to grow into the 1990s and beyond. These figures explicitly take into account rural to urban migration. Despite the scale of internal migration and the fact that the percentage of the workforce employed in agriculture has declined in all countries, the absolute numbers of the agricultural labor force rose in Turkey, Sudan, Iran, Saudi Arabia, PDR Yemen, and Yemen AR and declined in Jordan, Libya, Iraq, Lebanon, and Syria. Therefore, it is only in the latter set of countries that one might suspect that there are sectoral labor shortages.

The rates of growth of the labor force exceed any reasonable estimate of the rates of growth of demand for immigrant labor in the oil-exporting or industrialized countries. In 1984, before the sharp decline in oil prices, the World Bank estimated that the annual demand for immigrant labor of all types in Saudi Arabia would grow at 2.9% between 1985 and 1990 (Sherbiny 1984). During the same period, FAO estimated that the annual growth rates of the labor force in Jordan would be 4.3% and 3.1% in Yemen AR. It is unlikely that the growth of demand for labor in the oil-exporting countries will be able to absorb as large a proportion of the region's labor force in the late 1980s and 1990s as it did during the past 15 years.

Domestic demand for labor has been growing, but there is not much evidence to suggest that this growth has exceeded the rate of growth of the labor force for sustained periods. For some countries the rising level of investment needed to create a job retards the growth of employment. For example, in Morocco, the incremental capital to output ratio rose from 3 : 1 in the early 1970s to around 8 : 1 in the early 1980s (USAID 1986). High levels of protectionism and poor programming of public investment have contributed to this problem, although some current government policy shifts are intended to redress these problems. Similar difficulties have emerged in other countries. Employment creation is now severely hampered by the accumulation of external debt and by the slow growth of the international economy. In 1984, debt service payments as a percentage of exports of goods and services were 13.6% in Sudan, 26.6% in Yemen AR, 22% in PDR Yemen, 37.6% in Morocco, 31.9% in Egypt, 22.8% in Turkey, 24.4% in Tunisia, 14.8% in Jordan, 12.9% in Syria, and 33.6% in Algeria (World Bank 1987b).

However, there is considerable evidence that off-farm employment is of rapidly growing importance. In Morocco, some 40–50% of total rural employment is non-agricultural (USAID 1986), while in Algeria in 1976 only 40% of rural incomes came from agriculture, with 30% from off-farm employment and 30% from remittances from other family members in cities and abroad (Adler 1980). Some 50% of Jordanian farmers are not dependent on farming as their primary source of income (USDA 1986). In Tunisia, 50% of farm owners also had non-farm jobs; 300,000–400,000 rural people were employed in non-farm jobs, compared to a farm labor force of just over one million (World Bank 1982). In Yemen AR, nearly 40% of rural incomes came from remittances from family members working abroad (World Bank 1986b). The combination of rural, off-farm employment with remittances from family members has helped to increase the incomes of poorer farmers in the region. Off-farm employment may have reduced the supply of young adult males to agriculture and although the alternative jobs are not displacing labor directly, they seem to be changing domestic farm labor markets considerably.

The migration of young men out of agriculture is reinforced by the growth of rural education. Although rural illiteracy rates remain high, especially among women, nearly all boys of primary school age are now enrolled in Jordan, Tunisia, Turkey, Syria, Iraq, and Iran (El-Ghonemy 1984), and young men with some education seek to leave farming.

If the cultivated area and/or cropping intensity were rising quickly enough, the rate of growth of demand for farm labor could outstrip the rate of increase of supply. The ratio of the arable land area to the agricultural labor force (Table 5) is an indicator of the relative scarcity of land and labor. If this ratio increased dramatically, there may be a labor shortage in the sense of increasing scarcity of labor relative to land. However, at the sectoral level, there are marked differences among countries. In Jordan, Syria, Iraq, and Algeria there is evidence of an increase in the amount of land per farm worker. In the first three, this is not surprising, since the agricultural labor force actually fell, but in Algeria the increase in the cultivated area appears to have exceeded the increase in the farm labor force. In all of these countries, there may be sector-wide agricultural labor shortages. In other countries of the region, however, the amount of land per farm worker has fallen and for these countries, there is little evidence of sector-wide shortages.

A better indicator of labor shortages would be trends in real agricultural wages but most data are sparse and unreliable. Data for Turkey and Morocco are shown in Table 6. There is little evidence here of any marked increase in the scarcity of agricultural labor, and the data complement those on land/labor ratios.

Studies have been conducted in some countries to assess the balance of supply and demand for farm labor both at present and in the future. For

Table 5. Ratio of arable land to economically active population in agriculture (ha/man), (1960–85).

Country	1960	1965	1970	1975	1980	1985
Algeria	3.56	4.09	5.00	5.24	5.91	6.33
Libya	11.89	13.71	16.80	13.90	15.21	16.65
Morocco	3.38	3.23	3.21	3.20	3.08	3.14
Sudan	3.69	3.50	3.37	3.00	2.86	2.62
Tunisia	7.59	7.03	7.74	7.92	7.00	6.66
Iran	4.46	4.57	4.44	4.19	3.40	3.46
Iraq	4.92	4.72	4.44	4.93	5.00	5.45
Jordan	6.06	7.06	8.11	11.28	20.90	29.74
Syria	9.52	8.78	7.52	7.53	8.03	8.80
Turkey	2.48	2.37	2.43	2.42	2.55	2.47
Yemen AR	1.25	1.38	1.45	1.39	1.33	1.24
PDR Yemen	0.52	1.59	0.71	0.72	0.78	0.82
Egypt	0.55	–	0.57	–	0.44	0.42

Source: As in Table 1.

Table 6. Indices of real wages in agriculture for selected countries in the Near East.

Turkey			Morocco			Egypt	
Year	Index No. 1[1]	Index No. 2[2]	Year	Index No. 1[3]	Index No. 2[4]	Year	Index of real wages
1977	100	100	1975	92	87	1960	100
1978	106	106	1977	83	77	1974	110
1979	105	111	1979	89	85	1975	154
1980	84	92	1980	89	86	1976	174
1981	85	91	1981	95	90	1977	192
1982	85	89	1982	99	92	1978	208
1983	94	97	1983	112	105	1979	236
1984	84	82	1985	102	95	1980	208

Sources: Turkey: FAO data tapes; Morocco: Calculated from data in World Bank (1986); Egypt: ILO (1980).
[1]Agricultural daily wage deflated by C.P.I. for low-income residents of Ankara.
[2]Agricultural daily wage deflated by food price index for low-income consumers in Ankara.
[3]Agricultural minimum wage deflated by general C.P.I.
[4]Agricultural minimum wage deflated by general index of food prices.

example one study (World Bank 1986b) found that the average requirements for agriculture in Yemen AR amounted to about 60 person days for every 275 person days available, suggesting that no agricultural sectoral labor shortage existed. This is quite significant, since a larger percentage of the rural labor force of Yemen AR emigrated to work abroad than from any other country in the region. It is estimated that about 500,000 rural Yemeni men worked abroad in 1982, compared to an agricultural labor force of just over 1 million. In no other country is it believed that more than one third of the rural labor force has recently been absorbed in off-farm employment, whether at home or abroad. If the withdrawal of one third of the Yemeni labor force has not created sector-wide shortages, one might wonder how plausible such an argument is in other countries with a smaller percentage of the rural labor force leaving agriculture for off-farm employment.

A final consideration which casts doubt on the sector-wide shortage view is the dramatic increase in the area and production of labor-intensive crops like fruits and vegetables. For example, in Algeria value added in cereals stagnated between 1973/74 and 1986 while value added in vegetables grew at 7.4% per year and in fresh fruit at 4.3% per year (World Bank 1987a). Other countries show a similar pattern. This expansion suggests that if demand is sufficiently large, and if output prices are sufficiently favorable, labor supply does not constrain agricultural production.

The available evidence suggests that genuine sector-wide labor shortages in the region are limited to the Arab countries of the Fertile Crescent and perhaps Algeria. Other countries in the Middle East and North Africa have not shared that experience. Yet, those same countries have also experienced very rapid tractorization. There are two possible explanations for this: the sectoral level data used above are too gross to capture the specific, micro labor shortages which are inducing mechanization, and/or mechanization is spreading for reasons independent of labor shortages.

The exit of workers from agriculture may create labor constraints for agriculture even if there are no sector-wide labor shortages. Crops produced in marginal ecological zones may be affected by insufficient labor, especially if such crops face severe international price competition or are heavily taxed by national governments. For example, until recently, cereal production has stagnated in Jordan, Yemen AR, Tunisia, and Algeria. International prices are held down, partly because of subsidized exports from industrial countries, and governments have increased imports to ensure urban food security. In addition, many countries, such as Jordan, Tunisia, and Algeria have taxed cereal production, further reducing farmgate cereal prices. Where such policies exist, the increased cost of labor could not be covered by higher prices and farmers produced less. However, this would not necessarily have reduced the rate of substitution of machines for labor for those farmers who continue to produce cereals; if farmers take the cereal price as given and remain in operation, they will only look at the relative costs of inputs, assuming that some combination of inputs is cheap enough to produce either a profit or family subsistence. Further, many countries have recognized the problem, and producers' prices for cereals have been increased. Producers' prices now exceed international prices in Tunisia, Syria, Morocco, and Yemen AR. This policy change, combined with continued growth in the supply of farm labor in many countries, has improved the prospects for the expansion of basic food production in the region.

The difficulties with cereal production are especially acute in areas of low and fluctuating rainfall. In some cases, such as Jordan and Yemen AR, low output prices combined with increased labor costs have led farmers in marginal zones to abandon their farms. But because of the marginality and poverty of these areas, the departed labor has not been substituted by machinery. Further, the fragile ecology in these marginal zones suggests that widespread mechanization may at best afford only temporary benefits, and there is a considerable danger of creating "dust bowl" conditions if mechanization is indiscriminantly introduced. Such effects have been reported in Algeria (World Bank 1987) and Syria. Finally, in Yemen AR, the mountainous topography of marginal areas further impedes mechanization.

In most countries of the region, there are many agricultural labor markets. These markets are often poorly integrated and there may be shortages in one town and surpluses in another. Even within one village, there may be involuntary unemployment, and seasonal unemployment is widespread due to the cropping calendar. There is little evidence of widespread rural unemployment during peak seasons of demand, e.g., during harvests. Indeed, during these seasons farmers probably have greatest difficulty in finding sufficient labor. Further, mechanization of some operations can increase labor bottlenecks during the seasons when unmechanized operations occur. One response to this problem engenders involuntary unemployment during the off-season: farmers may be willing to pay workers higher wages than necessary (i.e., there are other workers who are willing to work for less) during the slack season, because these favored workers promise to work for the farmer during the peak season when it is hard to find labor. Such arrangements imply that labor markets do not automatically clear in every season.

Labor markets in the region are often very complex. Complicated transactions involving exchanges of labor, credit, animals, and other factors of production usually require the parties to have good personal knowledge of each other. At the same time, migration is often highly localized and one village may experience substantial out-migration or may have several flourishing rural non-farm factories or workshops, while a neighboring village may have much less of either. Yet the personalization of factor markets may make the surplus labor of the neighboring village a very imperfect substitute for the scarce labor of another.

A further difficulty is the exit of young adult males from agriculture resulting in an aging of the farm labor force in some countries. In Tunisia, for example, the average age of male farmers is now 55 (USAID 1982) and

Table 7. Female participation in the agricultural labor force in selected countries of the Near East region, 1960–85.

Country	Percentage of farm labor force which was female in			
	1960	1970	1980	1985
Iran	5.0	9.4	20.8	22.1
Iraq	2.0	5.3	41.0	41.0
Jordan	3.9	4.1	0.9	1.1
Lebanon	17.5	22.7	32.2	37.7
Libya	1.7	7.2	16.0	17.7
Morocco	2.8	10.4	14.1	16.0
Oman	3.1	2.9	2.9	2.6
Saudi Arabia	3.8	3.8	3.2	3.2
Sudan	23.3	23.8	24.1	26.4
Syria	9.2	13.4	27.4	27.5
Tunisia	1.5	5.5	19.9	20.8
Turkey	49.9	48.9	51.6	54.3
Yemen AR	4.6	5.5	9.4	10.2
PDR Yemen	11.2	13.9	13.0	13.6

Source: FAO (1986b).

most Algerian farm workers are over 50 years old with more than 100,000 expected to retire by the end of the decade (USDA 1986). The PDR Yemen reports a similar phenomenon. Such developments have also occurred in developed countries, such as Japan and USA, with no loss in agricultural output.

However, such developments have important implications for mechanization. Only more mechanized, and therefore more interesting, less onerous, and higher paying farm jobs can persuade young men to remain in the countryside. Also, more mechanized techniques may be necessary if older farmers are to maintain levels of output. This is likely to be especially true with power-intensive operations. For example, it is much easier for an older man to plow with a tractor than behind oxen.

The emigration of young adult males also causes the feminization of the farm labor force. The extent to which women participate in farm work varies widely from country to country and within regions of the same country. FAO data on female participation in the farm workforce are shown in Table 7. These figures are probably underestimates but they do show that female participation in the agricultural labor force has increased in many countries. This might be the result of emigration or increasing off-farm employment of males, but these increases may also partly reflect improved data collection. For example, women residing on family farms may perform various farm tasks along with domestic chores, and they might, or might not, be included in estimates of the farm labor force. The extent to which women are employed in off-farm activities seems to vary as widely within the region as their participation in agricultural work. They are extensively employed in such activities as the making of carpets, textiles, and household articles in Turkey and Iran, but such work is less common in Egypt.

Despite the data problems, there is little doubt that in certain villages and in certain regions relatively few men remain. The economic significance of this depends on the extent to which women's labor can substitute for men's labor. This substitutability varies by country and by task. In Oman, for example, the exit of males has not led to notable increases in women's participation in farm work. In the Egyptian Delta, women and children compete with men in cotton harvesting, but plowing and threshing are traditionally men's jobs. These tasks are now largely mechanized, suggesting that male labor shortages may have played a role even in the mechanization of power-intensive operations, and the percentage of total agricultural work done by women is increasing (Richards and Martin 1983). In Yemen AR, women do most of the work in cereal production, with the exception of plowing and threshing, while men usually do most of the work in cash crops, like coffee and qat. Significantly, the predominantly male tasks of plowing and threshing are also those which are the easiest to mechanize and which are the furthest advanced in the region. But in some areas of Yemen AR and PDR Yemen where mechanization is difficult, there is evidence that women are beginning to plow behind oxen (FAO 1986). Similarly in Middle Egypt, women's labor has substituted for men's in sorghum threshing (Hopkins 1987)

Even if women can perform all farm tasks, they will be even more overworked than previously. This may be detrimental to their health and possibly to the health of their children. In general, however, the extent to which women participate in farm labor is very much a function of income. Women from wealthy farm families throughout the region typically do not do farm work. Only if a family cannot afford to hire men, has insufficient male family labor, cannot pay for or obtain machine services, or lacks alternative sources of off-farm income, do women participate in farm tasks. Because of this, emigration of males may reduce female farm participation. When remittances come into a village, families are able to subsist without resorting to female farm labor. However, the pattern is highly complex, and in some cases women continue to perform certain farm tasks, especially those related to producing food for the family, while the remittances are used for other purposes.

It is possible that the increasing amount of farm work conducted by women may constrain agricultural production. Very little quantitative data are available on this issue but social customs, combined with high illiteracy among rural women, may pose certain difficulties for them as farmers. First, in those countries where subsidized inputs are allocated through government agencies, e.g., cooperatives, women may not have equal access to these inputs. Second, illiteracy impedes technological change and since female illiteracy is higher than male illiteracy in rural areas the feminization of the labor force may have unfortunate consequences for technological progress.

Third, the large majority of extension agents are men. Often, social customs make it difficult for these agents to reach women, although it is often unclear whether these agents have anything to offer farmers in the first place. Fourth, if women have limited experience in farm management, they may find it difficult to take over this task at first, but as they gain experience, this problem should become less important.

The relevance of these considerations varies from region to region within the same country and from family to family. Literacy is spreading among rural women in the region and there is considerable anecdotal evidence that women can successfully manage farms, often with assistance from male relatives. The role of women in agriculture in the region, for long underrated and understudied, is becoming increasingly important and technological policies should explicitly recognize this fact.

Remittances may cause local and temporal labor shortages, because they raise the reservation wage of labor. This is a major cause of labor shortages in Yemen AR, Egypt, Jordan, and certain areas of Sudan. These problems are most acute in areas with low productivity due to natural conditions, on farms for which there are few available technological possibilities of substituting machinery for labor, e.g., steep terraces in Yemen AR, and on farms producing crops which fetch low prices e.g., cereals.

Because labor shortages are peak-season shortages of adult male hired labor, they often affect larger commercial farms more than smaller, family-based operations. This has been observed in Egypt, Yemen AR, and Tunisia. Although it is common in many countries for even small farmers to hire some peak-season labor, a much larger percentage of farm work is conducted by hired workers on large commercial farms than on small subsistence farms. As landless agricultural workers are often among the first to seek construction jobs or other work in the cities, larger farmers must meet their labor requirements increasingly from the pool of smaller farmers. But if they grow the same crops as their less wealthy neighbors, the small farmers may for security reasons use their family labor first on their own fields, so larger farmers must offer even higher wages to attract labor, making mechanization even more attractive.

Large farmers are especially prone to adopt mechanized plowing at an early stage, because of timeliness problems and because of the high costs of supervising laborers working with animals. Larger farmers also have better access to credit, which encourages the early adoption of mechanization.

Finally, small farmers can often meet labor shortages in ways which are not usually available to large, commercial farmers, such as by various labor bartering agreements. In some cases traditional arrangements to share work animals have been extended to tractor services, as in the Moroccan *tuiza* system, increasing small farmer access to mechanization (USAID 1986).

An example of the differential impact of labor shortages by farm size may be found in the Yemen AR, where subsistence cultivation appears to have been relatively unaffected by the massive emigration of rural males. However, the development of commercial farming is inhibited by the combination of migration (and remittance) induced high wages and the lack of appropriate technologies for increasing labor productivity (World Bank 1986a). In Tunisia, large irrigated farms have been the most severely affected by the outflow of labor while the smaller dry farmers have been less severely constrained (World Bank 1982).

Conclusion

The Middle East and North Africa region offers an example of rapid economic growth, large-scale expansion of off-farm employment, widespread labor emigration from rural areas, and the most rapid rate of farm mechanization of any area of the developing world. Mechanization has followed a pattern very similar to that observed elsewhere. Power-intensive operations are now widely mechanized, while the mechanization of control-intensive operations is much more limited. The intermediate operation of grain harvesting shows considerable inter-country variation. In all countries, the relative costs of machine and animal power have provided powerful incentives and have changed because economic growth has accelerated the demand for livestock products and increased the opportunity cost of animal traction, and because governments subsidize mechanization. Agronomic considerations like timeliness and the advantages of plowing before the rains have also contributed.

In the countries of the Arab Fertile Crescent, agricultural labor may be increasingly scarce but elsewhere there is little evidence of sector-wide labor shortages. However, there may be seasonal shortages of young adult male labor which are more acute in some regions because farm labor markets are usually poorly integrated. Such shortages supply a further impetus to the mechanization of power-intensive operations and seem to have induced an increase in the proportion of farm work performed by women in the region. Such substitution has complex consequences. It is not obvious that labor is a principal constraint to increased agricultural production, although it does contribute to the problems of cereal production, especially in ecologically marginal zones and in mountainous regions.

There is little question that mechanization is irreversible, and there is every reason to suppose that mechanization in the region will proceed into grain harvesting and control-intensive operations. But there is little reason for governments to subsidize this process. Indeed, such subsidies are very dangerous from the point of view of national welfare and employment creation. If the rate of growth of industry and construction can be maintained, then market forces will be sufficient to encourage mechanization. Should regional growth falter, which is already occurring as a result of lower oil prices and accumulated international debt for many countries, there will be a need to create jobs in agriculture. Although the mechanization of power-

64

intensive operations may not, as yet, have actually displaced much labor, mechanization of grain harvesting and control-intensive operations would have this effect.

This does not suggest that agriculture can provide most of the new jobs which the region's rapidly growing labor force requires. But it does imply that policies which actually reduce agricultural employment are misguided. Governments and international agencies should continue to support research into increasing crop yields and, where appropriate, designing specific types of mechanization to overcome timing problems and other special agricultural difficulties. It is also important to improve the utilization of existing tractor power, through the expansion of repair facilities and access to spare parts. But sector-wide subsidies to farm mechanization are likely to exacerbate problems of national employment creation.

References

Adler, Stephen. 1980. Swallow's Children: Emigration and Development in Algeria. WEP-2-2/WP46, Geneva: International Labor Organization.

Berar-Awad, Azita. 1984. Employment Planning in the Sudan: An Overview of Selected Issues. Geneva: International Labor Organization.

Binswanger, Hans. 1986. Agricultural Mechanization: Issues and Policies. Washington, D.C.: World Bank.

El-Ghonemy, M. Riad. 1984. Economic Growth, Income Distribution, and Rural Poverty in the Near East. Rome: FAO.

El-Sherbiny, A.A. 1980. Food Security Issues in the Arab Near East. Beirut: UN Economic Commission for West Asia.

FAO (Food and Agriculture Organization). 1961a-1987a (Annual). Production Yearbooks 1960-1986. Rome: FAO.

FAO (Food and Agriculture Organization). 1986b. Worldwide Estimates and Projections at the Agricultural and Non-Agricultural Population Segments, 1950-2025. Rome: FAO.

FAO (Food and Agriculture Organization). 1986c. The Role of Women in Food and Agricultural Production in the Region. Paper presented at the 18th FAO Regional Conference for the Near East, 17-21 March, Istanbul, Turkey.

Hopkins, Nicholas. 1987. Agrarian Transformation in Egypt. Boulder, Colorado: Westview Press.

Khaldi, Nabil. 1984. Evolving Food Gaps in the Middle East and North Africa: Prospects and Policy Implications. IFPRI Research Report 47. Washington, D.C.: International Food Policy Research Institute.

Richards, Alan and Martin, Philip L. (eds). 1983. Migration, Mechanization, and Agricultural Labor Markets in Egypt. Boulder, Colorado and Cairo: Westview Press and the American University in Cairo Press.

Sherbiny, Naiem A. 1984. Expatriate Labor in Arab Oil Producing Countries. Finance and Development, December 1984: 34-37.

Stark, Oded. 1984. A Note on Modelling Labor Migration in LDC's. Journal of Development Studies 20(4): 318-322.

USAID (United States Agency for International Development). 1982. The Wheat Development Program. Appendix B. Institutions, Policy and Economic Influences of the Wheat Development Program in Tunisia: Impact Evaluation. Washington, D.C.: USAID.

USAID (United States Agency for International Development). 1986. Morocco: Country Development Strategy Statement. Washington, D.C.: USAID.

USDA/USAID (United States Development Agency/United States Agency for International Development). 1980. Syria: Agricultural Sector Assessment.

USDA (United States Development Agency). 1986. Middle East and North Africa: Situation and Outlook Report. Washington, D.C.: USDA.

World Bank. 1982. Tunisia: Agricultural Sector Survey. Report No. 3876–TUN.

World Bank. 1983. Turkey: Agricultural Development Alternatives for Growth with Exports. EMENA Projects Division, Report No. 4204–TU.

World Bank. 1986a. Jordan: Issues of Employment and Labor Market Imperfections. Report No. 5117–JO.

World Bank. 1986b. Yemen Arab Republic: Agricultural Strategy Paper. Report No. 5574–Yemen AR.

World Bank. 1987a. Staff Appraisal Report: Cheliff Irrigation Project Report No. 514–AL.

World Bank. 1987b. World Development Report. London and New York: Oxford University Press for the World Bank.

4. Household Labor Issues in West Asia and North Africa

DORENE R. TULLY

The International Center for Agricultural Research in the Dry Areas (ICARDA), P.O. Box 5466, Aleppo, Syria

Introduction

The term "peasant household" evokes the image of a man directing his own labor and that of his wife or wives and children, caring for stock or working the earth for subsistence, and sharing the burdens and rewards of family farming. Often, little more than this stereotype serves as the basic unit of analysis for micro-level economic studies and the diagnostic work of farming systems research. Simplicity is its attraction. The household is treated as a well-defined entity with a locus, assets, and a labor force. It is concrete; one can count households, list residents, record activities. Internal dynamics such as decision making are not as apparent, but the assumption that households are coresident units in which the members pool their resources to cooperate in production and consumption discounts the need to analyze the process in detail. It is enough to interview one competent adult member to understand household problems, strategies, and goals. As a consequence, much aggregated household data represent the opinion and knowledge of a single individual within each economic unit. Such data are presented as a powerful representation of local level economic processes and the basis for socioeconomic comparisons and policy recommendations across West Asia and North Africa (WANA).

This paper questions the value of this model in an analysis of farm labor management and the adoption of technologies which affect management strategies. It will be argued that an understanding of the internal dynamics of household organization is critical to a successful analysis of agricultural labor and technological change in the region.

The Domestic Cycle

Households exhibit a variety of demographic structures. Nuclear families are the core of the simplest household which is comprised of an unmarried person or a married couple with or without unmarried children. Extended households may be formed if the sons of a nuclear family marry and continue to reside with their parents together with their wives and children. Upon the

Dennis Tully (ed.), Labor and Rainfed Agriculture in West Asia and North Africa, 67–92.
© 1990 *ICARDA.*

death of the parents, each son may form his own nuclear household or the brothers may continue to live together in a joint household. Eventually brothers separate or die and joint structures give way to nuclear households which may begin the cycle anew. Therefore, as change is an integral part of the domestic cycle, so must it be of any analysis of local level economies.

Economic activities are organized in a variety of ways. Nuclear households may be physically and economically independent, but some maintain close ties with the husband's or wife's parents, sharing labor and resources freely (Bates and Rassam 1983). It is wrong to assume that husbands and wives necessarily establish a joint estate and pool production and consumption. Such sharing is frequently found in WANA (Sweet 1974), but so are other patterns. For example, consumption may be shared while production is not. Spouses may be primarily responsible for their own cultivation, maintain separate granaries, and use their production to meet individual economic obligations within the household (Tully 1987). Joint families may live together and share fields, animals, and domestic tasks (Ilaco, cited in Ehlers 1985), may merely inhabit the same house with food preparation, child care, and productive resources managed separately (Myntti 1979), or may live apart but share productive assets (Murdock 1979). Within extended households, several families may form a shared production and consumption unit managed by the eldest male (Bates 1982). Other extended households are structured as family firms in which responsibility for aspects of the farm is divided among the members, as are economic returns. One member may drive, repair, and lease out family-owned farm equipment, while another is a grain broker for the family and its neighbors, and a third works off-farm to provide cash inputs to shared enterprises (Kandiyoti 1975: Khattab and El Daeif 1982). Thus, the organization of production and consumption may differ significantly even between structurally similar households.

An important determinant of economic choice is household composition, particularly age and sex ratios. In the early stages of a nuclear family, children require more labor than they return. As they reach adulthood, their labor becomes a significant contribution to family maintenance. Labor supply reaches its lowest ebb when sons marry and the establishment of their own households coincides with parental aging (Hopkins 1980). Associated with this rise and decline in available family labor is the need for land. At certain points in the domestic cycle, a household may control more land than it can work and others may be allowed to rent or sharecrop the excess. With increased family labor, land may be added by rental, purchase, or the removal of tenants (Myntti 1984). Consequently, both land use patterns and family production levels can vary with household demographics.

The Life Cycle

Concurrent with household changes are individual cycles defined by social and biological events. Most people are born, marry, procreate, and die.

Thus, households are comprised of individuals of both sexes who are at various stages in this cycle. Position in the life cycle is an important determinant of economic status within a household. Commonly, children control few productive resources, but are helpers in production under the direction of a parent (Peters 1978; Saunders 1968). The effect of the maturation of children on household economic relationships is closely tied to family structure.

To establish a nuclear household, parents cede productive assets to their married offspring who take responsibility for both labor and management. Within an extended household, a newly married man continues to work for his father without gaining separate control over major productive resources. Resident adult sons allow their fathers to give up physical labor in favor of supervising farm tasks and increasing participation in political or religious activities (Sweet 1954). As the father ages, sons often take on more management responsibilities without receiving increased control over assets. A son brings his wife into his parents' domain where she labors for his mother, allowing her to reduce her workload in favor of supervision and less arduous economic activities. A new bride finds few opportunities to control economic resources within the household and little time to devote to economic tasks of specific benefit to her own family (Peters 1978). With the birth of children, a husband may be given more economic authority and his wife may shift some of her labor to her own nuclear family (Amin and Awny 1985). However, ultimate control of major assets remains with the husband's parents. Joint households that separate productive resources are similar to nuclear households, and those in which an elder brother and his wife control major resources follow the patterns of an extended household. As age, sex, and generation are closely linked with control over productive assets, economic responsibilities and opportunities are neither static nor equally distributed within a household.

Unequal access to productive resources can create inter-generational competition. Conflict may occur in nuclear families when sons reach adulthood and wish to establish their own households. This may contradict the parents' desire to maintain their assets intact and preserve control over their son's labor. Tensions arise in extended families after sons marry and produce children. Living space shrinks, financial resources stretch to support greater numbers, and loyalties are strained. If pressures become too great, a son may establish a separate household (Bates and Rassam 1983). His wife may encourage, or even instigate, such a schism to escape the control of her mother-in-law and establish her own economic base. Consequently, individual attempts to gain greater control of economic assets at the expense of other family members may split a household along generational lines.

Economic rights and obligations also differ between the sexes. Husbands and wives often have distinct responsibilities within a household. Commonly, women convert family production into consumable items such as clothing, food, and medicine; create monetary value by producing goods to sell; and reduce cash expenditures by processing farm produce, cooking for hired labor, and gathering fuel and water (Carapico and Hart 1977; Friedl 1981;

Hopkins 1985a). Men provide cash and the goods and services that it can provide.

To meet their obligations, husbands and wives may exploit different resources. Property may be separately owned by men and women. In parts of WANA, either sex can inherit or purchase land, animals, buildings, or machinery. If the property belongs to a woman, any income is hers and is neither jointly held with her husband nor incorporated into the household budget (Ilaco, cited in Ehlers 1985; Murdock 1979; Myntti 1979; Saunders 1968). Similarly, husbands and wives may retain separate control of incomes earned from their own labor. Women earn cash by selling handcrafted carpets, medicinal and culinary herbs they have gathered, hand-built fireplaces, or agricultural labor (Beck 1978; Sweet 1954). Men's incomes may be from tailoring, wage labor, or livestock sales (Tully 1987). Each sex receives the greatest return from investing cash and energy in their own enterprises. Therefore, either may resist providing labor to the other and attempt to divert available resources to their own activities. Even though husbands and wives cooperate in satisfying the needs of household members, there may be friction between them over the allocation of resources within the household.

Competition also characterizes the relationship between husbands and wives when their rights in productive resources are overlapping rather than individual. Husbands and wives may jointly cultivate a field and store the grain in a single granary. Both may have rights to the grain established by their labor, but these may be unquantified. This ambiguity results in attempts to gain control through internal transfers. A wife may brew beer to sell and through her labor transform shared grain into personal cash. A husband may arrange an agricultural work party to weed his personal field and pressure his wife to brew beer from jointly owned grain to pay workers. This results in an indirect transfer of the wife's labor to her husband's fields (Tully 1987). Once major family obligations are met, unassigned income may remain. Men or women may claim such income by careful investment. A man may spend money to court a second wife or obtain a political office and increase his local power. A woman may convert ambiguous funds into property, such as gold, which is accepted as her separate estate (Basson 1984). Thus, the decision to brew beer, work in a field, or invest profits may not be based on an analysis of the best household economic strategy, but of the best return to an individual.

Rights between spouses may be interlocking, with each controlling different aspects of the same asset. For example, a husband may buy and sell livestock or meat while his wife sells animal products from the same herd or flock. Each may control the proceeds from their own sales (Chatty 1978). Contrasting rights in resources mean that the costs and benefits of economic choices may be different for husbands than for wives. In the case of livestock, certain changes in husbandry techniques are advantageous to both parties, such as improvements in calf survival rates or increased carrying capacity of pastures. However, the possibility of improving either meat or milk production at the expense of the other through selective breeding results in competition. One spouse must give up production and income to the other. Therefore, the decision to keep general purpose animals or to specialize in particular production may be based on the resolution of a conflict between husbands and wives concerning their rights within the household rather than on the basis of what is most profitable for the household as a production unit.

The Wider Economic Context

All households contain people who have important economic interests in other households as a result of kinship or reciprocal exchanges. Nuclear households may be part of a wider system of productive relations in which parents assist their married children, providing advice, labor, and capital (Bates and Rassam 1983). In some areas, women inherit productive property from their parents. In others, they customarily surrender inheritance rights to their brothers. A brother also expects to receive a portion of the money or goods which his sister's proposed husband provides as gifts to the family before the marriage. In return, brothers provide economic help in the form of stock, food, or labor (Peters 1963; 1978). Thus, women may increase their economic options by investing their inheritances in the future labor of their brothers (Commander and Hadhoud 1986). Unrelated members of different households may also assist one another through reciprocal exchanges of draft animals, labor, or farm machinery. Clearly, household boundaries are permeable with significant and predictable economic transfers occurring across them. A determination of the economic feasibility of a course of action will be based on all of the resources available to an individual rather than just those which are part of the household of residence.

Economic strategies within the household are also affected by the world economy. For example, competition with cheap imported goods may decrease returns to occupations such as tailoring or the crafting of household goods, while wider demand for hand-made carpets may increase the value of that production (Beck 1978). This affects relative incomes within the household and may change investment strategies and labor allocation. Control over returns to labor may be affected by new marketing channels which shift sales from the home or village to market centers or government cooperatives. When products pass through formal channels, buying and selling are typically done by men, so they may receive cash for milk products produced by women or for grains produced in cooperation with their wives (Basson 1981; Beck 1978; Birks and Richards 1985; Sweet 1954). By reducing women's direct control over the returns to their labor, a new marketing structure may decrease their ability to meet household obligations and their commitment to certain farm enterprises.

The Organization of Family Labor

The majority of farm labor in the region is provided by unpaid family workers with the proportion of family to hired labor being greatest on small farms of 10 ha or less and in dryland areas (Crawford and Purvis 1986; Hopkins 1980; Rassam 1984). Task assignment within the family is based on biological and social attributes, primarily age, sex, and socioeconomic status (Myntti 1979; Tully 1987).

Division of Labor by Age

Throughout the region, children help with farm tasks from an early age (Richards and Martin 1983a; Sweet 1954; Tully 1987). Initially, directing a child's work may take as much of an adult's time as doing the task, but long before adulthood, children make significant contributions to family production. Children decrease farm losses to disease and injury through labor-intensive activities such as manual pest control for crops and livestock (Richards et al. 1983). By assisting adults, children improve the speed and timeliness of a wide variety of tasks such as planting, harvesting, and weeding. In addition, as children's labor has a lower opportunity cost than that of adults, they perform tasks, such as herding, which offer a low rate of return to labor. This frees adults for more lucrative enterprises while allowing families to engage in economic activities which are not cost effective with adult labor (Hogan et al. 1983; Tully 1987). Therefore, the work of children enhances the viability of family farming by improving productivity and increasing economic diversification.

As children grow to adulthood, their skill and experience increase, as does their physical strength. Young adults do most of the heavy domestic and agricultural work. At this stage, their participation in farm activities is primarily determined by the sexual division of labor rather than age. With greater age, both sexes tend to carry out less strenuous tasks (Myntti 1979), which is partly the effect of declining strength, but also reflects changes in economic status within the household. As individuals grow older, they gain control over economic assets and the labor of others which allows them to delegate arduous tasks to younger family members (Sweet 1954). In addition, the experience of older men and women makes them particularly suited to skilled activities such as supervising hired workers, selecting seed for the next year's planting, or negotiating the best price at livestock sales. The division of labor by age is based on an evaluation of physical strength, level of skill, opportunity costs, and control over resources. The presence of workers of all ages increases the flexibility of labor organization and improves farm viability.

Division of Labor by Sex

WANA is characterized by great diversity in the division of labor between men and women. Certain activities are usually assigned to a particular sex;

women cook and men shear sheep. However, many tasks are not consistently identified with one sex. For example, milking may be carried out by both sexes (Sweet 1954; Tully 1987), primarily by men (Bates and Rassam 1983), or solely by women (Carapico 1985; Khafagi 1983). Thus, an analysis of local labor patterns requires an understanding of task assignment in both the domestic and agricultural spheres.

Domestic responsibilities are associated with provisioning and maintaining a household. Some tasks are daily and are essential to household survival, such as providing water, processing and cooking foods, and child care. In most cases, they are the primary responsibility of women. In addition, necessary but occasional tasks include preparing staples such as rosewater, tomato or pepper sauces, or cooking oil; obtaining fuel; and repairing ovens. Extending the family house or granary is a typical male occasional activity. Finally, there are discretionary enterprises which benefit family members but are not vital to their survival. Women may create textiles to decorate and protect home furnishings or prepare time-consuming special foods. Male discretionary tasks are generally restricted to assisting women in their domestic work. Clearly, domestic production is disproportionately female, especially activities requiring daily labor inputs (Khafagi 1983; Myntti 1984; Sweet 1954; Tully 1987).

Most male domestic labor is not essential to the day-to-day functioning of the household. Men can postpone domestic tasks if other labor opportunities or demands arise. Consequently, male contributions to domestic labor can be met by hired labor or occasional visits from family members or nearby relatives. In contrast, the domestic labor of women is extremely difficult to replace. A man's female relatives may assist him temporarily, but they are seldom a long-term solution as they must provide for their own families. In some areas, it is possible to hire women for particular domestic chores such as cooking or child care, but it is not feasible to maintain a smallholding without unpaid female assistance (Peters 1978). When men are unable to hire female labor, single men cannot establish an independent economic unit, although women who maintain their own households are able to recruit male labor by offering lodgings in exchange for work (Peters 1978).

These differences have a significant impact on the economic options open to men and women. Because women's domestic work is both essential and constant, the amount of labor that women have available for non-domestic tasks and the flexibility of their work schedules are less than those of men. Thus, women may find it difficult to adopt innovations which are particularly sensitive to timing, demand heavy inputs of labor, or require long-term absences from home. This does not mean that women are unable to adjust their workloads, but the opportunity cost of rescheduling female domestic labor is high and replacement labor is limited. Returns to labor must be high enough to compensate for hardship.

In most of the region both sexes tend animals. Female labor is particularly important in the daily care of poultry, sheep, goats, and cattle (Cosar 1978; Crawford and Purvis 1986; Khafagi 1983; Rabo 1986). They collect water and fodder for animals stabled at home or grazed nearby (Carapico 1985; Crawford and Purvis 1986) and are often responsible for the intensive care of young stock, pregnant and lactating females, and sick or injured animals (Murdock 1979; Rabo 1986; Sweet 1954). Women produce cheese, butter and yoghurt (Khafagi 1983; Myntti 1979; Murdock 1979) and collect manure to make fuel and to fertilize cultivated land (Cosar 1978; Myntti 1979). Men slaughter, shear and clip, load stock into transport vehicles, and bring water and fodder by truck (Chatty 1978; Sweet 1954).

Male contributions to animal care increase with distance from the household (Ilaco, cited in Ehlers 1985). In fact, restrictions on male interactions with women who are not kin may preclude a man from tending stock maintained near other households even in his own village (Fazel 1977). Grazing patterns also affect the relative participation of men and women. A study in Syria showed that women contributed 83% of the labor to feeding animals held near their households, but only 12% when animals were moved to distant pastures (Rassam 1984). When animals are at home, women can coordinate their care with domestic tasks and children can assist in animal care. As returns to the intensification of labor are high in stock rearing (Irons 1972), livestock have a great capacity to utilize female family labor (Fitch and Soliman 1983). Careful feeding and watering, diligent removal of parasites, and prudent weaning and milking practices, all pay rewards in increased productivity. Moreover, women create value by processing milk products, family nutrition is improved, less money is spent to purchase dairy products, and if there is a marketable surplus, income is received (Fitch and Soliman 1983). When animals or their products enter the market, meat or live animals are usually sold by men (Fazel 1977; Fernea 1969), while women typically sell poultry, eggs, milk, and milk products to their neighbors (Rassam and Tully 1985; Sweet 1954). Thus, women play a major role in the production and sale of animals and their products whenever stock is home-based, care is intensive, and sales are made from the farm. Male participation in animal husbandry becomes particularly important when animals and their products leave the village sphere, either to graze or to enter the market.

In some areas, men do most of the agricultural work while women focus on domestic production and animal husbandry (Beck 1978; Hopkins 1977; Hopkins 1985a); in others, women manage domestic and agricultural production while men work outside agriculture (Kandiyoti 1977). It has been maintained that women do little agricultural work in WANA (Boserup 1970), but this is a misconception. What is characteristic is great variation in male and female participation. Compared to men, women's agricultural workload is high in Yemen (Myntti 1979), about equal in Syria (Rassam 1984), and low near the Red Sea (FAO 1986). Individual countries also exhibit considerable variation. In Egypt, Delta women do 30% of all field crop labor while women in Upper Egypt do little or no work in the fields (Richards and Martin 1983a). Clear differences may be found even within a single village. A woman may spend the morning cultivating her own field alone while her husband works off-farm and then earn wages in the afternoon working in the fields of elite men and women who do no agricultural work themselves (Beck 1978). Therefore, although sex is a good indicator of labor inputs to domestic activities, it is an inadequate index of participation in agriculture.

Most commonly, both sexes contribute to cultivation although participation may vary with task, crop, and technique (Davis 1978; Maher 1978; Peters 1978; Rassam 1984). Men prepare the land for cultivation, construct and maintain terraces and fences, broadcast seed, apply commercial fertilizers and chemical weed control, till, harvest, and thresh (Carapico 1985; Fazel 1977; Mundy 1985; Myntti 1979; Rassam 1984; Sweet 1954). Women assist in many of these tasks, but are primarily responsible for selecting seed, planting, fertilizing with manure, hand weeding, harvesting, cleaning grain, and winnowing (Carapico 1977; Cosar 1978; Glavanis 1984; Mundy 1985; Myntti 1979; Rassam 1984; Rassam and Tully 1985). The primary distinctions between male and female agricultural labor are based upon differences in technique and timing. Male tasks are frequently mechanized or animal-powered (Rassam 1984), so when machines are used to till, harvest, or thresh, the labor contribution of men is greater than that of women. Women, aided by children, provide intensive non-mechanized labor and the manual assistance required by most mechanical procedures. These distinctions are reflected in differing contributions to crops grown primarily for cash and those consumed by the family. Women commonly have less responsibility for the former and more for the latter than men do (Aswad 1978; FAO 1986; Myntti 1984). This is consistent both with women's primary responsibility for domestic production and men's greater involvement in the market economy (Friedl 1981; Rabo 1986). Crops which produce cash tend to receive more purchased inputs which are applied by men and greater use of farm equipment which is operated by men. The rhythm of male and female agricultural labor also differs. Men work in peak season bursts, to prepare land for cultivation or to harvest crops, while women increase their contributions during peak periods when timing is important and provide regular labor inputs throughout the agricultural cycle (Myntti 1984). Thus, even in areas where women are fully employed by their farm duties throughout the year, men may face chronic seasonal underemployment (Carapico 1985). Further, in rainfed systems women and children can often handle crop production with the help of a few men at peak periods (Myntti 1984).

Male and female labor may be integrated in a number of ways. Either sex may be able to carry out almost any agricultural activity or it may be possible to hire the few agricultural tasks which are sex-specific. Consequently, men or women may cultivate independently of one another (Peters 1978; Tully 1987). However, most family farms require coordinated male and female family labor. Men and women may go to the fields together and assist one another. For example, men may plant seeds which women cover (Peters 1978). Alternatively, production may be separate but complimentary. Responsibilities may be divided by crop, with women cultivating home gardens while men labor in distant fields, or by technique with men cultivating with machines and women by hand. Or women may specialize in livestock production and men in subsistence agriculture (Fazel 1977). Men and women may also provide sequential labor for the same production. For example, men may take animals to graze and then turn them over to women for milking or men may cut grain which women later bundle (Friedl 1981; Tully 1987). Whether they work separately or together, the functioning of the family farm depends upon coordinating the work of both sexes.

Division of Labor by Socioeconomic Status

Besides age and sex, socioeconomic status has an effect on the labor input of individuals and thus on the availability of labor within a household, village, or region. In some areas, certain families who belong to landowning lineages or who can trace their ancestry to particular notables, are set apart from other families. As one marker of their special status, elite men often do little or no physical labor, devoting themselves to politics, religion, or trade (Aswad 1967; Peters 1978; Shashahani 1986; Sweet 1954). The women of such families seldom cultivate (Myntti 1979), though they may supervise workers who tend their fields and provide other labor for the household (Shashahani 1986). Thus, in some villages certain households may be dependent upon non-family members for agricultural and domestic labor. In spite of assistance, the majority of elite women work long hours in domestic activities. As hospitality is expected from elite families, and feeding and entertaining clients and cronies is a necessary part of local politics, frequent entertaining increases the time women spend processing and cooking food. In addition, when wealthy men withdraw from daily supervision of their properties to pursue other interests, women's management responsibilities increase. Because elite families may control numerous properties and extensive herds and flocks, women may travel to distant pastures to oversee workers during sheep shearing or lambing (Aswad 1967; Sweet 1954; Tavakolian 1987). Thus, elite women's labor may be intensified by high levels of consumption, the demands of hospitality, and extensive management duties. There are similar changes in labor patterns in families that acquire sufficient income to hire workers. In such cases, family labor, particularly that of women, is withdrawn from agriculture (Aswad 1978; Crawford and Purvis 1986; Hopkins 1985a; Maher 1974).

Poor women also experience heavy workloads. Their domestic labor may be less than wealthy women due to low levels of hospitality and simple food preparations. However, women who must grind grains by hand, walk to central water supplies, and make more clothing than they purchase, face greater domestic labor burdens than their wealthy sisters. Certainly, their physical labor is greater whether in their own fields or those of elites who employ them. Therefore, in wealthy areas growing cash crops, women provide less labor to agriculture than they do in poorer areas, especially among families cultivating subsistence crops. Consequently, female agricultural labor is particularly important in poor families (Aswad 1978; Peters 1963). Thus, domestic and agricultural tasks may be allocated in different ways which reflect economic inequalities as well as sexual dichotomies.

Family Labor Availability

The number and age and sex ratios of individuals in a family's labor pool are largely the result of chance. Judicious marriages or combinations of families into larger household structures may temporarily adjust available labor. However, cyclical processes continually transform labor endowments independently of need. Newly established nuclear households usually consist of one adult of each sex and tasks requiring the cooperation of two or more men or women cannot be carried out with household labor alone. As children mature, this deficit is resolved (Glavanis 1984; Ilaco, cited in Ehlers 1985; Myntti 1979), but when sons and daughters establish their own households or accept long-term, off-farm employment, labor surpluses may become deficits (Bates and Rassam 1983; Hopkins 1985b). Therefore, the availability of labor changes as families expand and contract. Even when personnel remain constant, radical shifts may occur in labor availability. As young girls reach marriageable age, they may withdraw from cultivation to concentrate on domestic tasks and animal husbandry (Hammam 1986). Consequently, a family with numerous female children may possess abundant field labor for a few years and then require replacement workers. Or a family with an adequate number of workers may have an imbalance between the sexes. With many sons and few daughters, there is a greater demand for domestic work than labor to meet the need. Even a well-run farm undergoing no major changes in farming system experiences shifts in domestic or individual cycles which can result in a labor surplus or shortage.

Unlike hired labor, which may be brought in at need and dismissed at will, family labor is to a great extent a given. Family members may consume more than they produce. They may remain on the farm when their labor is superfluous and leave when it is vital. Nevertheless, they retain rights in household consumption and family production strategies must encompass these fluctuations. If a household has more labor than required, family members may seek off-farm employment, begin a business, or increase discretionary activities such as craft production or leisure (Hopkins 1983; Irons 1972). This situation is likely to be of short duration and adjustments to labor are more characteristic of family farms than is equilibrium. Because family labor is differentiated, labor shortages can be absolute or restricted to a particular type of labor. If the household does not have enough men for male tasks, it may have a labor shortage even if female workers are underemployed. Insufficient labor can be a problem throughout the year or may be limited to parts of the agricultural cycle. Thus, a household may support excess labor for 9 months of the year to have enough for peak demands at harvest or planting. Alternatively, family members may work off-farm on a seasonal basis which coincides with low labor needs at home (Bates 1982; Fernea 1969). Or they may commute to other jobs during the week and cultivate whenever they are free. The variety of possibilities which arise from a household's particular combination of demographic and employment choices means that some farms within a village or region may suffer chronic or seasonal shortages of family labor while others experience year long surpluses.

Family Labor Shortages

When families experience labor shortages, they can augment or reorganize labor and continue current economic practices or change the system of production to reduce labor requirements. Household labor may be increased by extending the number of hours worked (Amin and Awny 1985; Nelson 1973). Or family structure may be manipulated to secure greater access to labor. For example, parents in nuclear households may retain control of their children's labor by denying them the means to marry. However, this strategy may exacerbate household tensions and precipitate a split which ultimately reduces the labor supply. Extended families may encourage the marriage of sons to obtain the labor of daughters-in-law, but complex households may segment and face renewed labor shortages. Finally, friends, neighbors, or relatives may provide unpaid, reciprocal labor (Glavanis 1984; Tully 1987). However, wage employment has increased the opportunity cost of providing unpaid labor to one's neighbors and has contributed to male emigration from rural areas. Consequently, the availability of male cooperative labor has decreased (Glavanis 1984; Tully 1987). By contrast, women often turn to other women for assistance in their daily tasks, exchanging domestic chores to decrease boredom and to free their labor for work in the fields (Hopkins 1977; Murdock 1979; Myntti 1984; Rogers 1980). Moreover, in some areas women maintain networks to guarantee assistance when male labor is unavailable (Maher 1978; Rogers 1980). Thus, reciprocal labor is a viable option where male labor shortages leave agriculture in the hands of women.

Short-term labor needs may be met by reorganizing existing labor. During seasonal deficits, occasional and discretionary activities can be temporarily postponed until demanding activities are completed. When there is a shortage of a particular type of labor, substitutions may occur. If men work late in the fields, women take over male milking responsibilities (Murdock 1979). However, not all activities lend themselves to substitution. Male labor seldom replaces that of females when work occurs in locations identified with women (Friedl 1981; Peters 1978). Activities which are performed by groups of women, such as weeding, may also be closed to males. Women face similar restrictions. For example, herding or repairing terraces may be done by groups of men, far from home and continue late into the night. Any of these features might prevent female participation. Thus, the sexual division of labor is more often maintained when there are spatial and temporal restrictions on male or female behavior. Substitutions occur most frequently in activities which take place on the family farm, during daylight hours, and in the company of other family members (Hopkins 1985a). Further, as men are more constrained by the sexual division of labor, they are less flexible. A shortage of female labor may force a man to marry or join another household, but women who lack sufficient male labor may perform typically male tasks themselves, including plowing with animals, dipping stock, and transporting and spreading fertilizer (Amin and Awny 1985; Beck 1978; Glavanis 1984; Myntti 1979; Nelson 1973; Peters 1978). Shortages in farm labor often mean that women increase their labor, and when households possess minimum labor, that of women and children becomes disproportionately important.

When children replace adult labor or one sex takes over the economic activities of the other, there are costs as well as benefits. If family flocks are usually attended by an adult shepherd who seeks other employment, young children may take over this task. If the children would otherwise be watching younger siblings, part of the cost of their labor is an increase in the child-minding responsibilities of someone else. Further, young children may lack the ability to oversee animals properly, resulting in losses due to accidents, predation, or thievery. Similarly, when women take over a male task, families may be forced to buy staples which women no longer have the time to prepare, supplies of water or firewood may not be adequate to the family's needs, and cash incomes from the sale of eggs, crafts, and dairy products may be curtailed. Alternatively, changes in opportunity costs may encourage labor substitutions. For example, when cash is required, a man may take over some of his wife's domestic activities to free her to make carpets to sell.

In addition, substitutions affect the quality of labor. Men and women often develop distinct areas of expertise and even when they work together, they usually perform discrete tasks. Thus, certain activities are difficult to replace by workers of the opposite sex. Poorly executed broadcasting of seed or plowing produces an inferior crop. Shortages of skilled labor are often met by calling upon kin for assistance or hiring labor rather than by labor substitutions within the family (Maher 1974; Peters 1978).

During labor shortages farmers may change the way in which tasks are carried out. Work schedules can be altered and households may decrease the number of workers needed for labor-intensive operations by extending the time spent on these activities (Glavanis 1984). Alternatively, labor can be saved by decreasing the time allocated to certain tasks, dropping some operations, or employing labor-saving short-cuts. Similarly, livestock may have to subsist on reduced fodder and fewer waterings, herd size may be decreased, or capital investments intensified by purchasing feed (Bates and

Lee 1977). Fewer labor inputs often decrease productivity for both livestock and crops. Thus, there is a distinction between necessary and optimal labor inputs which may be exploited by labor poor farmers who maintain viability but decrease productivity. To offset these effects, farmers may substitute less labor-intensive crops, or capital may be used to purchase or rent farm equipment or hire laborers. The following sections will consider how these choices affect farm labor strategies.

Mechanization

Mechanization is widespread in WANA, but some spheres of the rural economy are more mechanized than others. The majority of domestic tasks are performed by hand (FAO 1986) and mechanization plays a limited role in animal husbandry. Cultivation is more extensively mechanized. Tractors, combine harvesters, and threshing machines make up the majority of mechanical inputs to rainfed agriculture. Consequently, land preparation, planting, and harvesting are more often performed by machines than are weeding, hoeing, winnowing, and post-harvest processing (Affan 1984; Birks and Richards 1985; Commander and Hadhoud 1986; Hopkins 1983; Tinsley and ter Kuile 1986; Tully 1986; Uner 1980). Thus, it is primarily seasonal, male agricultural operations which are mechanized while women's day-to-day economic activities remain manual.

When farmers employ mechanization to cultivate new lands or intensify agricultural practices on existing fields, more work is required of both sexes. More time may be spent in domestic tasks as women are forced to gather firewood, herbs, and vegetables farther from home (Affan 1984; Jaubert 1986; Mahhouk 1956; Mulayim 1985; Tully 1987; Uner 1980). When nearby grazing areas are planted to crops, animals may be moved to pastures which are distant and often marginal. If water and feed must be transported by truck, male contributions to livestock care increase. With expanded production, the demand for male agricultural labor intensifies and to meet peak season requirements, families may be forced to support surplus male workers throughout the year. Alternatively, underemployed men may migrate and leave the family seasonally short-handed (Hogan et al. 1984). Although mechanization may decrease the need for female or child labor in a particular task, increased production requires more labor from women and children, not only during peak periods but throughout the agricultural cycle for weeding, hoeing, and processing the harvest (Crawford and Purvis 1986; Hopkins 1985a; Mulayim 1985; Myntti 1979; Rabo 1986; Rassam and Tully 1985; USDA/USAID 1980). Consequently, mechanization may actually create a shortage of female and child labor.

If farms no longer keep draft animals, women's contributions to animal husbandry may lessen. However, work animals are often replaced by dairy and meat breeds which require more female labor to meet their greater demands for food and water and to process their products (Dyer and Gotsch

1983; Hopkins 1983; Hopkins 1985a). If machines are used to cultivate existing fields and yields remain relatively stable, family labor requirements may decrease. As it is male tasks which are mechanized, it is primarily male labor which is reduced. When this results in men leaving the farm, women and children often add remaining male responsibilities to their own workloads. Thus, mechanization may increase, decrease, or reschedule family labor needs (Crawford and Purvis 1986; Dyer and Gotsch 1983; Hogan et al. 1984; Hopkins 1985b; Tinsley and ter Kuile 1986).

A farm's level of mechanization is also determined by geographic, economic, and political factors beyond the control of the family. Appropriate equipment may not be available for some terrains (Hogan et al. 1984). For example, tractors are unable to work in the very hilly environments which characterize much rainfed country (Beeley 1985; Snobar and Arabiat 1984). Land which cannot be worked by machines has been abandoned in some areas with negative effects on grain production (Kopp 1985). In addition, government import policies control the availability of equipment, spares, and fuel and identify farmers who qualify for subsidies to purchase these products (Crawford and Purvis 1986; Hopkins 1983; Okazaki 1985; Uner 1980). Official priorities also determine where to develop necessary infrastructure, such as rural access roads or commercial centers. Government investment in areas offering a high rate of return has led to rapid mechanization in locations which can support intensive, irrigated cultivation (el-Ghonemy 1979; Hopkins 1985b). There has been a corresponding neglect of the rainfed sector where small, subsistence-oriented farms, lower incomes, and lack of nearby market centers are thought to limit the potential for increased production (Birks and Richards 1985; el-Ghonemy 1979; Hopkins 1985b; Snobar and Arabiat 1984). Finally, it is not possible to replace labor with capital without an income (Richards and Martin 1983b), so mechanization is correlated with off-farm employment (Crawford and Purvis 1986; Okazaki 1985) and the commercialization of agriculture (Bates 1982; Carapico 1985; Hopkins 1983; Uner 1980).

Mechanization has important implications for the management of family farms. Although some farmers buy equipment, most smallholders rent from custom operators. The dependence upon others for access to basic farm equipment is important in most agricultural decisions. Custom operators and government officials determine what equipment is available. If owners are unwilling to risk capital on necessary machinery, farmers who rent cannot employ equipment-dependent technologies (Hogan et al. 1983; Tully 1986). Moreover, owners may be unwilling to replace existing equipment even if it encourages erosion or loss of soil moisture (Hogan et al. 1984; Newberg 1982). Thus, risk minimization strategies of custom operators may block a farmer's ability to increase production or protect fields from degradation. Secondly, renters are not in control of the timing of mechanized agricultural operations. Custom operators who also farm will work their own fields first and give second consideration to their neighbors or the largest of nearby farms. Under such circumstances, farmers using animal traction may show more timely planting (Dyer and Gotsch 1983). As timing is particularly important in rainfed areas, such decisions may seriously decrease yields or force the planting of a less desirable crop (Ilaco, cited in Ehlers 1985). Finally, the economic interests of farmers and operators differ. Custom operators want to finish a job quickly to maximize returns so they may plow up and down slopes or set openings on combine harvesters which speed threshing but waste grain (Hogan et al. 1983). Thus, operators may maximize their profits at the expense of farm productivity (Tinsley and ter Kuile 1986). The necessity of integrating the economic interests of contract machine operators with those of the household blurs the borders of the household as a management unit.

For farmers that do mechanize, the choice often includes an implicit decision to commercialize production. Farms become more oriented toward the cash economy, growing crops which can be serviced by machines and sold on the market. Crops, such as lentils, which require manual labor may be replaced by more fully mechanized crops such as cereals resulting in a move towards monocropping and less diversity in production. As a result, farmers who mechanize increase their dependence upon economic and political forces outside of the local economy (Aswad 1967).

With mechanization, there are also changes within the household. When families transport animals by truck during migrations, women lose the opportunity to gather materials for their own productive activities (Beck 1978). Even among sedentary families, when trucks allow men to maintain animals separately from the household, women are deprived of the opportunity to make daily decisions regarding consumption, processing, and sale of animals and their products and they experience a loss of economic control to men. Similarly, with increased commercialization of crops or livestock, as men are the intermediaries between the farm and wider markets, they have final control over the income from sales. Thus, mechanization may decrease women's control over economic resources and their incentive to provide labor to farm activities.

Even families that do not mechanize find their economic choices structured by those who do. With the expansion of cultivation into former pastures and woodland, pastoral production may be jeopardized (Affan 1984; Jaubert 1986; Kopp 1985). It may be difficult to locate trained draft animals, skilled drivers, or animal-powered implements. Manure may be unavailable and families forced to accept lower yields or purchase chemical fertilizers (Dutton 1985; Okazaki 1985). Further, they may have to adjust the timing of agricultural activities to coincide with their neighbors. For example, if most farmers harvest early because of mechanization, non-mechanized farms must do so as well or face increased bird damage. As most farms in the region sell a portion of their production, farms that do not mechanize have trouble competing with those that do (Bates and Rassam 1983). The expense of technical packages associated with mechanization may force poorer households to cover deficits by selling land or preharvest crops at low prices (Dutton 1985; Okazaki 1985; Uner 1980). Thus, mechanization contributes to the economic differentiation of the rural population (Hopkins 1985b). This is clearly illustrated when tenants and sharecroppers are removed from the land when owners wish to increase cultivated area or find it unnecessary to share minimal risks and high profits to obtain labor (Beeley 1985; Hopkins 1985b; Okazaki 1985; Uner 1980). Mechanization devalues labor in favor of capital and may force capital-poor farmers, tenants, and sharecroppers off the land (Bates 1982; Hopkins 1977).

As many of those who are displaced enter the wage labor market, mechanization may increase the number of workers for hire. At the same time, because of the increased seasonality of mechanized agriculture, workers may leave rural areas if the hiring season is too short for them to earn a living (Crawford and Purvis 1986; Hogan et al. 1984) which may contribute to a shortage of peak season workers (Mulayim 1985). Labor shortages may also result when mechanization expands cultivated area and increases production (Crawford and Purvis 1986; Hogan et al. 1984; Hopkins 1983; Hopkins 1985a; Newberg 1982; Okazaki 1985; Richards and Martin 1983b). Therefore, the decision to mechanize and to hire labor are closely associated choices.

Wage Employment

Because of fluctuations in household composition which accompany the domestic cycle, most farm households employ labor, sell their labor off-farm, or both. The labor market therefore offers farm families a source of labor and income. The integration of these possibilities into farm economic strategies has important effects on household structure, labor organization, and farm management.

The labor market allows sons to establish independent nuclear households supported by wages and to withdraw their unpaid labor from the family farm (Crawford and Purvis 1986). Thus, off-farm income contributes to an increase in the proportion of nuclear families (Amin and Awny 1985; Bates and Rassam 1983; Mir-Hosseini 1987). The possibility of hiring labor means that productivity is less constrained by the size and composition of a family's labor pool and the viability of farms managed by nuclear households is improved, while the relative labor advantage enjoyed by extended families is reduced. At the same time, joint or extended families may be encouraged by market forces. The growing land market may inflate prices so sons may be unable to obtain sufficient land to farm separately, encouraging continued sharing of resources in joint or extended households (Bates 1982; Bates and Rassam 1983). Thus, household structure itself is dependent upon external economic factors.

The labor market also affects sex and age ratios within farm families. Although women in the region work as agricultural laborers, the majority of paid workers are men. Consequently, employment removes young, often

skilled, males during their prime working years, leaving agricultural operations to women, young children, and older men (Hopkins 1983; Khafagi
1983; Snobar and Arabiat 1984). Therefore, off-farm employment contributes to the aging and feminization of farm populations.

The other major contributing factor to this process is formal education.
School attendance removes children's labor from the farm (Friedl 1981) and,
more importantly, educated adults, particularly women, are less likely to
provide manual agricultural labor than uneducated adults (Crawford and
Purvis 1986; Shashahani 1986). Thus, education may permanently reduce a
family's farm labor force. Also, government jobs, increased industrialization,
and international and internal migration have increased opportunities for
educated adults to obtain full or part-time employment (Carapico 1985;
Glavanis 1984; Hopkins 1977). Thus, rural areas experience increases in
wage rates and shortages of labor due to migration to better paid, steadier
work (Hogan et al. 1983). Moreover, the opportunity to earn wages off-farm
becomes a part of farm economic calculations.

It is increasingly the case that farm households have an income source
other than agriculture (Hopkins 1983; Myntti 1984). Most often wage employment provides this alternative income. The decision for a family member
to seek off-farm work may be part of a household strategy to diversify the
farm's income base and provide cash for investment, or it may represent an
individual's decision (Horton 1961; Tully 1987). If a son, who has been the
family shepherd, chooses to work elsewhere and keep his wages for himself,
a labor and cash poor household may be forced to sell its flocks (Mir-Hosseini
1987). Therefore, wage employment may have different effects on the economic well-being of individuals residing in the same household.

The importance of income to households that are receiving remittances
varies. Often employment means unskilled labor entering temporary jobs
with no security. Such work supplements farm production but does not
replace it (Mir-Hosseini 1987). This is the primary role of off-farm income
for laborers who work for short periods and return to the farm in peak
seasons (USDA/USAID 1980). In other cases, regular, lucrative employment
reduces agriculture to a secondary source of income which may decrease
commitment to farm investment and maintenance (McLachlan 1985). In
either case, off-farm incomes increase the opportunity cost of family labor
and decrease its availability while raising the cost of hired labor (Khafagi
1983; Tully 1986).

Access to the labor market has important implications for family labor
organization. Male labor can be replaced to a significant degree by cash,
either to hire labor or to obtain farm equipment. This is not true for most
female labor, especially in the domestic sphere. Although both sexes expect
to work less when incomes increase, high male out-migration, labor scarcity,
and the increased cost of hired labor accentuate dependence on the family
labor of women and children (Myntti 1971; Rassam 1984). Thus, male access

to cash incomes may result in farms in which hired labor, machines, and women perform most productive activities (Maher 1974).

Hired labor is a different resource than family labor, and is usually reserved for peak season tasks. In rainfed areas, it is used primarily in harvesting, particularly non-cereals. Non-mechanized tasks with long, low-level labor demands are carried out by household labor on small farms. Hired labor is more specialized than family labor and the division of labor by sex is more rigid; even women who do male tasks at home may not provide them for wages (Amin and Awny 1985; Richards et al. 1983).

Expensive hired labor may make traditional manual methods of cultivation unprofitable (Snobar and Arabiat 1984). Therefore, hiring labor can affect farm technology, forcing the adoption of yield increasing techniques. Alternatively, in the face of decreasing male family labor, new technologies which require male labor may make hired labor a necessity.

Hired labor produces significantly lower quality work than does family labor in spite of greater supervision. This can decrease productivity. For example, in some areas wealthy households that use hired labor have lower yields than poor households where women do careful land preparation and continuous weeding (Myntti 1984). Hired labor may also reduce the availability of female family labor if women are unable to work with men who are strangers to them (Crawford and Purvis 1986; Maher 1974). Thus, hired labor is generally of lower quality than family labor and may have important impacts on agricultural operations and the use of family labor.

Farm Strategies

Households have access to variable endowments of land, family labor, and capital which fuel diverse economic strategies. The balance between opportunities for wage employment and agriculture differ with economic circumstances. Landless families or those with plots too small to provide subsistence must rent land to cultivate or sell their labor to survive (Bates 1982; Hopkins 1983; Mulayim 1985; Richards et al. 1983). On very small farms, household members migrate so that subsistence farming is possible for those remaining behind (Glavanis and Glavanis 1983). Others use the farm as a domestic base for educating children and sending out migrants who support the household. Farming may be carried out but good crops are considered windfalls (Hopkins 1980). The level of remittances determines whether such families live in poverty or comfort. Finally, some families rent their land to others and support themselves from rental income and wage labor (Bates 1982; Hogan et al. 1983). Thus, there is diversity in strategies even on plots which are too small to provide subsistence.

Larger farms offer more alternatives. The basic labor resource on farms of 5–10 ha is unpaid family labor, occasionally supplemented by neighbors or relatives. When this is insufficient, labor is replaced or supplemented by capital via mechanization or hired labor. These farms are labor-intensive enterprises utilizing few purchased inputs. These family operations, which make up the majority of smallholdings in rainfed areas, make rainfed farming economically feasible.

Subsistence is primary among farm goals. To meet household•needs, both livestock and crops are maintained (Uner 1980). Farm families keep livestock to provide meat and dairy products for their own use in spite of their heavy labor demands. The decision to keep animals also affects cultivation. Animal feed may be produced in environments more suited to other crops and crop yields may be affected by planting dates which are delayed to allow further grazing or cutting of fodder (Carapico and Hart 1977; Dyer and Gotsch 1983; Tully 1984). Nevertheless, the importance of livestock in consumption and as a buffer in times of low crop production outweigh these costs (Tully 1986). Cropping patterns also reflect a concern with subsistence needs. Most families produce a number of crops with differing maturation dates to provide variety in the diet and decrease the risk of long periods without food. Thus, small farms diversify into crops and livestock and pursue labor-intensive, low input agriculture. Family labor is the most important resource of smallholdings and the returns to labor-intensive techniques are reflected in high productivity per land or animal unit (Fitch and Soliman 1983).

There is a strong commitment to the integration of stock and crops, production for consumption, and the minimization of production costs. So, even if new crops are more appropriate, they will not be embraced if changing is costly or occurs at the expense of subsistence production (Affan 1984). Consequently, small and medium-sized farms benefit from labor-intensive crops (Hogan et al. 1983). Diversification is preferred to intensification as families are more reluctant to take risks with subsistence activities than to try and integrate something new such as wages (Hogan et al. 1983). Further diversification may be made into cash crops or wage employment to supplement income, but small-scale involvement in these activities seldom changes the basic orientation of these farms.

Farms above 10 ha are often too large to be worked by family labor alone so they are frequently mechanized and rely on hired labor. These patterns are associated with commercial agriculture and wage employment. Large, commercial farms are drawn into greater interaction with the state, becoming less self-sustaining, more cash-oriented, and less diversified (Beck 1978). Owners are frequently part-time managers with off-farm incomes (Hopkins 1985a). They may support livestock, but more commonly the greater share of their income comes from crops (Crawford and Purvis 1986; Hogan et al. 1984; Tully 1986). Monocropping of cultigens which can be handled by machines is common with increasing dependence upon purchased foods (Affan 1984). Larger farms are more mechanized and farming is a male activity. As labor is usually hired, it is often a constraint, being scarce or expensive (Hopkins 1985a; Tully 1987). Thus, large farms benefit from tractors, fertilizer, and insecticides rather than labor-intensive crops (Hogan et al. 1983). Crop choices may also be constrained with shifts to crops which can be serviced by machines (Aswad 1967).

Lacking adequate family labor, large commercial farms are dependent upon an unstable labor market and shortages can threaten their existence (Amin and Awny 1985; Peters 1978). When labor is scarce or expensive, capitalist farmers may be unable to work their farms profitably so in some areas labor shortages result in land being removed from production or commercial farmers returning to peasant management by tenants or sharecroppers (Peters 1963). Further, low returns to capital may result in diversification outside agriculture, resulting in fewer wealthy farmers operating capital-intensive operations (Affan 1984).

Thus, farms are differentiated by economic opportunity, size and composition of household, and access to resources such as land, labor, and capital. All of these factors influence farming strategies and determine where labor and capital will be invested to best secure family maintenance. The complexity of farm decision making calls into question studies which categorize farming systems and target populations on the basis of land size, participation in wage employment, or hiring of off-farm labor. Clearly, the impact of each of these factors on farm strategies is dependent upon the particular social, economic, and political context in which a household operates.

Conclusion

Clearly, the analytical utility of the peasant farm model which introduced this paper is questionable. Like a mirage, with close examination the sharp edges blur and the image of unity inverts. With increasing opportunities to earn cash off-farm, agriculture may not be the primary economic enterprise of rural households. Further, a focus on household residents will miss migrants who live elsewhere but make major contributions to farm management and income while including residents with limited participation in farm activities. Finally, management is not confined to the household, as the decisions of custom operators and the policies of local and international governments define certain economic possibilities. Thus, identifying the basic domestic economic unit with the activities and personnel of a household is inadequate even at a descriptive level.

Of equal importance, the assumption that households are undifferentiated and all members work together in pursuit of common goals is problematic. Households exhibit serious divisions between men and women and between children and parents. The widespread availability of off-farm wage employment allows individuals to withdraw their labor from the farm without considering the needs of other household members. Spouses may refuse to assist one another and may actually undermine each other's economic endeavors. One cannot assume that production and consumption objectives of individual family members are complimentary rather than conflicting, or that benefits and costs are equitably distributed. Rather, households are arenas in which individual strategies for resource acquisition come to terms with the constraints and opportunities of the wider social, political and economic context (Long 1984).

An adequate analysis of the household economy must consider both the internal dynamics of household organization and its relationship with the local and global economy. An ahistorical model cannot adequately represent this dynamic environment. Therefore, rather than focusing on the household as the major actor in the rural economy, it is best conceived as a nexus encompassing the strands of the social, economic, and political networks which are the context for economic action.

References

Affan, Khalid. 1984. Toward an Appraisal of Tractorisation Experience in Rainlands of Sudan. Development Studies and Research Centre Monograph Series No. 19. Faculty of Economic and Social Studies, University of Khartoum.

Amin, Galal A. and Awny, Elizabeth Taylor. 1985. International Migration of Egyptian Labour. Cairo: International Development Research Centre.

Aswad, Barbara. 1967. Key and Peripheral Roles of Noble Women in a Middle Eastern Plains Village. Anthropological Quarterly 40: 139–152.

Aswad, Barbara. 1978. Women, Class, and Power: Examples From the Hatay, Turkey. Pages 473–481 in Women in the Muslim World (Beck, Lois and Keddie, Nikki, eds). Cambridge: Harvard University Press.

Basson, Priscilla. 1981. Women and Traditional Food Technologies: Changes in Rural Jordan. Ecology of Food and Nutrition 2: 17–23.

Basson, Priscilla. 1984. Male Emigration and the Authority Structure of Families in North-West Jordan. Report to the Ford Foundation.

Bates, Daniel G. 1982. The Middle East Village in Regional Perspective. Pages 161–183 in Village Viability in Contemporary Society (Reining, Priscilla and Lenkerd, Barbara, eds). Boulder, Colorado: Westview Press.

Bates, Daniel G. and Lee, Susan H. 1977. The Role of Exchange in Productive Specialization. American Anthropologist 79: 824–841.

Bates, Daniel, G. and Rassam, Amal. 1983. Peoples and Cultures of the Middle East. Englewood Cliffs, NJ: Prentice-Hall.

Beck, Lois. 1978. Women Among Qashqa'i Nomadic Pastoralists in Iran. Pages 351–373 in Women in the Muslim World (Beck, Lois and Keddie, Nikki, eds). Cambridge: Harvard University Press.

Beeley, Brian W. 1985. Progress in Turkish Agriculture. Pages 289–301 in Agricultural Development in the Middle East (Beaumont, Peter and McLachlan, Keith, eds). Chichester: John Wiley and Sons.

Birks, J. Stace and Richards, Alan. 1985. Labor in Middle Eastern Agriculture. Pages 149–159 in Migrant Labor in Agriculture: An International Comparison (Martin, Philip L. ed). Berkeley: University of California.

Boserup, Ester. 1970. Woman's Role in Economic Development. New York: St. Martin's Press.

Carapico, Sheila. 1985. Yemeni Agriculture in Transition. Pages 241–254 in Agricultural Development in the Middle East (Beaumont, Peter and McLachlan, Keith, eds). Chichester: John Wiley and Sons.

Carapico, Sheila and Hart, S. 1977. The Sexual Division of Labor and Prospects for Integrated

Development: Report on Women's Economic Activities in Mahweet, Tawhile and Jihana Regions (Yemen Arab Republic). Report to USAID.

Chatty, Dawn. 1978. Changing Sex Roles in Bedouin Society in Syria and Lebanon. Pages 399–415 in Women in the Muslim World (Beck, Lois and Keddie, Nikki, eds). Cambridge, Massachusetts: Harvard University Press.

Commander, Simon and Hadhoud, Aly Abdullah. 1986. From Labour Surplus to Labour Scarcity? The Agricultural Labour Market in Egypt. Development Policy Review 4: 161–180.

Cosar, Fatma Mansur. 1978. Women in Turkish Society. Pages 124–140 in Women in the Muslim World (Beck, Lois and Keddie, Nikki, eds). Cambridge, Massachusetts: Harvard University Press.

Crawford, Paul R. and Purvis, Malcolm. 1986. The Agricultural Sector of Morocco: A Description. Country Development Strategy Statement, Annex C. Report to USAID.

Davis, Susan Schaefer. 1978. Working Women in a Moroccan Village. Pages 416–433 in Women in the Muslim World (Beck, Lois and Keddie, Nikki, eds). Cambridge, Massachusetts: Harvard University Press.

Dutton, Roderic. 1985. Agricultural Policy and Development: Oman, Bahrain, Qatar and the United Arab Emirates. Pages 227–240 in Agricultural Development in the Middle East (Beaumont, Peter and McLachlan, Keith, eds). Chichester: John Wiley and Sons.

Dyer, Wayne N. and Gotsch, Carl H. 1983. Public Policy and the Demand for Mechanization in Egyptian Agriculture. Pages 199–222 in Migration, Mechanization and Agricultural Labor Markets in Egypt (Richards, Alan and Martin, Philip L., eds). Boulder, Colorado: Westview Press.

Ehlers, Eckart. 1985. The Iranian Village: A Socioeconomic Microcosm. Pages 151–170 in Agricultural Development in the Middle East (Beaumont, Peter and McLachlan, Keith, eds). Chichester: John Wiley and Sons.

el-Ghonemy, Mohamad Riad. 1979. Agrarian Reform and Rural Development in the Near East. WCARRD, RNEA Paper No. 5. Rome: FAO.

FAO (Food and Agriculture Organization). 1986. Role of Women in Food and Agriculture in the Region. Eighteenth FAO Regional Conference for the Near East, Istanbul, 17–21 March 1986.

Fazel, G. Reza. 1977. Social and Political Status of Women Among Pastoral Nomads: The Boyr Ahmad of Southwest Iran. Anthropological Quarterly 50: 77–89.

Fernea, Robert. 1969. Land Reform and Ecology in Postrevolutionary Iraq. Economic Development and Culture Change 17: 356–381.

Fitch, James B. and Soliman, Ibrahim A. 1983. Livestock and Small Farmer Labor Supply. Pages 45–78 in Migration, Mechanization and Agricultural Labor Markets in Egypt (Richards, Alan and Martin, Philip L., eds). Boulder, Colorado: Westview Press.

Friedl, Erika. 1981. Women and the Division of Labor in an Iranian Village. MERIP Reports No. 95: 13–18.

Glavanis, Kathy R.G. 1984. Aspects of Non-capitalist Social Relations in Rural Egypt: The Small Peasant Household in an Egyptian Delta Village. Pages 30-60 in Family and Work in Rural Societies. Perspectives on Non-Wage Labour (Long, Norman, ed.). London: Tavistock Publications.

Glavanis, Kathy R.G. and Glavanis, Pandeli M. 1983. Trend Report. The Sociology of Agrarian Relations in the Middle East: The Persistence of Household Production. Current Sociology 31: 1–109.

Hammam, Mona. 1986. Capitalist Development, Family Division of Labor, and Migration in the Middle East. Pages 158–173 in Women's Work (Leacock, Eleanor and Safa, Helen I., eds). Massachusetts: Bergin and Garvey Publishers, Inc.

Hogan, Edward B., Furtick, William R., and Grayzel, John A. 1984. Morocco Increase in Cereal Production Project. Report to USAID.

Hogan, Edward B., Furtick, William R., and Hansen, Gary. 1983. Jordan Wheat Research and Production. Project Impact Evaluation. Report to USAID.

Hopkins, Nicholas S. 1977. The Emergence of Class in a Tunisian Town. International Journal of Middle East Studies 8: 453–491.

Hopkins, Nicholas S. 1980. Social Soundness Analysis of the Dryland and Irrigation Components of the Proposed Central Tunisian Rural Development Program (CTRD). Report to USAID.

Hopkins, Nicholas S. 1983. The Social Impact of Mechanization. Pages 181–197 in Migration, Mechanization and Agricultural Labor Markets in Egypt (Richards, Alan and Martin, Philip L., eds). Boulder, Colorado: Westview Press.

Hopkins, Nicholas S. 1985a. The Political Economy of Two Arab Villages. Pages 307–321 in Arab Society. Social Science Perspectives (Hopkins, Nicholas and Ibrahim, Saad Eddin, eds). Cairo: American University in Cairo Press.

Hopkins, Nicholas S. 1985b. The Social and Economic Impact of Agricultural Mechanization. Summary of a Conference, Amman, 18–21 May 1984. Amman: Mediterranean Research Cooperation Project.

Horton, Alan W. 1961. A Syrian Village in its Changing Environment. Doctoral Dissertation, Harvard University.

Irons, William. 1972. Variation in Economic Organization: A Comparison of the Pastoral Yomut and the Basseri. Pages 88–104 in Perspectives on Nomadism (Irons, William and Dyson-Hudson, Neville, eds). Leiden: Brill.

Jaubert, Ronald. 1986. The Semi-arid Areas of Syria: Farming Systems in Decline. Issues in Research Design. Pages 244–257 in Selected Proceedings of Kansas State University 1984 Farming Systems Research Symposium. Farming Systems Research and Extension: Implementation and Monitoring, Paper No. 9 (Flora, Cornelia Butler and Tomecek, Martha, eds).

Kandiyoti, Deniz. 1975. Social Change and Social Stratification in a Turkish Village. Journal of Peasant Studies 2: 206–219.

Kandiyoti, Deniz. 1977. Sex Roles and Social Change: A Comparative Appraisal of Turkey's Women. Signs 3: 57–73.

Khafagi, Fatma. 1983. Socio-economic Impact of Emigration from a Giza Village. Pages 135–156 in Migration, Mechanization and Agricultural Labor Markets in Egypt (Richards, Alan and Martin, Philip L., eds). Boulder, Colorado: Westview Press.

Khattab, Hind Abou Seoud and El Daeif, Syada Greisse. 1982. Impact of Male Labor Migration on the Structure of the Family and the Roles of Women. Regional Papers. Cairo: Population Council.

Kopp, Horst. 1985. Land Usage and Its Implications for Yemeni Agriculture. Pages 41–50 in Economy, Society, and Culture in Contemporary Yemen (Pridham, B.R., ed.). Beirut: American University of Beirut.

Lewis, Norman. 1953. Lebanon–the Mountain and Its Terraces. Geographical Review XLIII: 1–14.

Long, Norman. 1984. Introduction. Pages 1–29 in Family and Work in Rural Societies. Perspectives on Non-Wage Labour (Long, Norman, ed.). London: Tavistock Publications.

McLachlan, Keith. 1985. The Agricultural Development of the Middle East: An Overview. Pages 27–50 in Agricultural Development in the Middle East (Beaumont, Peter and McLachlan, Keith, eds). New York: John Wiley and Sons.

Maher, Vanessa. 1974. Women and Property in Morocco. Cambridge Studies in Social Anthropology 10. Cambridge: Cambridge University Press.

Maher, Vanessa. 1978. Women and Social Change in Morocco. Pages 100-123 in Women in the Muslim World (Beck, Lois and Keddie, Nikki, eds). Cambridge, Massachusetts: Harvard University Press.

Mahhouk, Adnan. 1956. Recent Agricultural Development and Bedouin Settlement in Syria. Middle East Journal 10: 167–177.

Mir-Hosseini, Ziba. 1987. Some Aspects of Changing Economy in Rural Iran: The Case of Kalardasht, a District in the Caspian Provinces. International Journal of Middle East Studies 19: 393–412.

Mulayim, Ziya Gokalp. 1985. Assessment of Rural Landless in Turkey. Rome: FAO.

91

Mundy, Martha. 1985. Agricultural Development in the Yemeni Tihama: The Past Ten Years. Pages 22–40 in Economy, Society and Culture in Contemporary Yemen (Pridham, B.R., ed.). London: Croom Helm.

Murdock, Muneera Salem. 1979. The Impact of Agricultural Development on a Pastoral Society: The Shukriya of the Eastern Sudan. Report to USAID.

Myntti, Cynthia. 1971. Women and Development in the Yemen Arab Republic. Eschborn: German Agency for Technical Cooperation.

Myntti, Cynthia. 1979. Women and Development in Yemen Arab Republic. Federal Republic of Germany: German Agency for Technical Cooperation (GTZ).

Myntti, Cynthia. 1984. Yemeni Workers Abroad: The Impact on Women. MERIP Reports No. 14: 3–10.

Nelson, Cynthia. 1973. Women and Power in Nomadic Societies in the Middle East. Pages 43–59 in The Desert and the Sown (Nelson, Cynthia, ed.). Research Series, No. 21. Institute of International Studies. Berkeley: University of California.

Newberg, Richard. 1982. Rainfed Agriculture Sub-sector Strategy. Morocco. Report to USAID.

Okazaki, Shoko. 1985. Agricultural Mechanization in Iran. Pages 171–187 in Agricultural Development in the Middle East (Beaumont, Peter and McLachlan, Keith, eds). Chichester: John Wiley and Sons.

Peters, Emrys L. 1963. Aspects of Rank and Status Among Muslims in a Lebanese Village. Pages 159–200 in Mediterranean Countrymen. Essays in the Social Anthropology of the Mediterranean (Pitt-Rivers, Julian, ed.). The Hague: Mouton and Company.

Peters, Emrys L. 1978. The Status of Women in Four Middle East Communities. Pages 311–350 in Women in the Muslim World (Beck, Lois and Keddie, Nikki, eds). Cambridge, Massachusetts: Harvard University Press.

Rabo, Annika. 1986. Change on the Euphrates. Villagers, Townsmen, and Employees in Northeast Syria. Stockholm Studies in Social Anthropology, No. 15. Stockholm: University of Stockholm.

Rassam, Andree Marie. 1984. Syrian Farm Households: Women's Labour and Impact of Technologies. MA dissertation, University of Western Ontario.

Rassam, Andree and Tully, Dennis. 1985. Gender Related Aspects of Agricultural Labor in Northwestern Syria. Pages 287–301 in Gender Issues in Farming Systems Research and Extension (Poats, Susan V., Schmink, Marianne, and Spring, Anita, eds). Boulder, Colorado: Westview Press.

Richards, Alan and Martin, Philip L. 1983a. Introduction. Pages 1–17 in Migration, Mechanization and Agricultural Labor Markets in Egypt (Richards, Alan and Martin, Philip L., eds). Boulder, Colorado: Westview Press.

Richards, Alan and Martin, Philip L. 1983b. The Laissez-faire Approach to International Labor Migration: The Case of the Arab Middle East. EDCC 31: 455–474.

Richards, Alan, Martin, Philip L., and Nagaar, Rifaat. 1983. Labor Shortages in Egyptian Agriculture. Pages 21–44 in Migration, Mechanization and Agricultural Labor Markets in Egypt (Richards, Alan and Martin, Philip L., eds). Boulder, Colorado: Westview Press.

Rogers, Barbara. 1980. The Domestication of Women. London: Tavistock.

Saunders, Lucie Wood. 1968. Aspects of Family Organization in an Egyptian Delta Village. Transactions of the New York Academy of Sciences, Series II 30: 714–721.

Shashahani, Soheila. 1986. Mamasani Women: Changes in the Division of Labor Among a Sedentarized Pastoral People of Iran. Pages 111–121 in Women's Work (Leacock, Eleanor and Safa, Helen I., eds). Massachusetts: Begin and Garvey Publishers, Inc.

Snobar, Bassam A. and Arabiat, Suleiman M. 1984. The Mechanization of Agriculture and Socio-economic Development in Jordan. Dirasat (Agricultural Sciences) 11: 159–200.

Sweet, Louise. 1954. Tel Toqaan: A Syrian Village. Museum of Anthropology Papers No. 14. University of Michigan.

Sweet, Louise. 1974. In Reality: Some Middle Eastern Women. Pages 379–397 in Many Sisters (Matthiasson, Carolyn J., ed.). New York: Free Press.

Tavakolian, Bahram. 1987. Sheikhanzai Women: Sisters, Mothers, Wives. Ethnos 1/2: 180-199.

Tinsley, R.L. and M. ter Kuile. 1986. Mechanization in Small Farm Systems. Pages 349–363 *in* Selected Proceedings of Kansas State University 1984 Farming Systems Research Symposium. Farming Systems Research and Extension: Implementation and Monitoring, Paper No. 9 (Flora, Cornelia Butler and Tomecek, Martha, eds).

Tully, Dennis. 1984. Land Use and Farmer's Strategies in Al Bab: The Feasibility of Forage Legumes in Place of Fallow. Research Report No. 13. Aleppo: ICARDA.

Tully, Dennis. 1986. Rainfed Farming Systems of the Near East Region. Discussion Paper No. 17. Aleppo: ICARDA.

Tully, Dennis. 1987. Culture and Context in Sudan: The Process of Market Incorporation in Dar Masalit. New York: State University of New York Press.

Uner, Sunday. 1980. Effects of Migration on the Rural Economy of Turkey. Employment and Manpower Planning Studies. Rome: FAO.

USDA/USAID and State Planning Commission Syrian Arab Republic. 1980. Syria. Agricultural Sector Assessment, Vol. 1. Summary Report.

5. Perspectives from Labor Studies in South Asia

JAYATI GHOSH

Centre for Economic Studies and Planning, Jawaharlal Nehru University, New Mehrauli Road, New Dehli 110 067, India

The problems of agricultural labor in many developing countries tend to be broadly similar but it is difficult to make a comparison between two such vast regions as South Asia, and West Asia and the Mediterranean. Also, it is difficult to describe the salient features of rural labor markets in South Asia, primarily because of substantial regional variations in patterns of growth of output and farm investment, property relations, tenurial conditions, labor use, and the labor process.

Some of the main similarities and differences between the above two regions are discussed in this paper, using India as a reference point although some trends in other South Asian countries are discussed.

Property Relations and Land Reform

In the West Asian and Mediterranean countries discussed in this volume, there appear to be broadly two types of systems of landholdings. One is characterized by concentration of landownership with a broadly dichotomous structure: very large capitalist holdings with widespread use of farm labor and/or machinery, and a mass of small holdings in which farming is more subsistence-oriented. The latter holdings tend to be based on dryland farming with low income and high risk. Wealthier holdings tend to be clustered in the irrigated sector, where most of the agricultural investment is also concentrated. This dualism seems to prevail, for example, in Iraq, Tunisia, and Turkey. The second system is more equitable and is dominated by relatively small holdings, often (as in Morocco) as the result of post-independence state land reform. These small farms may be run on entirely capitalist lines, with diversified farming, profit orientation, and reliance on mechanized techniques (e.g., Cyprus). Alternatively, they may remain peasant in character and orientation, with more limited farm investment and operating both as cultivators and suppliers of labor to neighboring countries (e.g., Yemen AR).

In all cases, two features stand out. One is the simplicity of land tenure. The other is the significance of the state, either in perpetuating a particular form of dualism through selective agrarian policies, or in dramatically altering landownership and tenurial patterns through successful land reforms.

Dennis Tully (ed.), Labor and Rainfed Agriculture in West Asia and North Africa, 93–99.
© 1990 *ICARDA*.

The situation in rural South Asia, and particularly India, is far more complex. Property and production relations range from areas in which large, semi-feudal holdings dominate to areas characterized by smallholder subsistence farming and areas where capitalist farming based on hired labor and/or heavily mechanized techniques is common.

Generally, land and other assets are highly unequally distributed, with landholding concentrated in the upper categories of holding size and a substantial proportion of the rural population being landless. Most cultivators have holdings of 0–5 ha, very small by West Asian standards. Land reforms have had little or no effect on this structure, and the operation of credit and input and product markets tends to reinforce inequality. Rural market interlinkage has been increasingly noted in recent years and contracts in particular markets are used to influence or control the outcomes in other markets. Labor-tying, through rental or credit contracts, is one obvious example, but a much wider range of alternative forms has been catalogued and analyzed (Bharadwaj 1974; Bhaduri 1981; Bardhan 1984).

Forms of production relationship include capitalist farming, small-scale peasant proprietorship, and other, more oppressive "semi-feudal" forms of tenure. The agriculture of the northwestern region has been the most dynamic, largely because of the tradition of small peasant proprietors and also because of public investment in irrigation. The eastern region remains the most backward, because of inadequate public investment and also the constraining effects of property relations.

The situation of tenants in many parts of rural India remains precarious and unprotected, while for agricultural laborers, real incomes have stagnated or declined and employment remains insecure. Legislation on minimum wages and land reforms have had very little impact. The abolition of intermediate, rent-receiving, non-cultivating landlords in a number of states, especially in the northern provinces in the 1950s, paved the way for peasant proprietorship. Attempts to provide tenant security have been sporadic and, with few exceptions, largely unsuccessful. Thus tenant legislation in Punjab in the 1960s, which also attempted to regulate rents, primarily increased the rate of eviction. In West Bengal, the systematic registration of sharecroppers in the late 1970s and early 1980s has provided limited security. In a number of states tenancy has been legally banned, but this has usually increased the insecurity of *de facto* tenancy contracts.

Ceilings have been placed on landownership in most states but they have had a nominal impact and patterns of land distribution continue to be similar to the past. Less than 1 million ha have been released for redistribution, giving an average of only 0.5 ha per beneficiary, even with a very restricted group eligible to benefit (see Joshi 1975; Haque and Sirohi 1986). Consolidation of holdings has been more successful primarily because landownership is not threatened and the existing distribution can be more efficiently organized.

In general, in rural India, the combination of continued population pressure on land, concentration of landholdings, and prevailing institutional and property relations has constrained agrarian growth and innovation, and allowed the continued widespread prevalence of rural poverty and underemployment. Thus, unlike West Asia, the ineffectiveness of land reforms and the resistance of particular production relations have meant slower growth as well as the continuation of the backward and oppressive conditions of agricultural laborers. They have also meant that the nature of the migration within and from rural areas (of a "push" variety) has been very different from that witnessed in the West Asian and Mediterranean countries described in this book.

Patterns of Labor Use and Mechanization

Differences in rural labor use stem from different patterns of property relations. In South Asia there is a long tradition of inequitous ownership and fairly rigid demarcation of the rural population, with production relations tending to perpetuate traditional forms of cultivation. This has affected agricultural labor relations and labor use. In contrast, in many West Asian countries, settled cultivation has evolved and coexisted with nomadic pastoralism and has been affected by colonial and post-colonial policies.

The resource boom of the 1970s has been the crucial factor affecting labor markets and labor processes in the Middle East in the recent past. The demand for labor in oil-producing countries has caused marked shortages of skilled labor and a significant depletion of the rural workforce in many countries, sometimes causing a gender shift in the labor force as more women take over previously male tasks. Only Morocco appears to be facing increasing population pressure on scarce land, similar to South Asia. Migration has also affected macro-economic variables, through its effects on the balance of payments, sectoral distribution of income, and patterns of labor use. More resources have been generated for domestic industrialization and there has been significant growth of the services sector. Jordan is the most striking example of this; agriculture accounts for only 8% of national income. Expansion of services has generated more employment and greater demand for agricultural products. The response to the decline in labor supply and the need to increase output has typically involved falls in gross cultivated area, increases in labor-substituting mechanization (e.g., Turkey, Tunisia, Iraq, Yemen AR, and Cyprus), and rises in rural real wages.

In South Asia external migration and its impact on domestic labor markets have been more limited than in West Asia and Mediterranean countries. Migration to oil-rich countries for employment has been mainly from Pakistan, Bangladesh, and in India, coastal Kerala and Gujarat. In these areas, Pakistan in particular, there have been increases in income and consumption flows due to remittances. Elsewhere in India, migration is related to the push forces created by a growing population dependent on a stagnating agricultural sector.

Rural-urban migration has changed little in the past 2 decades, although circular or seasonal rural-rural migration has increased. This reflects regional inequalities in agricultural growth: inter-regionally as labor from the stagnating rural areas of the eastern states moves to the more dynamic northwest ("Green Revolution") areas; and intra-regionally as the irrigated areas with more intensive cropping and labor use attract labor from drier, more traditional regions. The availability of rural employment in different regions has determined the choice of destination, while migration is more affected by declining living standards in places of origin. In all regions, despite occasional and skill-based shortages, labor surplus predominates. Rural population pressure on land has been high, and exacerbated by slow industrial employment growth and limited opportunities for migration.

In many of the less dynamic agricultural regions, there is a pattern of seasonal migration to more productive areas, which has been instrumental in maintaining household survival at subsistence levels (Bharadwaj 1988; Breman 1985). Such migration has not caused any tightening of labor markets in source areas, and has restrained wage increases in labor-intensive and productive regions. In all states except Punjab and Haryana, the two states with fastest output growth rates, real wages of agricultural laborers have increased only slightly since 1970/71 (Jose 1988). In this environment, the demand for labor-substituting mechanization is limited.

In the 1950s and 1960s, investment in mechanization was related primarily to irrigation. The recent spread of groundwater irrigation (mainly tubewells) has increased the geographical scope of such investment. Since the 1970s the use of tractors, threshers, harvester combines, and other labor-substituting machines, primarily in wheat-growing "Green Revolution" areas, has grown (Hanumantha Rao 1975), in response to the need for greater labor control and supervision and greater seasonality of labor demand in the new HYV seed-based strategy, as well as potential pressures for rising wages (Binswanger and Rosenzweig 1984; Sen 1981a; 1981b). Increases in output per hectare in "newer" Green Revolution areas, such as parts of eastern Uttar Pradesh, Bihar, Gujarat, and Madhya Pradesh, have not been associated with any rises in labor use per hectare, which has remained unchanged due to labor-substituting mechanization (Bhalla 1987).

In South Asia, mechanization of agriculture has taken place in a context of relative labor surplus and land scarcity, implying that output increases need not involve greater labor absorption in agriculture. Thus over time mechanization substantially increases rural unemployment and underemployment, which are compounded by a low demand for labor in the industrial sector, especially in India. Industrial employment in the formal sector has stagnated over the last decade in India, while in the growing informal sector employment growth in manufacturing and services has lagged behind output growth. This has been related to falls in absolute income per capita, especially in the rural informal sector, and continued inefficiency of labor use (Sen 1988), reflecting broader macro-economic trends which have entailed the neglect of agriculture in terms of investment and pricing policies.

The Impact of Government Policies

In the countries discussed in this volume, the trend has been towards mechanization, often actively supported by the state, as in Turkey and Cyprus. In South Asia, state strategies have focused on high-yielding seed varieties, fertilizer, pesticides, and irrigation in an attempt to increase yields. These strategies, most prevalent in Pakistan and northwest India, have neglected dryland farming and concentrated on irrigated areas. Thus in India the highest growth rates for output and yield since independence have been in the irrigated northwest region while in some rainfed regions, particularly in the east, per capita agricultural output has stagnated or declined in the past 2 decades.

The use of labor-substituting mechanization, while not actively supported and subsidized by the state (except to some extent in Pakistan), has nonetheless been encouraged by the provision of credit, particularly to wealthier cultivators who have invested in mechanization. Institutional credit has increased substantially in rural India, but essentially remains very unequally distributed (Dasgupta 1980). Thus policy has reinforced patterns of inequality based on farm size as well as regional inequalities stemming from inherent differences in productivity.

Inter-sectoral terms of trade have tended to move against agriculture in the past decade and a half, not only in world markets but also within all of the national economies considered here. This has also been true in South Asia, with some regions severely affected, although at the macro level there has been no noticeable deceleration in agricultural investment or output growth (Ghosh 1988). This contrasts with the countries experiencing a resource-led boom. As Commander and Burgess point out, windfall resource gains mainly accrued to the non-agricultural sectors and in several countries led to a rapidly widening food deficit. Despite a bad drought in 1987/88, India has moved from a food deficit to a food surplus situation, but this shift at the aggregate level is based on a pattern of income and consumption distribution in which rural poverty and malnutrition continue to be widespread.

Concluding Remarks

One striking common feature of the Middle East and South Asia is the continued subsistence orientation of dryland cultivation which has been adversely affected by a resources squeeze, diversion of funds into sectors other than agriculture, and by negative movements of terms of trade. In West Asian and Mediterranean countries, the income problem in the rural economy has been largely mitigated by labor migration and remittance inflows; the other side of the coin has been chronic labor shortages. In South Asia, the absence of employment alternatives on a sufficiently large scale has meant continued labor surpluses, widespread underemployment, and generally stagnating incomes in a backward agriculture. As the resource boom plays itself out in the Middle East, pressure on land may increase, resulting in some of the problems currently faced by laborers in South Asia.

The extent to which these problems can be avoided depends on the inter-sectoral allocation of investible resources and patterns of employment growth outside agriculture. Indeed, one basic lesson which emerges from these different country studies is that the conditions facing agricultural labor are fundamentally affected by changes within agriculture (tenurial patterns, land distribution, technological changes) and by both national and international macro-economic processes. These processes affect the employment and income alternatives for agriculturalists and the potential for greater income generation within agriculture.

References

Bardhan, Pranab. 1984. Land, Labor and Rural Poverty: Essays in Development Economics. Delhi: Oxford University Press.

Bhaduri, Amit. 1981. The Economic Structure of Backward Agriculture. London: Academic Press.

Bhalla, Sheila. 1987. Trends in Employment in Indian Agriculture, Land and Asset Distribution. Indian Journal of Agricultural Economics 42: 537–560.

Bharadwaj, Krishna. 1974. Production Conditions in Indian Agriculture. A Study Based on Farm Management Surveys. Occassional Paper No. 33, University of Cambridge. London: Cambridge University Press.

Bharadwaj, Krishna. 1988. On the Formation of the Labor Market in Rural Asia. Mimeo. Geneva: International Labor Office (EMP/RU).

Binswanger, Hans P. and Rosenzweig, Mark R. (eds). 1984. Contractual Arrangements, Employment and Wages in Rural Labor Markets in Asia. New Haven: Yale University Press.

Breman, Jan. 1985. Of Peasants, Paupers and Migrants: Rural Labor Circulation and Capitalist Production in West India. Delhi: Oxford University Press.

Dasgupta, Sipra. 1980. Class Relations and Technical Change in Indian Agriculture. New Delhi: Macmillan.

Ghosh, Jayati. 1988. Terms of Trade, Agricultural Growth and the Patterns of Demand. Social Scientist, May.

Haque, T. and Sirohi, A.S. 1986. Agrarian Reforms and Institutional Changes in India. New Delhi: Concept Publishing Company.

Hanumantha Rao, C.H. 1975. Technological Change and the Distribution of Gains in Indian Agriculture. New Delhi: Macmillan.

Jose, A.V. 1988. Agricultural Wages in India. Working Paper. New Delhi: ILO-ARTEP.

Joshi, P.C. 1975. Land Reforms in India: Trends and Perspectives. New Delhi: Allied Publishers.

Sen, Abhijit. 1981a. Market Failure and Control over Labor Power: Towards an Explanation of 'Structure' and Change in Indian Agriculture. Part 1. Cambridge Journal of Economics 5: 201–228.

Sen, Abhijit. 1981b. Market Failure and Control over Labor Power: Towards an Explanation of 'Structure' and Change in Indian Agriculture. Part 2. Cambridge Journal of Economics 5: 327–350.

Sen, Abhijit. 1988. A Note on Employment and Living Standards in Unorganised Sectors in India. Social Scientist, April.

Vaidyanathan, A. 1986. Labor Use in Rural India: A Study of Temporal and Spatial Variations. Economic and Political Weekly 21(52): A130–A146.

PART TWO

Country Review Papers

6. Agricultural Labor and Technological Change in Turkey

HALUK KASNAKOGLU, HALIS AKDER and A. ARSLAN GURKAN
Department of Economics, Middle East Technical University, Ionu Bulvari, Ankara, Turkey

Introduction

Turkey is a large country with a wide range of temperature, rainfall, and soil quality (Figure 1), and its agricultural sector has many diverse farming systems. The country is divided into 67 provinces, which are grouped into nine geographical regions for agricultural statistics. These roughly define economic regions, but we have devised another classification which is more relevant to agriculture and includes consideration of land use. Predominantly rainfed provinces were divided into three categories according to aridity. The first two categories can be considered dryland farm areas, while the third includes the wetter Black Sea provinces. Provinces with over 20% of the cropped area irrigated are a fourth category (Figure 2).

In this paper the discussion is focused on the first two categories for comparison with other rainfed countries of West Asia and North Africa. Particular attention is given to Central Anatolia, an important component of the rainfed area, made up of regions I and IX.

On the whole, disequilibrium between labor supply and demand is not the most important issue in Turkish agriculture, but it is important in certain regions. In this paper, we have identified some issues on which detailed analyses could fruitfully be carried out. We have concentrated on collating and analyzing the empirical evidence rather than surveying views or assertions not well supported by facts. Most of the information presented here is the result of the authors' on-going research, involving substantial reworking and analyses of the raw data.

Historical Background

Until the second half of the 19th century, the agricultural pattern in semi-arid Central Anatolia changed little, and it may be considered typical of Anatolia generally. Farm size was limited by the available draft power (i.e., a pair of oxen), use of a steel-tipped wooden plow, and climatic conditions (Herrman 1900). Holdings were divided into small plots scattered throughout the village, probably to avoid the risks of having all land on a single type of soil at the same location. After summer fallowing, the soil was cross-plowed several times and wheat seed from the previous harvest was broadcast and

Dennis Tully (ed.), Labor and Rainfed Agriculture in West Asia and North Africa, 103–133.
© 1990 *ICARDA.*

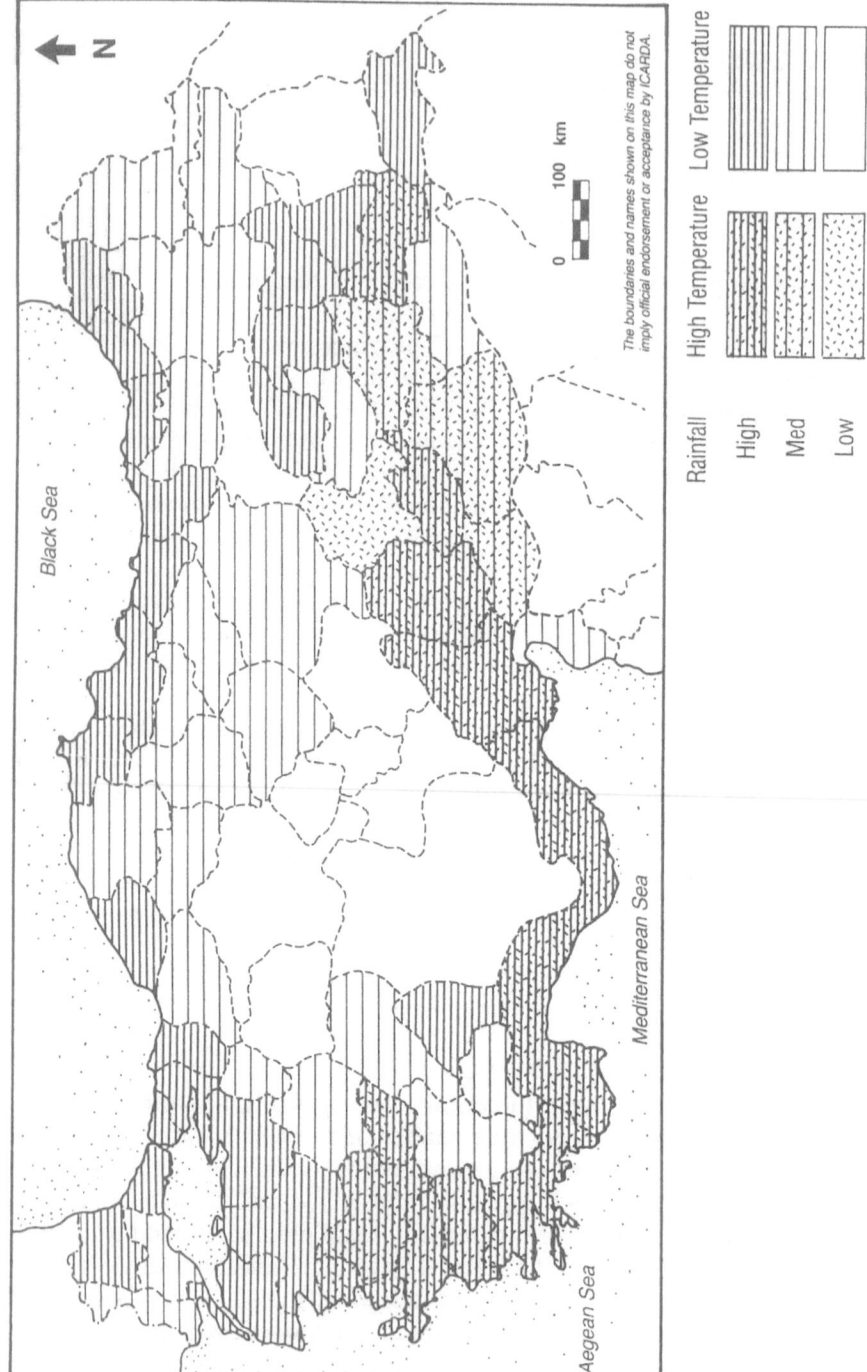

Legend text within figure:

Rainfall

High Temperature   Low Temperature

High

Med

Low

The boundaries and names shown on this map do not imply official endorsement or acceptance by ICARDA.

0   100   km

N

Black Sea

Aegean Sea

Mediterranean Sea

Fig. 1. Distribution of rainfall and temperature in Turkey. *Source:* United Nations Maps No. 3088 Rev. 2 (1986), No. 3063 Rev. 1 (1986).

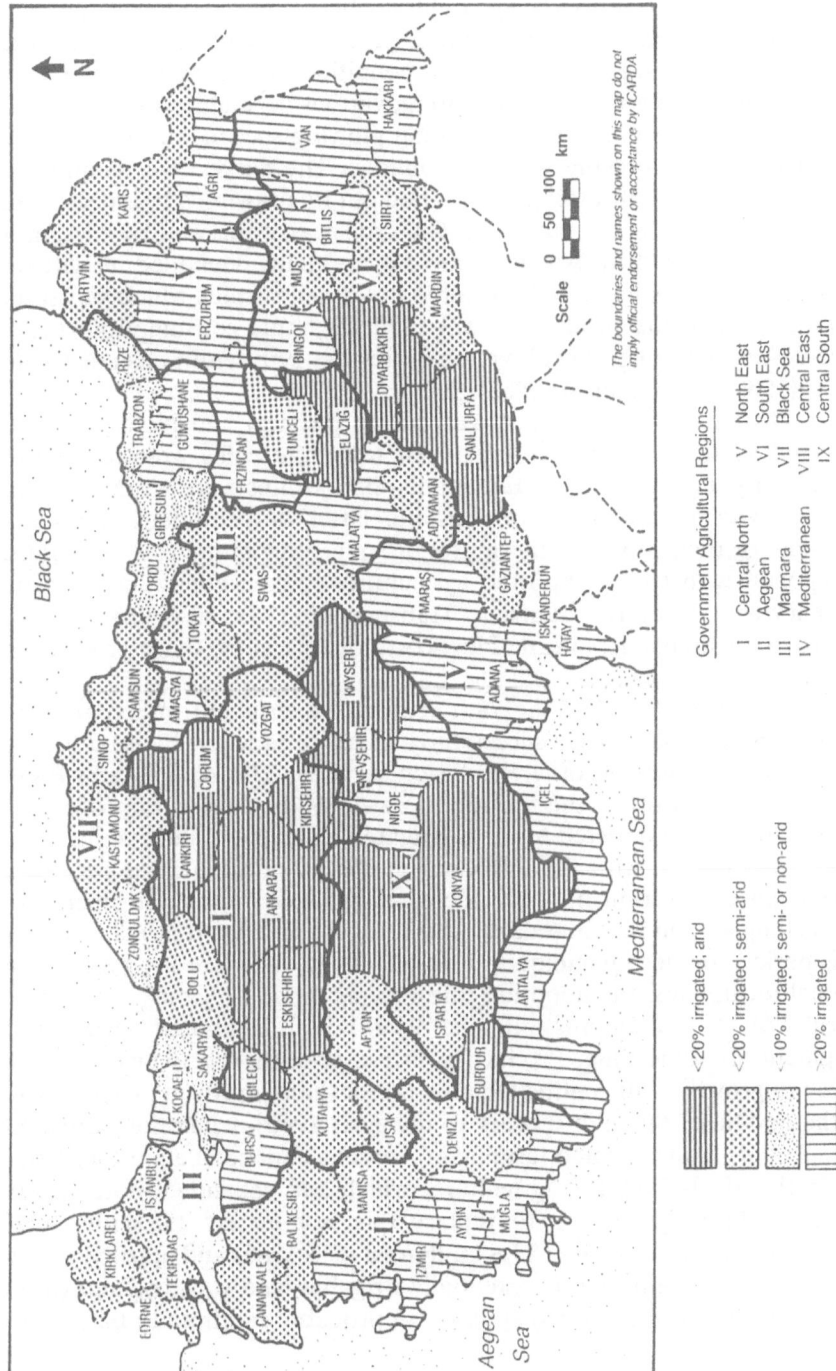

Fig. 2. Moisture/land use categories in Turkey. *Source:* Adapted from Gurkan (1986); United Nations Maps No. 3088 Rev. 2 (1986), No. 3063 Rev. 1 (1986).

covered by plow. Autumn plowing could not begin until the first rains, and was brought to an end by winter; the period when plowing was possible could be as short as 2 weeks. Under these conditions, one man with two oxen could plow approximately 3–5 ha. In spring, various vegetables were planted, usually for home consumption. The spring plowing season was also limited to a brief period, before the onset of summer. The harvest was completed about mid-August and threshing was carried out by drawing a board studded with sharp flints across the cut crop. The crop was winnowed by tossing it into the air, and the straw was used as forage. Yields depended heavily on the weather. Good yields usually resulted in a small surplus which was sold, but a large part of the production was used for home consumption. More detailed descriptions of grain cultivation during the 19th century can be found in Issawi (1980) and Helburn (1955).

The persistence of this pattern was due to physical and biological constraints and socioeconomic factors. In Anatolia, the cities were located in areas with irrigated agriculture. Economic relationships were weak between the rainfed villages and the towns as the latter did not depend on the countryside for their food. Furthermore, urban centers in the coastal regions were supplied with cheap imported agricultural commodities, due to trade agreements with Western countries. Transportation was also difficult. All these factors resulted in a pattern of low-cost, low-risk subsistence production in rainfed areas.

During the second half of the 19th century, government intervention in agriculture increased; policies successfully raised production by establishing schools, model farms, and other extension services. The Agricultural Bank was also founded during this period (Issawi 1980). Major improvement schemes were mainly directed towards irrigated agriculture, but rainfed agriculture also benefited with the introduction of new crops such as potatoes, support for tobacco production, and increases in wheat production (Herrman 1900). As long as land was abundant, which was the case in Anatolia throughout this period, modern techniques diffused slowly and were used alongside traditional techniques, the adoption rate varying among regions.

World War I, the Depression, and World War II limited agricultural development for some time. Production increases in this period were essentially due to growth of the cultivated area rather than improved productivity. From the middle of the 19th century to the early 1960s, when the cultivated area reached its limits, rainfed farming in Anatolia expanded almost continuously. In the 1960s irrigated agriculture, which had expanded more slowly than rainfed farming, started to replace this system. By 1985, the total cultivated area in Anatolia was 23.9 million ha (SIS 1985) and approximately 37.9% of the 8.5 million ha of potentially irrigable land was irrigated (SPO 1985). Nevertheless, the rainfed system will probably continue to be dominant.

In the 1950s, responding to increased demand for wheat exports and the opportunities of Marshall aid, economic policies favored imports of mechanical equipment, highway construction, and general government support to agriculture. The most notable changes occurred in mechanization. Tractor imports increased from under 2,000 to 40,000 in the early 1950s. For a short period, imports stopped, due to a lack of repair services and spare parts as well as foreign exchange constraints. However, since 1963 tractor imports and domestic production have increased continuously and the stock is now around 584,000 (SIS 1960d-85d).

Increased mechanization was controversial during the 1950s. Mechanization was blamed for releasing labor from agriculture which exacerbated unemployment and migration to cities. Also, until the 1950s, German scholars in the agricultural faculties in Turkey expressed the view, based on Western European experience, that improvements in land productivity through adoption of modern biological techniques were a precondition for mechanization. They also pointed out the advantages of traditional machinery and equipment. For example, steel-tipped, traditional plows reduced erosion compared with many modern tractor plows. Wooden equipment was easier, faster, and cheaper to repair than tractors.

Arguments for mechanization stressed the disadvantages of traditional technology. The primary source of draft power in traditional rainfed farming was, and still is, a pair of oxen. To feed them, a very large area of cultivable land must be reserved for pasture. Moreover, the time available for plowing by animal power in spring and autumn is very short. With tractors and harvesters work is completed in a timely manner, and both labor and land productivity are increased.

Whether mechanization led to unemployment and rural migration during the 1950s is uncertain. During this period, cultivable land was still available, so even if tractors replaced sharecroppers on large estates, they could have acquired new land. Besides, many of those who migrated to the cities were from families with small landholdings, who were more common than sharecroppers. Also, during the 1950s the total number of tractors was only 50,000, while during 1970–80 the number increased from around 100,000 to 400,000 and at the same time the rural population and cultivated land remained almost constant, with no proportionate increase in the number of landless farmers.

Regions with relatively large farms became mechanized much more rapidly than those with small farms and sharecropping gradually disappeared. The demand for seasonal workers increased, especially at peak periods and in regions where cash crops for export dominated, while total labor hours employed were reduced.

After the 1960s, mechanization expanded rapidly and the number of draft animals and related equipment declined. There were still regional differences in mechanization and there was no satisfactory pooling system, which could have reduced the existing excess capacity and helped the mechanization of small farms. Many tractors were used for non-agricultural purposes, for example for transportation and construction.

Complementarity between different modern techniques may have been important to the rapid diffusion of mechanization when labor was abundant. During the 1960s, commercial fertilizer, pest control, and improved seeds became more widely used. These could be applied in a better and more timely manner with mechanization. Another factor was the availability of very generous government subsidies on agricultural credits for machinery, fuel, and other inputs.

Agriculture in the Economy

From 1950 to 1980, the share of agriculture in GNP declined, both overall and per capita, and the gap between the agricultural and non-agricultural sectors grew until 1980. However, from 1975 to 1980 the gap suddenly narrowed and the growth rate of non-agricultural income fell, and per capita income declined. This was probably due in part to the social turbulence that characterized the period. The contribution of agriculture to GNP growth, calculated by the method of Kuznets (1964), declined steadily from 1948 to 1977, but increased in the period 1978–82. Agriculture's contribution to the growth of GNP per capita also increased during 1978–82 (Kasnakoglu and Gurkan 1987; Gurkan and Kasnakoglu 1986).

From 1950 to 1980 the agricultural share of the adult, economically active population declined but agriculture still employs more than 60% of the population (Table 1). The agricultural sector has been the major source of labor for the urban areas, through migration.

Crop production currently constitutes the major part of total agricultural production, followed by livestock, forestry, and fishing. The share of livestock has declined since 1960 while other sectors have increased. Within the crop category, cereals are the major product, but fruits and vegetables are increasing (Figure 3). Although there has been a steady decline in agricultural exports as a proportion of the total since 1950, in 1985 agriculture was still the most important source of foreign exchange earnings (Table 2). Crops,

Table 1. Economically active population (15+).

Census years	Active population (x1,000)		
	Total	In agriculture	Other
1950	12,929	8,969	3,959
1955	12,205	9,446	2,759
1960	12,993	9,737	3,255
1965	13,558	9,750	3,808
1970	14,051	9,281	4,770
1975	16,050	10,458	5,591
1980	17,220	10,489	6,730

Source: SIS (1950b–85b).

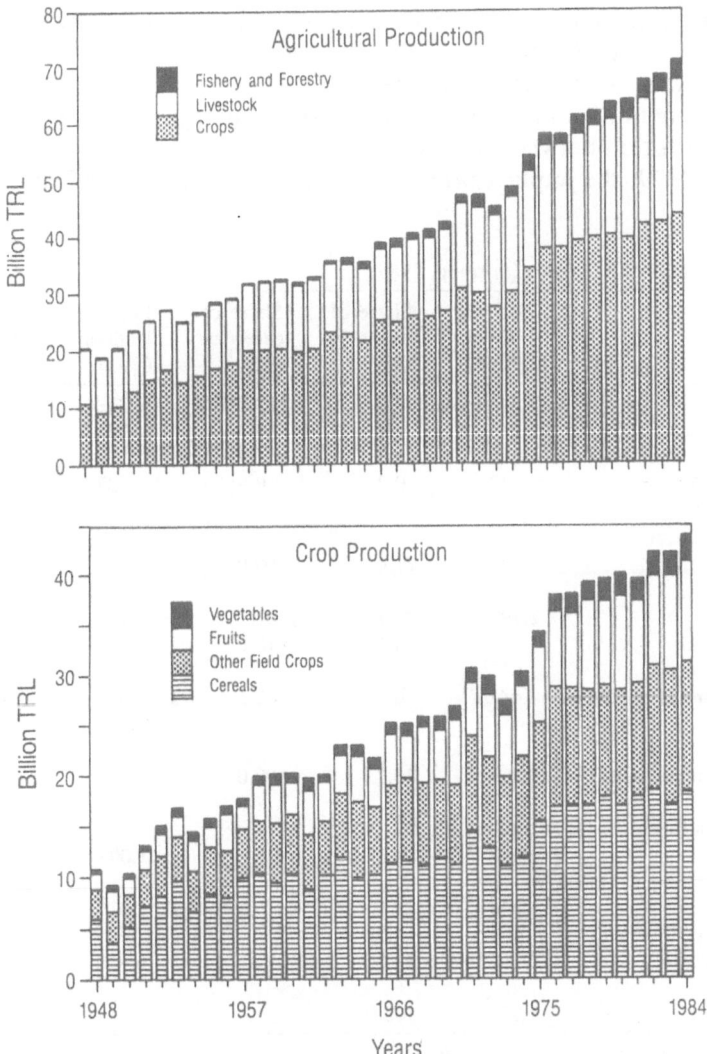

Fig. 3. Value of agricultural and crop production (1968 prices). *Source*: Gurkan and Kasnakoglu (1986).

primarily tobacco, hazelnuts, and cotton, formed the vast majority of primary agricultural exports and textiles constituted the major part of the agriculturally based industrial exports.

Since 1948, rates of growth in output, area, and yield have declined for most agricultural products. After 1960, the rate of growth of cereal production (and as a result, crop and total agricultural production) was significantly below the rate before 1960, and after 1976 it was lower still. Growth rates for other crops, forestry, and fishing declined at other times during 1956–70, while the growth rate of livestock output actually increased. Prior to 1960,

growth in crop area was the most important source of output growth in
agriculture (Table 3), but after 1960 increases in yields accounted for most
output growth. Although the limits of arable land have been reached, the
area component continues to make a contribution of just over 10%, implying
some substitution between alternative land uses.

Before 1979, the implicit price indices of both crop and agricultural prod-

Table 2. Composition of foreign trade, 1985.

Sector	Exports (million USD)	Share of total (%)	Share of category (%)	Imports (million USD)	Share of total (%)
Field crops	1,441	17.1	83.8	242	2.1
Animal products	244	3.1	14.2	105	0.9
Fishery products	21	0.3	1.2	1	0
Forestry products	13	0.2	0.8	27	0.2
Total agricultural exports, raw	1,719	21.6	100.0	375	0.2
Textiles	1,790	22.5	61.3	146	1.3
Leather goods	484	6.1	16.6		
Other processed agricultural products	647	8.1	22.1	487	4.3
Total agricultural exports, processed	2,921	36.7	100.0	633	5.6
Total agricultural exports	4,640	58.3		1,008	8.9
Mining	244	3.1		3,626	32.0
Petroleum products	372	4.7		290	2.6
Other products	2,702	34.0		6,420	56.6
Total exports	7,958	100.0		11,344	100.0

Source: Turkiye Is Bankasi A.S. (1986).

Table 3. Sources of growth in crop output for various periods.

Period	Growth in output (%)	Components of growth			
		Area (%)	Yield (%)	Cropping pattern (%)	Yield-crop interaction (%)
1950–60	45.42	94.50	5.65	1.84	−1.99
1960–70	41.77	13.06	69.88	20.93	−3.87
1970–80	36.88	12.45	94.49	6.03	−12.98

Source: Gurkan and Kasnakoglu (1986).

ucts relative to the implicit price index of non-agricultural commodities remained above 100, with only few exceptions. From 1968 to 1977 there was a sharp improvement in the internal terms of trade in favor of agriculture, but after 1977 it deteriorated, remaining well below 100 after 1978/79. A similar pattern is observed for the relative prices of agricultural outputs and inputs (Figure 4), although the ratio remained above 100 even after 1978. This meant that the agricultural populace lost more as consumers than as producers.

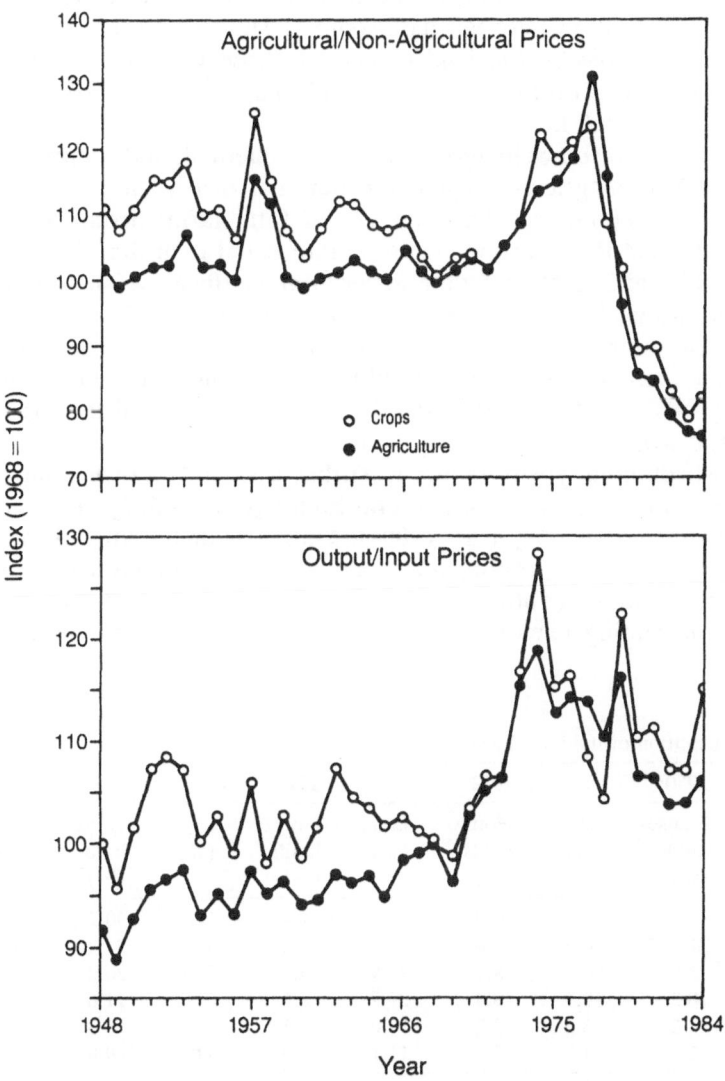

Fig. 4. Terms of trade, 1948–84. *Source*: Kasnakoglu and Gurkan (1987).

Land and Income Distribution in Agriculture

It is very difficult to combine the information on land distribution and tenure systems available from various sources. These results, obtained from a number of studies, should be interpreted with care. The distribution of households and landownership by farm size is unequal, with over 60% of households cultivating around 20% of the area and 5% of households cultivating 30–40% of the area (Table 4). (Gini coefficient=0.6–0.7). Most farms are less than 5 ha. During the period 1950–80, there was little change in the distribution of farm size. Minor changes are probably due to differences in coverage and definition between data sources. However, there has been an increase of about 6% in the land owned by those with over 20 ha and a decrease in the area of farms under 5 ha. This may represent a trend towards consolidation of small holdings.

Distribution of incomes in agriculture is also unequal, and is more unequal than both the non-agricultural and the national distribution. One fifth of agricultural households receive only 3% of total farm income (Table 5). However, the distribution of incomes is more equal than the distribution of landownership, as suggested by the lower Gini coefficients. The distribution of landholdings is clearly related to income (Table 6).

Almost two thirds of total agricultural land in Turkey is owned by the state, village communities, or other public institutions and around one third is privately owned. Most of the state- or community-owned land is natural pastures and woods which farmers can use without cost for grazing.

The most common type of tenure in Turkey is small landownership. There are over 3.5 million private agricultural holdings, excluding those engaged solely in animal husbandry (SIS 1980a). Only 3% of holdings do not own any land; among owners, renting additional land is not common and is mainly done by large holdings (Table 7). After the 1950s the extent of sharecropping declined and renting expanded, although both are relatively unimportant

Table 4. Distribution of holdings by size.

Holding size (ha)	1950[1]		1963		1970		1980	
	House-holds (%)	Land (%)	House-holds (%)	Land (%)	House-holds (%)	Land (%)	House-holds (%)	Land (%)
0–2.0	30.6	4.3	40.9	7.0	44.4	8.0	30.2	4.1
2.1–5.0	31.5	14.3	27.8	17.3	28.2	18.7	31.9	15.9
5.1–10.0	21.9	20.7	18.1	23.9	15.7	22.0	20.2	21.3
10.1–20.0	10.3	19.3	9.4	23.7	7.7	21.5	11.6	23.8
20.1–50.0	4.2	16.6	3.2	17.0	3.1	17.9	5.3	22.9
>50	1.5	24.8	0.6	11.1	0.9	11.9	0.8	12.0

Source: SIS (1950a–80a).
[1]Autumn sample of the 1950 census.

Table 5. Income distribution in Turkey and in agriculture.

Percentile of families	Turkey (% of income)			Agriculture (% of income)		
	1963	1968	1973	1963	1968	1973
0–20	4.5	3.0	3.5	6.0	4.0	3.0
21–40	8.5	7.0	8.0	9.2	6.0	5.5
41–60	11.5	10.0	12.5	13.8	10.0	9.5
61–80	18.5	20.0	19.5	21.4	19.0	20.0
81–100	57.0	60.0	56.5	49.6	61.0	63.0
Gini	0.55	0.56	0.51			

Source: SPO (1976); Kasnakoglu (1975).

compared to small landownership, and most rented or shared holdings are small. Sharecropping has sometimes been considered a remnant of feudalism and renting an indication of expanding capitalist production relations in agriculture. Further research and careful analysis have shown that this distinction is misleading (Varlier 1978; Aricanli and Somel 1979). Both types of tenure are found in well developed regions of Turkey as well as in regions where traditional relations still exist.

Sharecroppers can no longer be regarded as a potential supply of non-agricultural labor, but further labor-saving technologies may release labor from small holdings. Even if such farmers continue to cultivate their own land, the amount of time they can offer for seasonal work is likely to increase.

Approximately 23% of all agricultural holdings in Turkey, excluding those engaged solely in animal husbandry, are located in Central Anatolia (Regions I and IX) (SIS 1980). The average holding size is well above the national mean. Holdings of less than 5 ha constitute 48% of all the holdings in the region compared to 61% of those in Turkey. However, the predominant farming system is rainfed and soil productivity is well below that in the coastal regions so the area required for subsistence is larger. Relatively large farms, above 50 ha with land rented in, are more widespread in this region than in Turkey in general, but the absolute number of such holdings is small (SIS 1980a).

Ownership of Agricultural Machinery

Apart from tractors and harvesters, the majority of farmers own their animals and equipment. According to the 1980 census over 60% of all households cultivating over 50% of agricultural land used tractors; of these, a quarter owned the tractors used, 5% shared ownership, and the remaining 70% rented tractor services. Tractor owners hold 46% of the land and, in general, the farmers with large holdings own tractors, while the smaller farmers rent tractor services (Table 8). Fewer combine harvesters are owned; of the

Table 6. Distribution of holdings by income groups (%).

Holding size (ha)	Income group (TRL)						
	0–5,000	5,001–10,000	10,001–15,000	15,001–25,000	25,001–50,000	50,001–100,000	>100,000
0–2.0	32.1	26.3	13.5	13.4	10.8	3.2	0.8
2.1–5.0	10.4	21.6	22.1	26.3	15.4	3.3	1.0
5.1–10.0	2.4	11.1	15.0	30.9	24.6	13.9	2.1
10.1–20.0	0.0	3.8	7.9	20.8	38.7	21.5	7.4
20.1–50.0	0.0	7.9	0.0	7.9	31.1	36.8	16.3
50.1–100.0	0.0	0.0	0.0	0.0	19.4	0.0	80.6
>100	0.0	0.0	0.0	0.0	0.0	44.9	55.1

Source: Varlier (1978).

Table 7. Land tenure in Turkey and dry areas by holding size (% of holdings in each size category).

Holding size (ha)	Farming own land only	Owning and renting in	Renting land in and out
	Turkey		
0–2.0	93.1	4.0	0.1
2.1–5.0	91.6	7.1	0.2
5.1–10.0	89.3	9.1	0.2
10.1–20.0	88.2	10.1	0.1
20.1–50.0	83.0	13.2	0.7
50.1–100.0	80.8	17.6	0.9
>100	76.7	19.7	2.1
Total	90.6	7.4	0.2
	Dry areas (Regions I and IX)		
0–2.0	97.4	1.4	0
2.1–5.0	97.1	2.0	0.2
5.1–10.0	95.9	3.5	0.1
10.1–20.0	93.3	6.1	0
20.1–50.0	85.0	12.1	1.3
50.1–100.0	66.4	33.5	<0.1
>100	56.2	38.2	5.2
Total	94.6	4.4	0.2

Source: SIS (1980a).
Note: Holdings without owned land (<3%) not shown.

Table 8. Access to agricultural equipment by farm size (% of households).

Holding size (ha)	Tractors			Combine harvesters		
	Owned	Shared	Rented	Owned	Shared	Rented
0–2.0	8.7	2.6	88.7	0.3	0.0	99.7
2.1–5.0	15.0	3.6	81.4	1.0	0.6	98.4
5.1–10.0	26.9	6.8	66.3	2.9	1.1	96.0
10.1–20.0	39.8	7.3	52.9	3.3	2.0	94.7
20.1–50.0	48.0	9.7	42.2	7.4	4.4	88.3
50.1–100.0	57.4	10.4	32.2	8.9	6.2	84.9
>100	76.8	8.2	15.0	18.2	4.6	77.2
Total	24.7	5.5	69.8	3.5	2.0	94.5

Source: SIS (1980a).

families who use harvester services, 3.5% are owners, 2% are joint owners, and 95% rent harvester services. In Central Anatolia, only 1.7% of households own a combine but they own 9% of the land (SIS 1980a).

Composition of Income and Expenditure in Rural Areas

Over 45% of the total income in rural areas is earned directly from agriculture. Wages constitute the second largest component of income in rural households followed by trade (Table 9), with most of the wage earnings probably coming from non-agricultural activities. The share of payments in kind is little over 5% of total income, indicating the dominance of cash markets in rural areas. The share of income from agricultural production decreases as farm size and income increase, whereas the shares of income from trade, services, and rent, which are basically non-agricultural activities, increase with farm and income size (Table 10).

Most expenditure is on food, followed by clothing and housing (Table 11). Expenditure on wheat products accounts for about 20% of food spending in most groups. Food and housing form smaller shares of total expenditures of larger farms.

Production for one's own household's consumption, or home production, is an important component of disposable income in agriculture, amounting

Table 9. Sources of rural household income (% of total) by income groups.

Income groups (TRL)	Production cash	Trade cash	Services		Wages	
			Cash	Kind	Cash	Kind
0–5,000	51.4	6.0	5.8	2.2	21.6	1.8
5,001–10,000	51.1	6.1	5.8	2.0	23.3	1.4
10,001–15,000	48.4	6.7	6.4	1.7	24.0	1.5
15,001–25,000	47.2	7.7	5.5	1.9	25.7	1.0
25,001–40,000	44.6	12.2	7.6	2.1	20.1	1.1
>40,000	34.0	29.9	11.5	1.7	12.1	0.5
Total	45.8	11.5	7.0	1.9	21.6	1.1

Income groups (TRL)	Rent		Transfers		Interest
	Cash	Kind	Cash	Kind	Cash
0–5,000	1.6	0.4	6.9	1.5	0.9
5,001–10,000	1.1	0.5	7.3	1.1	0.4
10,001–15,000	1.6	0.5	7.5	1.3	0.4
15,001–25,000	2.0	0.2	7.4	1.0	0.4
25,001–40,000	2.1	0.2	8.8	1.0	0.3
>40,000	3.3	0.3	6.5	0.2	0.2
Total	2.0	0.3	7.5	1.1	0.4

Source: SIS (1973/74).

to over 20% of the total (Table 12). The value of home consumption in low income groups is nearly one third of total disposable income, but is much less in groups with more land. This observation is consistent with the marketing ratios for wheat and barley. Farms with 2 ha or less sell only 7% of their wheat and 5% of their barley, compared to 41 and 23% for farms with 10–20 ha and 76 and 65% for farms over 100 ha (Varlier 1980).

Composition and Distribution of Production

Cereals occupy 79% of the area sown to field crops (Figure 5). Of the cereals, wheat, barley, and maize account for 70, 22, and 4%, respectively, of the area sown. Lentils, chickpeas, and vetch are the major pulses, while cotton, sugar beet, and tobacco constitute 46, 27, and 17%, respectively, of the area

Table 10. Sources of rural household income (% of total) by holding size.

Holding size (ha)	Agricultural	Non-agricultural	
		Kind	Cash
0–2.0	48.9	5.0	46.1
2.1–5.0	47.8	4.7	47.6
5.1–10.0	45.1	4.2	50.7
10.1–20.0	42.6	3.0	53.5
20.1–50.0	39.7	3.5	56.9
50.1–100.0	36.0	3.0	61.0
>100	37.0	2.7	63.4

Source: SIS (1973/74).

Table 11. Expenditure shares (%) by holding size in rural areas.

Expenditure categories	Holding size (ha)						
	0–2.0	2.1–5.0	5.1–10.0	10.1–20.0	20.1–50.0	50.1–100.0	>100.0
Food	64.0	63.4	61.7	60.5	58.9	58.6	55.9
Clothing	12.7	12.9	13.1	13.3	13.2	12.0	12.6
Housing	9.8	9.2	8.7	8.3	8.0	8.2	7.8
Others	14.3	14.8	16.5	17.8	19.8	21.3	23.7
Wheat products	12.4	11.8	11.1	10.6	10.3	10.0	9.5
Barley products	1.8	1.8	1.8	1.8	1.7	1.8	1.7
Cotton products	12.0	12.5	13.1	13.3	13.2	12.0	12.6
Tobacco products	3.1	3.1	2.9	2.8	2.6	2.5	2.3
Sugar products	3.9	3.6	3.2	2.8	2.7	2.5	2.4
Mutton products	1.0	1.1	1.3	1.5	1.7	1.9	1.9

Source: Kasnakoglu and Gurkan (1986).

Table 12. Components of disposable income as percentages of total income.

Disposable income groups (TRL)	Income in cash	Income in kind	Home production		Furnishings	Clothing	Change in stock
			Food	Food processing			
Turkey							
0–1,999	48.6	3.8	23.8	3.5	2.3	2.3	15.8
2,000–5,999	53.8	2.9	25.8	5.3	1.5	1.5	9.3
6,000–9,999	63.4	3.3	21.3	4.0	1.0	1.0	6.1
10,000–19,999	72.0	3.5	15.0	3.1	0.6	0.6	5.3
20,000–39,999	79.2	3.5	10.1	2.3	0.3	0.3	4.2
>40,000	83.5	2.3	6.8	3.5	1.1	0.2	2.7
Total	71.9	3.3	14.8	3.3	0.8	0.7	5.4
Dry areas (Regions I and IX)							
0–1,999	48.5	1.8	26.3	4.7	2.9	3.0	12.9
2,000–5,999	57.6	3.2	22.6	5.9	1.4	1.5	7.8
6,000–9,999	68.5	3.1	17.9	3.3	0.9	0.9	5.4
10,000–19,999	73.0	4.0	13.4	2.9	0.6	0.6	5.5
20,000–39,999	81.8	2.1	9.5	2.7	0.4	0.3	3.3
>40,000	77.0	1.4	4.8	9.8	4.8	0.2	2.0
Total	73.2	2.9	13.2	4.1	1.2	0.7	4.8

Source: SIS (1980a).

119

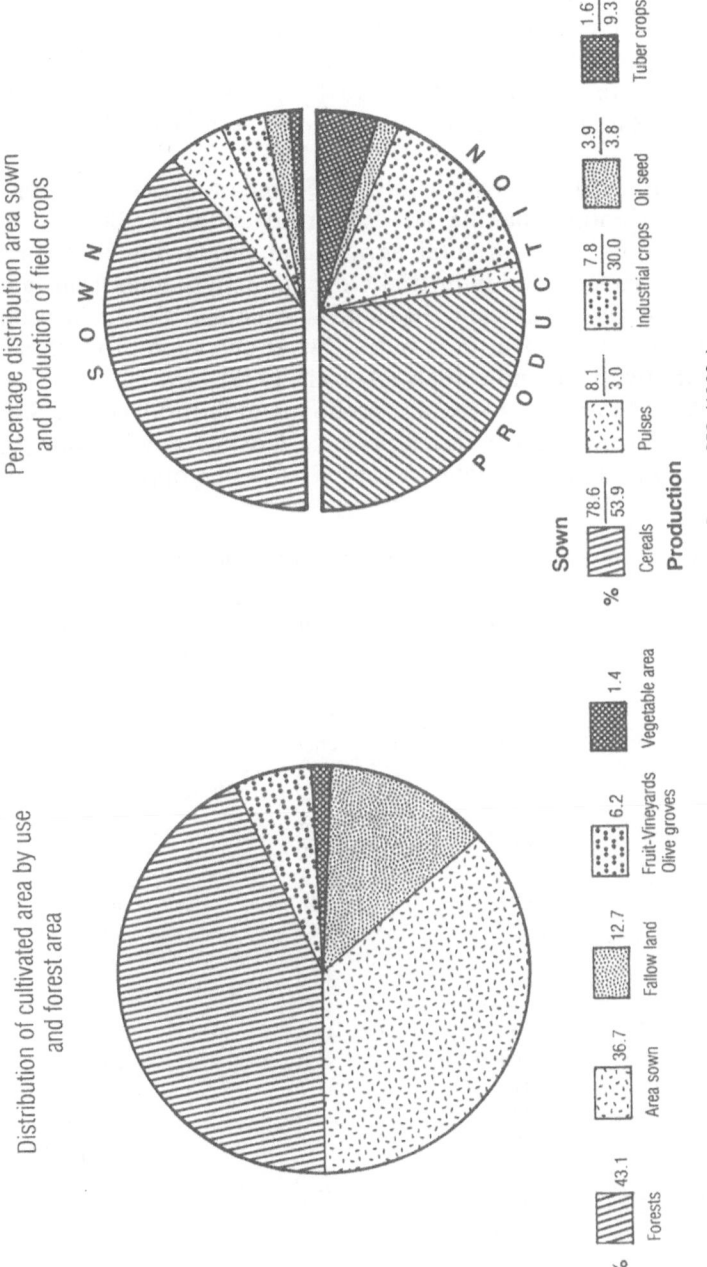

Distribution of cultivated area by use and forest area

Percentage distribution area sown and production of field crops

Fig. 5. **Land use and cropping patterns.** *Source:* SIS (1983c).

sown to industrial crops. The remaining major field crops are sunflower, onions, and potatoes.

Hazelnuts, tea, apples, peaches, pistachios, citrus fruits, olives, cherries, figs, grapes, melons, cabbages, beans, spinach, leeks, tomatoes, eggplants, and green peppers are the major tree and vegetable crops. Sheep constitute nearly 60% of the livestock, followed by goats (20%) and cattle (17%). Donkeys, mules, horses, and camels make up less than 3% of animal numbers.

The country can be divided into three major agricultural regions: the coastal areas, Central Anatolia, and Eastern Anatolia (Table 13). The coastal areas account for 50% of total agricultural production and specialize in horticultural products and industrial crops, although other crops and animal husbandry are also significant. In Central Anatolia, the main products are grain, livestock, sugar beet, and pulses, while the main products in Eastern Anatolia are livestock and cereals (Aresvik 1975).

Turkey's sheep and tobacco are predominantly produced on holdings of less than 10 ha, and larger farms produce most of the remaining products (Table 14). On average, wheat occupies 35–40% of cropped areas irrespective

Table 13. Regional distribution of agricultural production in Turkey in 1958 (% of national totals).

	Coastal regions	Central Anatolia	Eastern Anatolia
Wheat	34.3	54.2	11.5
Barley	29.2	55.5	15.3
Pulses	45.0	38.4	16.6
Tobacco	88.5	8.3	3.2
Sugar beet	31.9	63.2	4.9
Cotton	95.1	2.1	2.8
Hazelnuts	95.7	3.0	1.3
Grapes	52.8	34.1	13.1
Citrus	100.0	n.s.	n.s.
Olives	98.7	n.s.	1.3
Total crops	54.0	36.9	9.1
Sheep	29.2	42.4	18.4
Goats	48.8	26.8	24.4
Cattle	41.8	33.7	24.5
Milk	43.9	28.9	17.2
Meat	60.0	27.1	12.9
Total livestock	44.0	35.2	20.8
Crops and livestock	50.0	36.4	13.6

Source: Aresvik (1975).

n.s. = not significant.

Table 14. Distribution of production by product and holding size (% of area or number).

Holding size (ha)	Product					
	Wheat	Barley	Sugar beet	Cotton	Tobacco	Sheep
0–2.1	2.0	1.1	0.7	2.3	6.3	18.7
2.1–5.0	13.3	11.2	2.3	14.3	28.5	22.8
5.1–10.0	20.7	19.6	23.9	19.2	29.8	17.5
10.1–20.0	24.8	25.0	31.2	25.6	20.5	11.4
20.1–50.0	27.7	30.9	34.4	25.3	12.4	23.6
50.1–100.0	8.4	9.2	5.6	5.8	1.2	5.6
>100	3.1	3.0	2.0	7.6	1.2	0.4

Source: SIS (1980a).

of holding size, but most other products show differences related to holding size (Table 15). For holdings of 0–10 ha, wheat, barley, maize, tomato, and tobacco constitute over 75% of the area planted. In the next two groups (10–50 ha), the share of barley increases substantially, and chickpeas, sunflower, and lentils become major crops. In farms with 50 ha or more, cotton and sunflower are relatively important and, with wheat and barley, make up nearly 80% of the total area planted.

In Central Anatolia (zones I and IX), the shares of cotton, maize, and tobacco are substantially lower than the national average while wheat, barley, and potato areas are higher. The composition of production by holding size is similar to the national pattern.

Costs and Profitability of Crops

In intensive systems, the value of variable inputs is relatively high in relation to the value of fixed inputs, mainly land, while in extensive systems, the total cost of variable inputs is relatively low. Cereals and sunflower are extensive dryland crops, while vegetables, fruits, pulses, and tobacco may be considered intensive (Table 16). In general, intensive crops have a higher demand for all of the variable inputs; thus, at this level, labor and tractors appear as complements to each other rather than as substitutes. Production activities can also be classified according to relative shares of individual inputs in variable costs, with no reference to their overall input intensities. Tree crops, vegetables, pulses, and tobacco are the most intensive users of labor relative to other inputs.

The rate of return to variable inputs is higher in intensive, irrigated production activities, and lower in cereals and industrial crops. The rates of return of traditional export crops, such as cotton, tobacco, and hazelnuts, are very low, indicating that the rents from export subsidies accrue to the exporters rather than the producers. The higher returns in irrigated production activities are partly due to highly subsidized water charges.

Table 15. Areas of major crops by holding size (% of area).

Holding size (ha)	Wheat	Barley	Food legumes	Sugar beet	Cotton	Sunflower	Maize	Potato	Tomato
Turkey									
0–2.0	36.1	13.3	4.7	1.0	2.9	1.5	15.0	5.1	8.8
2.1–5.0	35.0	19.7	6.6	2.1	2.7	2.5	11.2	3.8	6.0
5.1–10.0	26.8	16.8	7.1	2.5	1.8	3.4	4.9	2.4	4.7
10.1–20.0	34.6	23.1	10.0	3.5	2.6	5.4	3.8	2.8	5.5
20.1–50.0	34.2	25.4	12.9	3.4	2.2	4.6	1.4	3.1	5.4
50.1–100.0	40.3	29.4	10.8	2.1	2.0	3.5	0.7	1.7	3.3
>100	39.5	25.0	6.0	2.0	7.0	8.0	2.0	1.0	3.4
Total	32.6	18.7	7.4	2.3	2.4	3.2	8.2	3.4	5.9
Dry areas (Regions I and IX)									
0–2.0	40.7	20.6	6.0	1.9	0.3	0.7	5.0	9.5	9.1
2.1–5.0	38.3	27.3	8.1	2.1	0.2	2.1	3.8	6.1	6.2
5.1–10.0	35.7	27.3	11.7	3.1	0	2.4	2.6	4.8	6.5
10.1–20.0	36.5	29.0	11.2	3.8	0	2.2	1.3	3.9	6.0
20.1–50.0	33.6	27.7	12.7	4.4	0	3.7	0.4	4.6	5.4
50.1–100.0	36.4	31.3	6.7	4.3	0	3.6	0	3.6	3.5
>100	41.7	40.0	1.8	7.6	0	1.0	0.5	1.6	0.9
Total	40.2	29.3	10.8	3.3	0.1	2.5	2.9	6.0	6.4

Source: SIS (1980a).

Table 16. Economics of major crops.

Crops	Rainfed (R) or Irrigated (I)	Input shares (% of cost)				Total cost	Yield (t/ha)	Forage yield	Revenue (USD/ha)	Rate of return (t/ha)
		Labor	Tractor	Fertilizer	Seed					
Wheat	R	6	63	10	21	147	1.6	1.9	247	1.69
Wheat, F	R	6	70	7	17	86	1.0	1.2	160	1.85
Wheat	I	13	64	7	15	193	3.4	4.1	543	2.82
Maize	R	55	30	7	7	158	1.7	3.4	343	2.17
Maize	I	58	37	2	2	485	4.7	9.4	948	1.96
Barley	R	23	61	4	12	236	2.3	2.8	333	1.41
Barley, F	R	5	76	7	13	76	1.5	1.7	209	2.73
Sunflower	R	43	50	4	3	178	1.2	0	329	1.85
Chickpea	R	28	32	2	38	294	1.0	1.1	620	2.11
Lentil	R	32	49	1	17	260	1.0	1.1	574	2.21
Onion	R	79	18	2	1	653	9.2	0	1,609	2.46
Tomato	I	65	21	1	13	1,681	32.4	0	6,535	3.89
Cucumber	I	64	23	2	11	1,063	16.7	0	4,240	3.99
Cotton	I	59	36	3	2	991	0.9	0	1,119	1.13
Potato	I	36	26	2	37	1,081	13.9	0	3,512	3.25
Tobacco	R	74	17	0	8	1,154	0.9	0	1,343	1.16
Water melon	R	44	34	1	20	492	10.4	0	2,148	4.37
Olive	R	41	47	1	11	219	0.9	0	624	2.85
Grape	R	60	25	1	14	662	4.0	0	1,871	2.83
Pistachio	R	48	41	0	11	465	0.4	0	952	2.05
Hazelnut	R	88	2	3	8	641	0.9	0	867	1.35
Citrus	I	79	9	3	9	1,311	22.8	0	5,589	4.26
Apple	I	42	39	1	18	525	5.9	0	1,941	3.70
Apricot	I	49	31	1	19	783	4.0	0	1,498	1.91
Cherry	I	60	27	1	12	1,464	4.7	0	1,984	1.36
Strawberry	I	50	3	0	47	2,417	4.4	0	5,701	2.36

Source: Le-Si et al. (1983).
Note: F: with fallow (crop-fallow rotation).

124

Technology Use and Farm Size

In 1980, less than 14% of total cultivated land was irrigated and 28% was fallow (Table 17). Irrigation is more common on small farms, while fallowing and the use of tractors, fertilizers, and pesticides increases with land size. This dichotomy can be explained by the fact that irrigation in very small holdings can be carried out with little capital investment, using wells or nearby water resources, whereas in larger holdings it requires substantial investment by the farmer if it is not provided by the government. The use of modern inputs, on the other hand, can be achieved at a lower unit cost in larger holdings, as they benefit from scale effects.

Decision Making in Agriculture

Agricultural decisions may be considered from two directions: who decides among the given choices, and who determines the choices. Behavior is the outcome of decisions taken at many levels, not just the farm. Thus in a general way, decisions in agriculture are determined by the family, village, government, commerce and industry, and international relations. The most important decisions are related to consumption, input and technology, marketing, and labor supply and demand.

The ultimate decision-making unit is the family. In rural Turkey there are two types of family, extended and nuclear. Both are decision-making units, particularly for decisions related to consumption and labor requirements. Most rural families are nuclear households with small landownership. Large landownership seems to be more characteristic of patriarchal-extended (three generation) households (Timur 1972). The number of nuclear families is

Table 17. Agricultural practices by holding size.

Holding size (ha)	Agricultural practices (% of area)				
	Irrigation	Fallowing	Tractor cultivation	Use chemical fertilizers	Use pesticides
0–2.0	28.4	9.7	31.1	45.3	32.2
2.1–5.0	17.9	18.7	40.5	47.3	31.3
5.1–10.0	14.7	25.2	50.0	46.9	33.0
10.1–20.0	12.9	29.3	57.5	48.3	36.3
20.1–50.0	9.7	32.9	62.9	49.9	40.6
50.1–100.0	7.0	39.2	64.0	50.9	37.8
>100	14.7	35.8	72.4	56.3	54.1
Total	13.6	27.7	54.3	48.6	36.3

Source: SIS (1980a).

increasing, encouraged by the equal division of inherited land among children. In both types of family, the father or oldest brother is the most important decision-maker although in nuclear families women may have more to say about consumption and probably production as well.

Sociologists in Turkey prefer to perceive the village rather than the family as the primary unit of analysis and in fact, many production activities involve common operations, such as the use of commonly owned pasture. The village coffeehouse is seen as an important institution where decisions are taken or discussed and where information is shared.

International relations rarely exert a direct influence on households but usually have an effect through the mediation of government, particularly through development policies. Trade policies have a strong influence on marketing decisions while technology transfer, especially as foreign aid, affects technology and input systems. Another type of international influence is through demonstration effects on consumption patterns. Observation of higher consumption levels tends to increase the demands of the observers. The latter has been especially important for Turkey because of rural migration to urban centers and western Europe; returning migrants bring new consumption patterns.

The government's input and output price supports are its main instruments for determining the constraints on choices open to the farmer. The widespread use of modern inputs (commercial fertilizers, improved seeds, mechanical equipment, etc.) even on small farms is probably due to government efforts to make them available and promote new technologies. The influence of the industrial sector is usually felt via government policies but the commercial sector has more direct access to the farmer. Private and semi-government buyers have a major influence on the marketing of agricultural produce.

Agricultural Support Policies

Output support prices, input subsidies, credits, quotas, tariffs, taxes, land distribution, investments in infrastructure, extension services, etc. have been employed to achieve a range of objectives, such as income and price stability, stimulating output and income, satisfying demand, and improving the balance of payments (Kasnakoglu 1986). Five ministries and about 20 semi-autonomous agencies, i.e., state enterprises, state monopolies, and unions of cooperatives (sales, credit, or both), are involved in the formulation and administration of agricultural price policies. Apart from fresh fruits and vegetables, almost all major agricultural products are under government support schemes, constituting over 90% of the total value of agricultural production, and state institutions purchase a substantial proportion of marketed production (Table 18). Similarly, most modern inputs are either produced, distributed, or priced by the government. Infrastructural investments in roads, irrigation schemes, land development and conservation, and exten-

Table 18. Shares of support purchases in agricultural production.

Commodity	% of Total production purchased by SEE[1] or ASC[2]	% of Total production marketed	% of Marketed production purchased by SEE or ASC (Range 1970–83)
Wheat	5–19	35	14–54
Barley	0.3–14	30	1–47
Oats	0.01–2	30	0.3–7
Maize	0.07–0.6	35	0.2–2
Chickpea	10–15	40	25–38
Lentil	5–10	40	13–25
Tobacco	13–91	100	13–91
Sugar beet	100	100	100
Cotton	49–100	100	49–100
Sunflower oil	1–56	100	1–56
Olive oil	0.01–0.6	90	0.01–0.7
Hazelnut	44–79	90	49–88
Sheep	6–35	50	12–70
Cattle	34–50	50	64–100

Source: Kasnakoglu (1986).
[1]SEE: State Economic Enterprise or Monopoly.
[2]ASC: Agricultural Sales and/or Credit Cooperative.

sion services also provide inputs to agriculture free of charge or at highly subsidized prices (Kasnakoglu 1986).

The goals of support price policies as stated in the 5–year development plans (SPO 1963–87) have consistently included or implied reducing inflation, encouraging exports, preference for input subsidies rather than subsidized purchases to support farm incomes, and avoiding direct income transfers. In addition, one must consider the enduring overall goals of the planners: industrialization, improved nutrition, low consumer prices, and reduction of regional inequalities.

The planners have understood that price support policies carry with them the danger of inflation, and therefore have attempted to identify non-inflationary interventions. One strategy has been a preference for input price subsidies which reduce the cost of production to the farmer and encourage input use. Resulting increases in land and labor productivity can be expected to increase output and farm incomes, make agricultural products more competitive in world markets, and keep food prices low, thereby improving nutrition. Other policies to increase farm productivity will also serve these goals. These include improved extension services and better agricultural statistics and forecasting, more efficient design of production plans, prompt payments to farmers, and infrastructural development, including irrigation schemes. However, the pressing need for foreign exchange may lead to policies which conflict with the goal of fighting inflation, specifically the export of commodities whose support prices are above world market prices.

Support policies may also be applied to promote non-agricultural policy goals. For example, to further industrialization, supports for raw material crops such as cotton, wool, and tobacco may be stressed. However, since they are not related to land size, agricultural price policies have generally increased inequalities in income distribution within rural areas.

The support price policies may have considerable influence on regional income distribution, partly because of the geographic distribution of crops and partly because the same crop, for example tobacco, is supported with different prices in different regions, mainly because of varying quality. Effects of input subsidies may also differ according to region. For example, irrigation schemes with price supports may improve regional equity if situated in relatively poor areas, as is the recent Southeast Anatolia Project.

The most important effect of agricultural policies is their impact on the allocation of resources, in both the short and the long term. Government subsidies of input costs and infrastructural investments have been expansionary, resulting in more and better quality land and fertilizer use, improved irrigation practices, increased mechanization, and less fallow. These changes have resulted in increased demand for labor.

Government intervention in output markets has no doubt also had allocative effects. Several studies on Turkish agriculture (see Kasnakoglu 1986) have shown that producers are responsive to market signals in general and to output prices in particular. Support prices have, to a large extent, determined relative output prices (Kasnakoglu 1986). Relative farmgate prices from 1938 to 1982 generally favored cash and export crops and, as these are more labor-intensive, this may have created a positive pressure on labor demand.

The effects on output and labor demand of distorted domestic agricultural output prices relative to world prices, through an overvalued exchange rate, export and import restrictions, and supported domestic prices, are more complicated. Kasnakoglu and Gurkan (1986) found that, except for cotton and sugar beet, output price policies have resulted in absolute output losses in all products considered, including wheat, barley, and tobacco. The average output gain per year in cotton during 1960–80 was 0.4% and for sugar beet 2%, while output losses due to intervention in wheat, barley, and tobacco were 15, 8, and 3%, respectively. Resulting decreases in labor demand due to reduced output of the latter crops more than offset the labor demand gains due to expansion of cotton and sugar beet.

Thus, in general, interventions in the output markets have tended to curtail labor demand in Turkish agriculture, while interventions in input markets have increased labor demand. Furthermore, the expansionary effects of input policies have in the past more than offset the contractionary effects of output policies resulting in a net expansion in agricultural labor demand.

Structure of the Agricultural Labor Force

The nature and structure of labor demand in Turkey varies among physical, social, and economic regions. To illustrate this, we have grouped the provincial data of Gurkan (1986) into four subregions based on those of Gurkan (1985): predominantly irrigated, high rainfall coastal areas, Central Anatolia, and Eastern Anatolia (Table 19).

The labor variables show interesting differences. The average labor force in agriculture per ha in irrigated areas is nearly twice as high as in the dry farming regions, despite the fact that the latter include the most underdeveloped provinces which rely almost solely on agriculture for subsistence. Hired labor also makes its greatest contribution relative to the active agricultural population in the irrigated areas. The proportion of females and elderly males in the workforce is smallest for the developed, non-irrigated coastal regions, with irrigated regions occupying second place, and highest in Eastern Anatolia which has the higher rate of rural emigration (cf. Cerit 1986).

Comparison of other variables indicates that the intensity of labor usage tends to be related to the intensity of farming. Of the rainfed areas, the coastal areas have the smallest average farm size, greatest use of fertilizers and mechanical tillage, and the lowest proportion of fallow, all of which are consistent with their high income per ha. Therefore, the fact that these

Table 19. Selected regional indicators aggregated from provincial data by farming systems.

	Irrigated	Rainfed			Turkey
		Coastal	Central Anatolia	Eastern Anatolia	
Farm income/ha (TRL)	2,612	2,668	1,908	1,483	2,268
Farm income/capita (TRL)	6,563	6,136	6,211	5,639	6,257
Tractor power (hp/ha)	91	168	81	100	103
Mechanical/draft power ratio	8.2	28.3	14.4	4.7	12.4
Fertilizer use (TRL/ha)	32.1	53.3	30.7	20.8	33.4
Mean holding size (ha)	3.6	5.3	8.2	5.7	5.3
Irrigated area (%)	22	9	8	9	15
Fallow area (%)	32	18	64	56	41
Cereal area (%)	19	31	37	28	27
Vegetable area (%)	27	12	7	6	16
Net rural emigration (%)	5.6	5.3	6.0	7.4	6.0
Labor force in agriculture/ha	1.14	0.65	0.48	0.73	0.83
Hired laborers/holding	0.046	0.031	0.021	0.009	0.032
Female and elderly proportion of labor force(%)	51	48	52	54	51

Source: Based on Gurkan (1986).

provinces have low emigration, a high proportion of adult males in the agricultural labor force, and employ more labor than the Anatolian provinces is not surprising. Eastern Anatolia is at the opposite extreme in being a relatively underdeveloped labor-surplus area.

Labor Availability, Rural Out-Migration and Productivity

We have used provincial labor availability, net rural out-migration, and land productivity as the dependent variables in a system of three regression equations to ascertain whether they are influenced in a consistent way across different provinces by variables selected to represent the essential features of their farming systems. In addition to the usual variables, climatological differences and crop mixes have been included. Details of the analysis may be found in Kasnakoglu et al. (in press).

In the equation for income per ha, 8 of the 13 variables included have statistically significant coefficients at the 5% level. The variables for per ha fertilizer use, tractor hp, and labor force are all significant and positive, and the first two are the first and third most important variables, respectively, contributing to income variation. Two variables representing the cropping pattern, proportion of cereals and proportion of vegetables, are statistically significant and the former is the second most important variable. The variables representing aridity and the number of frost days are also important, while the extent of irrigation and of fallow, used as measures of different farming systems, have no statistically significant influence on income per ha.

In the equation for labor availability, only four variables make significant contributions to the explained variation. Increased irrigation and precipitation are associated with greater labor availability per ha as is increased inequality in land distribution. Labor availability is negatively related to proportion of land under cereals, as cereals are mostly found in extensive farming systems.

Finally, net migration is positively associated, implying less emigration, with greater availability of tractors, a larger educated populace, and greater incomes per ha and per capita. Thus a labor displacing effect of increased mechanization is not evident in the period chosen (1975–80). Perhaps those provinces that are mechanized are now demanding more labor, having previously lost their surplus labor. Of the important dry farming features, only one is significantly associated with migration; the larger the area of fallow land, the greater the out-migration from rural areas.

Seasonal Employment

There has been seasonal employment in Turkish agriculture for many years. For example, at the turn of the century labor was insufficient in cotton-growing regions of the south but labor was available from the east and southeast due to a complementary agricultural labor cycle (Bruck 1919). Agaoglu (1936) differentiated between long- and short-term seasonal employment, subdividing the latter into hoeing and harvesting, and the former into summer work (harvesting) and winter work (cultivation). Fertilizer application and pest control are new types of seasonal work. Some seasonal workers resided in the region in which the work was performed but many came from other provinces. In the 1930s the lack of efficient transportation and competing demand from the hazelnut-growing regions along the Black Sea and Aegean coasts were considered to be the most important constraints on the availability of seasonal labor in the cotton-growing southern regions (Agaoglu 1936). The village coffeehouse and middle-men, "elcibasi", were the principal means of recruitment of seasonal labor and both continue to fulfill essentially the same function today.

In 1980, one third of all agricultural holdings employed seasonal workers. The total number of family workers, i.e., the total working population in rural areas, was approximately 14 million in 1980, assuming four persons over 12 years old per holding. Although the number of seasonal workers given in Table 20 is too high because of double counting we estimate that at least 6 million persons in rural areas were engaged in seasonal work, the most widespread being harvesting and hoeing (Table 21). Assuming that those employed in harvesting also performed other work (except animal husbandry), then the average seasonal worker probably worked less than 20 days per year.

In Anatolia, the type and duration of work vary according to agricultural region. On average, 30% of all agricultural holdings employ seasonal workers. In the Marmara (northwest), Aegean (west), and Mediterranean (south) regions the rates of employment are well above the national average (48, 40, and 38%, respectively), while in all other regions they are below the national average. Southeastern and Eastern Anatolia, with only 13 and

Table 20. Seasonal employment according to holding size.

Holding size (ha)	Total no. holdings	Hiring seasonal workers (× 1,000)	Seasonal workers (× 1,000)	Days worked (× 1,000)	Average workers/ holding
0–2.0	1,102	202	1,031	5,230	5.1
2.1–5.0	1,165	351	2,881	15,476	8.2
5.1–10.0	738	277	3,497	19,952	12.6
10.1–20.0	422	176	3,527	22,178	20.1
20.1–50.0	195	89	1,996	17,499	22.3
50.1–100.0	26	13	622	3,746	48.2
>100.0	3	3	208	3,216	72.2
Total	3,651	1,111	13,763	87,296	12.4

Source: SIS (1980a).

Table 21. Seasonal workers and type of seasonal work.

Type of work	Workers (× 1,000)	Days worked (× 1,000)	Average days/ worker
Cultivation	1,555	6,447	4.1
Fertilization			
and pest control	803	3,968	4.9
Hoeing	4,773	26,371	5.5
Harvesting	6,128	36,530	6.0
Animal husbandry	116	13,104	113.4

Source: SIS (1980a).

20% of all farms employing seasonal labor, are the main suppliers of seasonal workers to other regions, so it appears that dry farming regions supply seasonal workers to irrigated regions. This is supported by the fact that cultivation and harvesting of cereals, typical of rainfed agriculture, are mechanized whereas harvesting of the main cash crops, especially cotton, is not. The large irrigation projects planned for dry farming regions may upset the existing balance by increasing local employment opportunities for seasonal workers, mainly in the Mediterranean region.

Concluding Remarks

Although the introduction of modern inputs into Turkish agriculture dates back to the turn of the century, their use has only become widespread since the 1950s. Adoption of new inputs has required new cultivation techniques and these in turn have affected labor input needs. Land-saving techniques, such as fertilizer application, improved seeds, and irrigation, have generally increased labor demand whereas labor-saving techniques, such as mechanization, have reduced labor demand. The introduction of new varieties of certain crops, such as open-ball varieties of cotton in place of the traditional closed-ball varieties, has at times caused the rescheduling of labor activities. The effects of these new techniques cannot be easily separated because they are usually adopted as parts of a package. Therefore efficient adoption of land-saving techniques has required complementary adoption of mechanization, a labor-saving technology. Moreover, the adoption of new techniques was encouraged by government policies, and thus did not necessarily arise out of the inherent needs of the farmer.

Given the rich variety of farming systems in Turkey, such developments seem to have resulted in the seasonal labor demand in coastal regions, where irrigated agriculture prevails, being met by the labor surplus of Central, East, and Southeast Anatolia where rainfed agriculture is dominant. However, this pattern will probably change because further changes in the intensity of input utilization are likely. Comparing the present level of input consumption per

132

ha with that in developed or neighboring developing countries with similar geographical conditions, for example Greece and Bulgaria, there is clearly potential for further intensification in Turkish agriculture.

The large-scale irrigation projects about to be undertaken in Turkey, such as the Southeast Anatolian Project, are likely to have an impact. They may reduce the movement of seasonal workers between regions, so cotton harvesting may become mechanized or production may shift to Southeast Anatolia. A lack of labor in the potentially irrigable areas may hinder the adoption of the desired techniques and crops.

In the near future, labor will not become a constraint on the adoption of new techniques in rainfed agriculture. However, the adoption of more labor-intensive technologies, or technologies which might reschedule labor needs in rainfed agriculture, will affect labor demand of irrigated agriculture. More intense use of labor-saving technologies in rainfed agriculture, on the other hand, may release more workers who might be absorbed by the expanding irrigated agriculture. However, without conscious coordination and effort it is highly unlikely that the labor surplus of rainfed agriculture will be employed effectively in irrigated farming.

References

Agaoglu, S. 1936. Memleketimizde Mevsimlik Ziraat Iscisi Meselesi. Ulku 7: 273.

Aresvik, O. 1975. The Agricultural Development of Turkey. New York: Praeger Publishers.

Aricanli, T. and Somel, K. 1979. Turkiye'de Toprak Dagiliminda ve Tarimda Degisimin Niteligi Uzerine Gozlemler. Middle East Technical University, Studies in Development 6: 91–110.

Bruck, W.F. 1919. Turkische Baumwollwirtschaft. Jena: Verlag von Gustav Fischer.

Cerit, S. 1986. Turkiye'de Iller Arasi Gocler. The Turkish Journal of Population Studies 8: 81–103.

Gurkan, A.A. 1985. The Regional Structure of Agricultural Production in Turkey: A Multivariate Perspective. Middle East Technical University, Studies in Development 12: 27–47.

Gurkan, A.A. 1986. A Statistical Data Base for Turkish Agriculture, Ankara. Mimeo.

Gurkan, A.A. and Kasnakoglu, H. 1986. Tarimin Turk Ekonomisinin Gelisimine Katkisi: Bugun ve Yarin. ENKA Science and Arts Awards, Ankara.

Helburn, N. 1955. A Stereotype of Agriculture in Semiarid. Turkey. Geographic Review 45: 375–384.

Herrman, R. 1900. Anatolische Lanwirtschaft. Leipzig. Publisher not known.

Issawi, C. 1980. The Economic History of Turkey 1800–1914. Chicago: University of Chicago Press.

Kasnakoglu, H. 1986. Agricultural Price Support Policies in Turkey: An Empirical Investigation. Pages 131–157 in Food, States and Peasants: Analyses of the Agrarian Questions in the Middle East (Richards, A., ed.). Boulder and London: Westview Press.

Kasnakoglu, H., Akder, H., and Gurkan, A.A. In press. Agricultural Labor and Technological Change in Turkey. Aleppo: ICARDA.

Kasnakoglu, H. and Gurkan, A.A. 1986. The Political Economy of Agricultural Pricing Policies: The Case of Turkey. Project Report No. 3. Prepared for the World Bank, Ankara.

Kasnakoglu, H. and Gurkan, A.A. 1987. Past Trends and Future Prospects of Turkish Agriculture. Yapi Kredi Economic Review 1(2): 23–32.

Kasnakoglu, Z. 1975. Distribution in Turkey: A Case Study on the Determinants of Male Earnings Differentials in 1968. PhD thesis, University of Wisconsin.

Kuznets, S. 1964. Economic Growth and the Contribution of Agriculture: Notes on Measurement. Pages 102–119 *in* Agriculture in Economic Development (Eicher, C. and Witt, L., eds). New York: McGraw-Hill.

Le-Si, V., Scandizzo, P. and Kasnakoglu, H. 1983. Turkey: Agricultural Sector Model. AGREP Division Working Paper No. 67. Washington, D.C.: World Bank.

SIS (State Institute of Statistics). 1950a-80a (Annual). Census of Agriculture. Ankara: SIS.

SIS (State Institute of Statistics). 1950b-85b (Annual). Census of Population. Ankara: SIS.

SIS (State Institute of Statistics). 1960c-83c (Annual). Agricultural Structure and Production. Ankara: SIS.

SIS (State Institute of Statistics). 1960d-85d (Annual). The Summary of Agricultural Statistics. Ankara: SIS.

SIS (State Institute of Statistics). 1973/74. Income Distribution and Consumption Expenditures in Rural Areas. Ankara: SIS.

SPO (State Planning Organization). 1963–87 (Every 5 years). Five-Year Development Plan. Ankara: SPO.

SPO (State Planning Organization). 1976. 1973 Gelir Dagilimi. Ankara: SPO.

SPO (State Planning Organization). 1985. 1985 Programi. Ankara: SPO.

Timur, S. 1972. Turkiye'de Aile Yapisi. Haccettepe University Publication No. 15. Ankara: Haccettepe University.

Turkiye Is Bankasi A.S. 1986. Economic Indicators of Turkey, 1981–85. Ankara: Economic and Research Department.

Tutuncu, M.B. 1984. Turkiye'de Destekleme Politikasinin Uygulamasi Hakkinda Not. KFB/KKID-TKS, Ankara. Mimeo.

Varlier, O. 1978. Turkiye Tariminda Yapisal Degisme, Teknoloji ve Toprak Bolusumu. SPO Publication No. 1636. Ankara: State Planning Organization.

7. Agricultural Labor and Technological Change in Cyprus

GEORGES PANAYIOTOU

Agricultural Research Institute, Nicosia, Cyprus

Historical Background

The economic life of Cyprus, based mainly on agriculture, has been closely linked to that of its colonizers and the influence of the last two, the Ottomans and the British, remains today. The Ottomans inherited a feudal system from the Venetians and restructured it to suit their purposes for administration and taxation. The governorship of the island, with rights of taxation, was purchased and governors tried to maximize their revenues by heavy taxation mainly of agricultural products (Hill 1972). Economic stagnation, heavy taxation, and corrupt administration, coupled with a series of droughts and natural calamities, badly damaged agriculture and reduced the population.

Feudalism was the dominant type of land tenure until the middle of the 19th century, when land became state property with private rights of use (usufruct) distributed to the peasants, who were allotted individual plots in return for a tax or fee amounting to 10% of production (Christodoulou 1959). To increase production, and therefore state revenue, the state provided security of possession of land in 1857 through the Ottoman Land Code. With this legislation, usufruct rights were registered and could be inherited, transferred through sale, exchange or gift, and mortgaged, but were lost if land was left uncultivated for a specified period.

Through these developments, a form of widespread private landownership was institutionalized for the first time (Christodoulou 1959; Camelaris 1977). This system was maintained by the British, with minor improvements in land registration, until after the Second World War. In 1945 the British gave those possessing land full and absolute *de jure* ownership rights in place of their usufruct rights. With independence in 1960 private ownership rights were safeguarded by the constitution and they continue to predominate.

Agriculture has always been an important part of the economy, particularly in its contribution to export earnings. In the early days of British rule the chief exports were wine, silk, carob, raisins, hides, wool, cereals, and cotton. At the end of the 19th century, world cereal prices fell sharply as cheap grain became available from the New World. Wine and carob prices also fell. Prices improved during World War I, mainly due to the disruption of agricultural production in Europe, but the depression that followed led

Dennis Tully (ed.), Labor and Rainfed Agriculture in West Asia and North Africa, 135–161.
© 1990 *ICARDA.*

to a fall in export prices and widespread poverty in rural areas, with heavy indebtedness and forced sales of farms.

During the first 10 years after the Second World War, the economy grew rapidly, but economic conditions deteriorated during the armed struggle of 1955–59. Between 1950 and 1959 agriculture grew at 5.2% per annum compared to 8.8% for the whole economy, and the share of agriculture decreased from 25.5 to 21.2%. In 1958 about 58% of total exports were minerals, and only 33% were of agricultural origin. Cyprus, which was traditionally a net exporter of food and agricultural products, became a net importer, with exports worth only about 35% of imports. At independence in 1960, the island faced stagnation of incomes, a trade deficit, an unstable political environment, flight of capital, and increasing emigration (Apostolides 1980).

The new government embarked upon a vigorous development effort to mobilize all productive resources and restore economic growth. Agriculture was given first priority and was gradually restructured to increase exports. Citrus groves were established in the Famagusta and Morphou areas and oranges and grapefruits became important export crops. Potatoes and vegetables also started gaining significance in the export markets along with the traditional carob, wine, and grape products.

Agricultural productivity rose tremendously due to improvements in infrastructure and irrigation, increased use of mechanization, fertilizers and chemicals, and improved farming methods. By 1972, agricultural products made up 55% of total exports and the value added by the sector was 48.4 million CYP, 17.7% of GDP (current factor cost) (Planning Bureau 1972). In the same year, agriculture employed 95,600 people or 34.8% of the economically active population. Thus, despite improvement, agriculture's contribution to GDP was small compared to the number of people it employed (SRD 1961a–86a).

Standards of living and social services were improving and the modernization of industry and agriculture continued until July 1974, when progress was abruptly stopped by the Turkish invasion and occupation of nearly 40% of the country. The displacement of 200,000 people, about a third of the total population, and the loss of the best agricultural land together with the majority of investments in industry and tourism caused a severe shock to the economy. Per capita GDP fell by 19% in real terms in 1974 and a further 16% in 1975.

During 1974–86 four Emergency Economic Action Plans emphasizing the development of agriculture, tourism, and industry, were successfully prepared and implemented. By 1978, full employment was achieved for the first time since 1974. A major aim of agricultural policy is to increase marketed output by expanding irrigated land, introducing higher yielding varieties of crops and improved breed stock, and through credit and extension programs. Rural income support, to discourage movement from remote villages to towns for employment, is another important objective and is pursued through integrated rural development programs, subsidies, and structural reform (Planning Bureau 1982).

Extensive investments in agriculture and interventionist agricultural policies resulted in an increase in GDP from agriculture, from 68.8 million CYP in 1976 to 78.6 million CYP in 1984 (1980 market prices). As a percentage of total GDP, however, the contribution of agriculture fell from 13.7 to 8.8% (SRD 1961a–86a) due to even faster development in other sectors of the economy.

Agriculture is now well integrated into the economy and there are no substantial social or economic differences between rural and urban areas. There is close contact between rural and urban populations and the amenities of the towns can also be found in the villages. Many town-based people operate agricultural enterprises and many farmers work off-farm in the towns in almost every sector of the economy.

Overview of Rainfed Farming Systems

Of the total area of Cyprus, only 46.8% is cultivated. Another 16.7% remains uncultivated because of low fertility, erosion, steep slopes, abandonment, or other reasons, while the remainder is forests (18.7%), communal grazing land (0.6%), and urban dwellings and roads (8.5%) (Papachristodoulou 1979). After 1974, about 38.5% of the land, including the most fertile agricultural land, came under Turkish occupation (MANR 1984).

For purposes of agricultural policy the government divides the area under its control into four regions based on soil and land suitability, climatic conditions, and cropping patterns (Figure 1). These divisions were used as the frame for agricultural censuses in 1977 and 1985 (SRD 1979; 1986b).

In 1985 the total crop area was about 162,000 ha, of which 78.4% was rainfed and 21.6% was irrigated, mostly in the Coastal and Mountain Zones. About 60% of the crop area was under annual crops and 40% under permanent crops. The majority of cereals, legumes, and fodder crops are grown in the Coastal and Dryland Zones where they account for over 50% of the total crop area (Table 1). Vegetables and citrus are grown mainly in the irrigated land of the Coastal Zone. In the other two zones, permanent crops account for three quarters of the cultivated area, with almost exclusively vines in the Vines Zone and both vines and deciduous fruits in the Mountain Zone. The prevailing farming systems in the four agroeconomic zones are described by Papayiannis and Papachristodoulou (1982).

In the Coastal Zone, irrigated production of cereals, potatoes, vegetables, and tree crops predominates, combined with off-farm employment. The rainfed farming system of this area combines cereal and fodder production with livestock, either sheep and goats or cows. The zone includes 49% of all sheep, 29% of all goats, and 31% of all dairy cows. All cows' milk and about

138

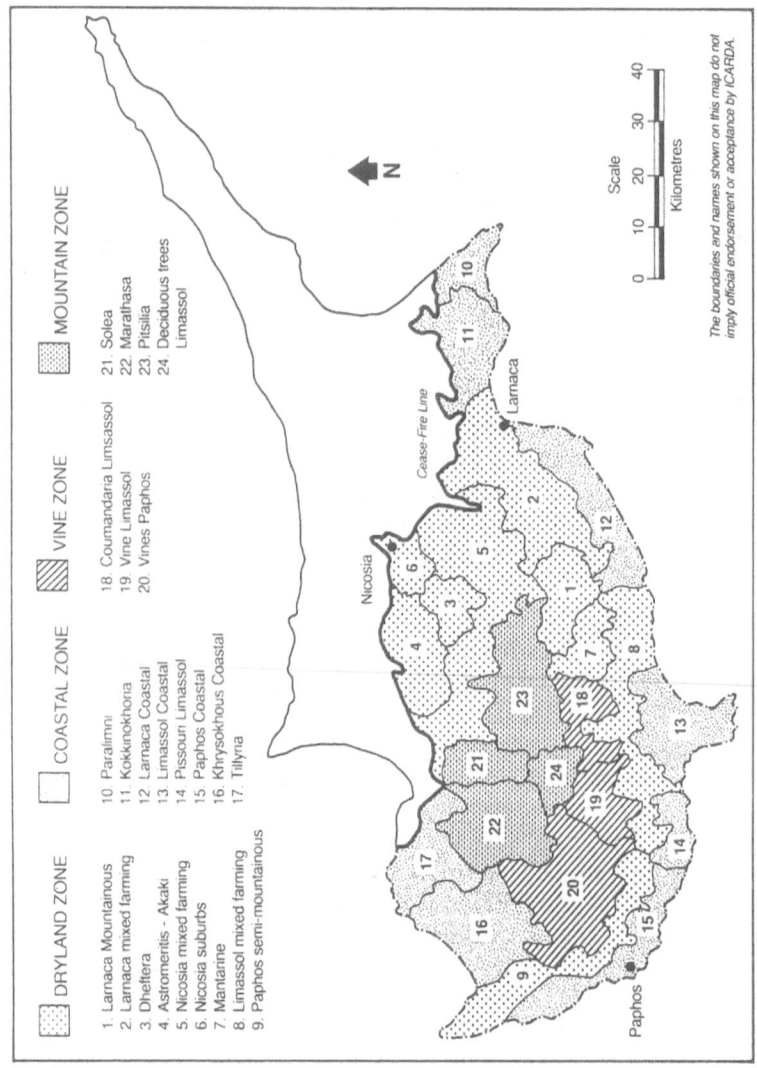

DRYLAND ZONE

1. Larnaca Mountainous
2. Larnaca mixed farming
3. Dheftera
4. Astromeritis - Akaki
5. Nicosia mixed farming
6. Nicosia suburbs
7. Mantanne
8. Limassol mixed farming
9. Paphos semi-mountainous

COASTAL ZONE

10. Paralimni
11. Kokkinokhoria
12. Larnaca Coastal
13. Limassol Coastal
14. Pissouri Limassol
15. Paphos Coastal
16. Khrysokhous Coastal
17. Tillyria

VINE ZONE

18. Coumandaria Limsassol
19. Vine Limassol
20. Vines Paphos

MOUNTAIN ZONE

21. Solea
22. Marathasa
23. Pitsilia
24. Deciduous trees Limassol

Scale

0 10 20 30 40

Kilometres

The boundaries and names shown on this map do not imply official endorsement or acceptance by ICARDA.

Fig. 1. Agroeconomic regions of Cyprus. *Source:* Philippides and Papayiannis (1983).

Table 1. Cropping pattern by zone, 1985.

Crop	Zone				Cyprus
	Coastal	Dryland	Vines	Mountain	
	% of crop area				
Cereals	31.2	42.7	17.2	4.0	34.7
Legumes and fodder	21.7	24.9	8.1	3.1	18.2
Vegetables	21.0	3.2	0.6	4.9	8.3
Vines	11.0	7.3	62.6	54.7	20.1
Citrus	6.6	3.0	0.2	0.8	4.0
Fruits	1.4	1.0	1.7	19.2	2.6
Dryland trees	7.1	17.9	9.6	13.3	12.1

Source: SRD (1986a).

80% of the sheep and goats' milk is sold to pasteurizers or cheese factories, while producers in remote areas process their milk into halloumi cheese or other products for sale. Green fodder or hay is produced on the farm, but large quantities of subsidized grain are usually bought for supplemental feeding. All farm operations are mechanized, including milking. In this zone there is one tractor for every 3.1 holdings or each 9 ha of land, one combine harvester for every 100 ha of cereals, one mower for every 50 ha of green fodder, and one pick-up for every 7.7 holdings (SRD 1986b).

In the Dryland Zone the prevalent system again combines dryland annual crops (cereals, legumes, and fodder) with livestock, as well as rainfed trees and vines in some areas. The characteristics of the system are the same as in the rainfed part of the Coastal Zone.

The Dryland Zone has the largest concentration of animals and includes 45, 51, 87, 68, and 67%, respectively of the sheep, goats, pigs, poultry, and dairy cattle. A number of commercial farms in this region specialize in dairy cows, pigs, and poultry, using more or less modern methods of production and management. They use standardized feed rations, are mechanized, and regularly use hired labor. These farms use resources efficiently and have high returns to labor and capital and a well-organized market for their products.

The Vines Zone is dominated by three rainfed farming systems, all of which have wine grapes, which demand little labor, as their most important component. Wine grapes are the only crop in the first system, which has the largest percentage of farmers with part-time, off-farm employment. These farmers used to produce raisins or wine and spirits and market the final product themselves, but during the past 10–15 years they have sold directly to the wineries. There is widespread underemployment among these farmers, especially the older ones and the female population. Draft animals are still used due to lack of access roads for machinery.

In the second system, found in Commandaria villages, a cottage industry for processing wine grapes is important, as well as tree crops. The grapes are sun-dried and crushed at local cooperative wineries to produce a concen-

Table 2. Value of crop and livestock production by zone, 1979.

Crop	Zone				Cyprus
	Coastal	Dryland	Vines	Mountain	
Total value of production (million CYP)	44.1	37.9	11.2	5.5	98.7
Cereals (%)	5.2	10.8	4.5	1.3	7.1
Legumes and fodder (%)	3.8	8.7	3.3	2.0	5.6
Vegetables (%)	47.1	6.7	1.4	10.9	24.5
Vines (%)	6.4	3.1	63.3	14.6	12.0
Citrus (%)	10.6	5.7	0.3	0.2	6.9
Fruits (%)	2.5	1.2	3.3	34.0	3.8
Dryland trees (%)	1.6	8.3	6.6	18.2	6.0
Livestock products (%)	22.8	54.4	17.3	18.8	34.1

Source: Philippides and Papayiannis (1983).

trate which is sold to factories in Limassol that produce Commandaria dessert wine. The farmers' wives have a small number of goats to supply the family with milk, milk products, and meat. All operations are carried out by family labor and draft animals are still used.

In the third system, viticulture is combined with cereals, fodder crops, and livestock and most grapes are sold directly to the winery. Although there is considerable underemployment throughout the year, hired labor, some from other villages, is required during grape harvesting. Farming practices are more mechanized than in the other two systems mainly due to a favorable landscape and plant spacing.

In the Mountain Zone, the farming systems are based on irrigated deciduous fruits, plus rainfed trees or vines and part-time employment off-farm.

Farm Economics

The Coastal and Dryland Zones accounted for most of the total value of production in 1979 (Table 2). In the Coastal Zone vegetables are the most valuable crop, followed by livestock and citrus, while in the Dryland Zone livestock contribute over half of the total value. The Vines Zone gets 63% of its value from wine grapes and 17% from livestock, while the Mountain Zone gets 34% of its value from deciduous fruits, 19% from livestock, 18% from dryland trees, and 15% from vines.

In 1979 the Coastal Zone had consistently better productivity indicators than the other zones mainly due to the larger proportion of irrigated land, better soil fertility, and the larger area of the holdings (Table 3).

The most profitable crop is olives, followed by barley, wine grapes, almonds, wheat, and vetch (Table 4). However, olive trees need 5 or more years to bear much fruit; in the short term barley is far more profitable due

Table 3. Productivity indicators in Cypriot agriculture, 1979.

Zone	Value added (CYP) per					
	Holding	Ha of cultivated land	Gainfully employed person		Man-month's work	
			From crops	From livestock	From crops	From livestock
Coastal	2,108	75	2,009	1,173	248	83
Dryland	834	27	1,425	1,515	152	108
Vines	1,159	34	1,010	884	123	57
Mountain	492	47	570	1,273	71	88
All zones	1,209	45	1,401	1,330	166	94

Source: Philippides and Papayiannis (1983).

Table 4. Profitability of the main rainfed crops, 1981–83.

Crop	Gross revenue	Variable costs	Fixed costs	Gross profit	Net profit
	CYP/ha				
Wheat	195	115	64	80	16
Barley	213	92	55	121	66
Vetch	256	94	149	162	13
Wine grapes	440	160	235	280	45
Olives	380	68	164	312	148
Almonds	288	55	212	233	21

Source: Papayiannis et al. (1982–84).

to the high price paid by the government. This subsidy distorts resource use and as a result farmers plant barley instead of wheat. Most livestock keepers growing barley sell it at a high price and buy barley for feed at one third of the sale price (CGC 1986).

Both locally produced and imported barley is heavily subsidized to encourage livestock production and keep livestock product prices low to combat inflation. The low price of feed grain has led to wasteful substitution of roughage by barley. As a result, much of the grazing land is no longer utilized, while animals with very little roughage in their diets have developed digestive diseases (Economides 1985). Seasonal surpluses and marketing problems have also been created.

The production of wine grapes has also been distorted by price policy. Subsidies encouraged producers to continue expanding until, finally, the government banned vine expansion except under the Vine Replanting Scheme. The disposal of wines, spirits, and other vine products has been a serious problem for many years because grape production is very large for such a small country (180,000–200,000 t per year, of which over 90% must be exported) (SRD 1985) and almost one third of the agricultural population depends on grapes for a living.

Olive production has expanded rapidly during the past 10 years due to its relative profitability and to the extensive promotion of olive culture by the

Department of Agriculture. Thousands of seedlings have been given to farmers and in 5 years Cyprus is expected to become self-sufficient in olives.

Social Organization of Agriculture

The first half of the 20th century was a difficult period for Cypriot agriculture. In the years of poor harvests farmers had to borrow money from moneylenders or merchants, at interest rates of up to 50%, to buy agricultural inputs and maintain their families (Christodoulou 1959). As one bad crop followed another, due to droughts and the low productivity of land, many farmers became heavily indebted to moneylenders and had to sell their land to pay their debts. Between 1920 and 1926 there were 16,559 compulsory sales of land which considerably increased the number of landless peasants (Georgallides 1979). Many farmers turned to mining for employment and others moved to the towns or emigrated to America or Australia.

Eventually, moneylenders became the dominant landowners in every village. Most farmers were either landless or possessed only marginally productive land so they had to rent land. Payment was in kind, sometimes as much as 50% of the harvest (Christodoulou 1959; Lanitis 1944). Repeated efforts by the colonial government to solve this problem were unsuccessful due to the lack of alternative sources of credit and it was only with the establishment and growth of the cooperative societies that the problem was solved.

Absentee farmers very rarely have a role in farming in Cyprus. If they do not intend to cultivate the land part-time, they lease it on a long-term basis or give relatives the right to farm however they think proper. With tree plantations, they usually demand good management in return for the annual income from the crop.

Custom operators are found in all zones but are more common in the Vines and Mountain Zones where farmers cannot afford machinery. Custom operators usually own a tractor and undertake cultivation, seedbed preparation, and harvesting on a contract basis. They also undertake chemical application by high pressure sprayers (for citrus) or harvesting with specialized machinery (groundnuts). In all zones some farmers, both full- and part-time, may undertake limited custom work in the fields of relatives and friends. Many of the older operators, especially in the dryland areas, lease their plots on a short-term basis to tractor owners who put them under cereal cultivation.

Many town-based, part-time farmers have share arrangements with full- or part-time farmers residing in the village. These vary from land leasing to full sharecropping. Sharecropping is more common between farmers living in the same village whose properties are adjacent. Sometimes, share arrangements include not only farm operations but also marketing of agricultural produce and new investments in machinery, commercial farm equipment, etc. However, according to Papayiannis et al. (1982–84), less than 1% of agricultural land is operated under the sharecropping system.

The Role of Cooperatives

The first cooperative credit society was established in 1908. The movement grew slowly until 1937 when the Department of Cooperative Development and the Central Cooperative Bank were established. The number of cooperatives of all types increased from 219 in 1938 to 861 in 1982, while credit societies increased from 112 in 1938 to 477 in 1958. Marketing cooperatives were established for fruits and cereals. In 1946 SODAP, the largest winery in Cyprus, was established as a cooperative enterprise. The number of consumer cooperatives also increased, from 16 in 1940 to over 300 in 1958, and the value of purchases went up from 54,000 USD to 9 million USD (Angastiniotis 1965).

The first collective farm was established in 1946 by 28 ex-servicemen who subscribed their gratuities and resettlement grants as capital. They learned modern practices and demonstrated how unpromising, derelict land could be turned into a model farming unit. Their example was followed by others (Christodoulou 1959).

The emergence of cooperatives combined with certain other developments improved the farmer's lot. According to Christodoulou (1959) the most far-reaching development in the post-war years was the alienation in 1947 by government of the Paphos chiftliks (large estates) which belonged to absentee landlords living in Constantinople. The land was used for experiments and demonstrations or leased to farmers for long terms (25–33 years). During the same period the church, the largest landowner on the island, sold a large part of her properties. Many were purchased by cooperative societies who sold parcels to members on an installment basis.

With the increased productivity of agriculture following independence, the cooperative movement expanded and by the 1970s it was engaged in all forms of economic activity and dominated the economic life of the island. Its activities included banking to assist credit and other cooperatives, retail and wholesale trade including export of agricultural products and imports of agricultural inputs, and transport, processing, and marketing of agricultural products. In this period the agricultural marketing societies enjoyed a very large share of the market and accounted for 70, 95, 84, 65, and 94% of citrus, carob, vegetable, wine, and olive exports, respectively (Andreou 1972).

For over a decade the cooperative enterprises functioned well. However, in the early 1980s, heavy investment in a series of diverse enterprises undertaken without feasibility studies led to the failure of most of these projects and the loss of millions of pounds. As a result, almost all these companies went bankrupt and closed down and, more important, the contribution of the cooperatives to the economy was greatly reduced. The remaining cooperatives operate alongside similar ventures in the private sector, which is now expanding. Credit, marketing, and other services are offered to farmers by individuals and institutions operating on a competitive basis.

Technological Changes

Cypriot agriculture is currently market oriented, with emphasis on exportable products because the local market is very small. A number of technological changes have helped to transform the pre-war subsistence-oriented farming into today's modern agriculture.

Mechanization

The introduction of the tractor soon after the Second World War had a great impact on farming operations. Cultivation was better and more timely and cultivated area was expanded. Coupled with the use of fertilizers and pesticides, these changes resulted in surplus production for the first time in this century. The threshing machine was introduced next and like the tractor it was rapidly adopted by cereal farmers. Today, the tractor has displaced draft animals altogether and the combine harvester has displaced the standing thresher.

Mechanization has taken place in other farming systems but to a lesser extent. Small tractors were introduced in areas with citrus and deciduous orchards, and two-wheeled, hand-braked tractors were eventually adopted in the vine growing areas where the four-wheeled tractor was not suitable due to the landscape and lack of roads. With the introduction of the two-wheeled tractor, the vine acreage has increased and the new vineyards have been planted in rows. Similarly, with the introduction of a harvesting machine for groundnuts, the area has increased from 130 ha in the 1970s to 900 ha in 1985 and production from 300 to 3,400 t.

Improved Irrigation Systems

The introduction of improved, water-saving irrigation systems such as sprinkler, mini-sprinkler, and drip irrigation has led to an increased area under irrigation, with better quality products and higher yields. Despite the high initial costs, these systems were adopted by farmers because they could get credit from cooperatives. Nowadays, with the exception of deciduous fruit in the Mountain Zone and some small projects in the Dryland and Vines Zones, the majority of irrigated areas use improved water delivery systems.

New Varieties and Crops

Higher yielding varieties of durum and bread wheats, lentils, fodder crops, and almonds have been introduced, as well as more marketable varieties of wine grapes, citrus and deciduous fruits, potatoes, melons, and vegetables suitable for greenhouse production. Other innovations include citrus rootstock resistant to Tristeza virus, local production of potato seed, and new crops such as avocado, mango, lotus, and jojoba.

Modernization of the Livestock Sector

The majority of the small ruminant population currently consists of improved breeds. About 75% of sheep are Chios purebreds and 20% are other improved races, while about 15% of goats are Damascus purebreds and 70% are improved crosses. Dairy cows represent about 97% of the cattle population and are all Holsteins (SRD 1986b). Due to advances in breeding, nutrition, and management, livestock production and productivity have increased rapidly and milk production is now 150 kg/ewe, 250 kg/goat, and 5 t/cow (Papachristodoulou et al. 1987). Pig and poultry production are now totally commercialized and production meets local demand.

Agricultural Research

The Agricultural Research Institute has played a key role in the introduction, development, and dissemination of technological changes in Cypriot agriculture. It was founded in 1962 as a joint project between the Government of Cyprus, the United Nations Special Fund (UNSF), and FAO. The Institute carries out applied agricultural research to address problems facing plant and livestock production, evaluate recent scientific findings in other countries under local conditions, and determine how the agricultural potential of Cyprus may be fully developed.

The extension service of the Department of Agriculture disseminates improved technology to farmers and provides feedback to the research unit. Bonuses in kind or subsidies are sometimes used to facilitate adoption. The rate of adoption of new technology, with the exception of the older and poorer farmers in remote villages, is relatively high in Cyprus. The farmer must be convinced only that the new technology will be appropriate and profitable for his farm. Many farmers do not wait for the extension service to bring new technology, but actively seek it by visiting experimental stations and consulting the scientists working on new developments. Private firms are also important in disseminating new technology as they carry out demonstrations in farmers' fields. Funding by credit from the cooperative societies supports adoption.

Table 5. Average area of agricultural holdings.

Year	No. of holdings	Total area (ha)	Average holding area (ha)
1946	60,179	432,800	7.2
1960	69,445	431,066	6.2
1977	43,807	200,093	4.6
1985	47,248	178,504	3.8

Source: SRD (1949; 1963; 1979; 1986b).

Land Consolidation

The land tenure system, combined with laws of inheritance which provide for equal division of property among children, and sentimental bonds with the land, have led to an increasing number of landowners and extreme fragmentation of agricultural holdings. The number of holdings has increased while cultivated area has declined, as more land is taken up by roads and buildings. The average size of holding was 3.8 ha in 1985 compared to 4.6 ha in 1977 and 6.2 ha in 1960 (Table 5). Each holding consists of up to 24 parcels, often widely dispersed, with an average area between 0.65 and 1.24 ha (SRD 1986b).

This extreme fragmentation does not permit economies of scale, precludes introduction of mixed farming or other improved systems, and wastes scarce resources. To solve this problem, the government enacted the Land Consolidation Law in 1969. The main objectives are to consolidate holdings with good access and enlarge small holdings to establish viable farm units. Because the law provides for voluntary acceptance of land consolidation, the process will be slow. For the time being, land consolidation is practised only in the project areas which are usually irrigated. The first results are very satisfactory and it is expected that rainfed areas will soon be included.

The Effects of Off-Farm Employment and Migration

The Labor Force

The economically active population of Cyprus, which was 235,500 in 1960 and represented about 41% of the total population, reached 280,000 or 44% of the total population just before the Turkish invasion in 1974. With the displacement and emigration that followed, the economically active population fell to 203,600 in 1976, then started to rise again and in 1984 was 248,000 persons or 46.4% of the total population (Figure 2).

The population growth rate was 0.7% in the 1960s, declined to 0.2% in the 1970s, but showed signs of recovery in 1984–85. The drop in the growth rate was the result of high rates of emigration and declining birth rates. As pointed out by Chappa (1982), the climate of uncertainty resulting from the

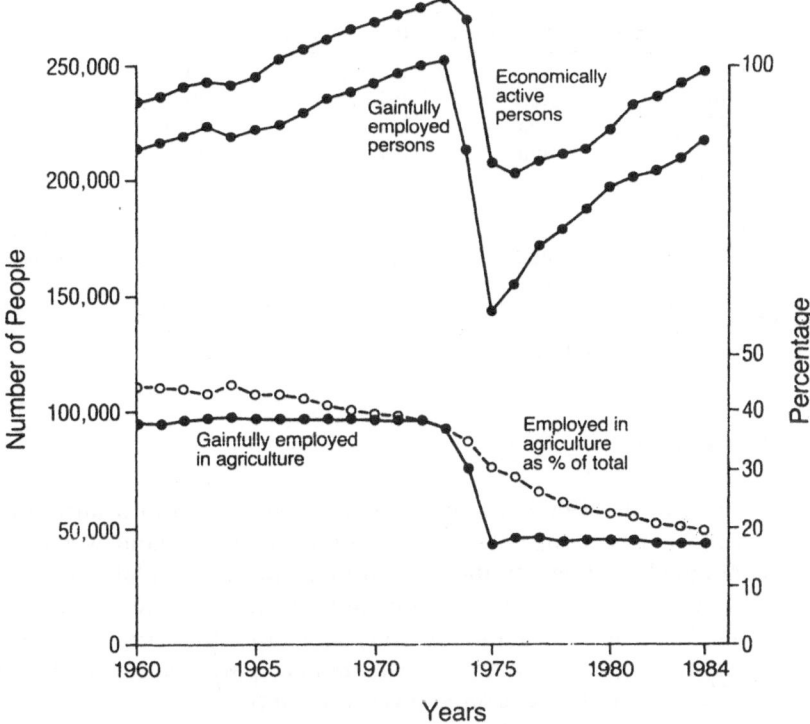

Fig. 2. Labor force and employment in agriculture, 1960–84. *Source*: SRD (1961–86a). *Note*: As from mid-1974 data refer to the government controlled areas only.

political situation has been at times a very strong "push" factor for emigration. During the past 25 years, emigration has been slight (about 0.5%) in periods of political calm and economic development and greater in periods of political upheaval and armed struggle. Peaks in emigration occurred during 1960–62 soon after independence, in 1964 because of intercommunal fighting, and during 1974–76 as a result of the Turkish invasion. During the past few years net emigration has gradually declined and has now stopped completely.

The steady decline of the birth rate, which is clearly manifested in the aging of the population, has been attributed largely to increased urbanization (Chappa 1982). The urban population rose from about 36% of the total in 1960 to 42% in 1973. After the displacement of nearly 200,000 people from their homes in 1974, urban residents outnumbered the rural population, and by 1985 they represented 66.3% of the total population (Table 6).

In 1982 10% of the population was aged 65 and over compared to only 6.4% in 1960, while the percentage of children below 15 years dropped to 24.7% from 36.4% during the same period (Demetriades 1984a). This trend is expected to continue in the next few decades.

148

Table 6. Urban and rural populations, 1981–1985.

Year	Urban population	% of total	Rural population	% of total	Total
1891	38,695	18.5	170,591	81.5	209,286
1901	44,103	18.6	192,919	81.4	237,022
1911	51,973	19.0	222,135	81.0	274,108
1921	61,512	19.8	249,203	80.2	310,715
1931	76,351	21.9	271,608	78.1	347,959
1946	115,808	25.7	334,306	74.3	450,114
1960	205,983	35.9	367,583	64.1	573,566
1973	266,803	42.2	364,975	57.8	631,778
1985	358,900[1]	66.3	182,200[1]	33.7	541,100[1]

Source: SRD (1961b–86b).
[1]Refers to government controlled area only.

The Labor Market

According to Matsis (1982), the labor market developed in a fairly orderly manner in the period before 1974. Registered unemployment was only 1% of the economically active population, but there was no real labor constraint. There was considerable unregistered unemployment, especially among Turkish Cypriots, and the very large agricultural labor force continued to be a source of labor for other sectors. Also, female participation in the labor force was very low and represented a reserve supply.

After 1974, the labor market underwent major changes. The fast reactivation of the economy, loss of the Turkish labor force, human losses during the invasion, and permanent and temporary emigration reduced the domestic labor supply and accounted to a great extent for the shortage of labor after 1978.

The labor constraint has been accompanied by structural imbalances because the supply of labor did not respond sufficiently to the changing demand pattern and the training institutions could not cope with the new demand structure. Cyprus, having no university, cannot adequately control education beyond the secondary level so the overall labor shortage has been accompanied by surpluses in some occupational groups, especially graduates of tertiary education.

In the agroeconomic surveys of the Agricultural Research Institute (Panayiotou et al. 1981; Panayiotou and Papachristodoulou 1983), family labor availability is calculated by assuming that the family head and all other adult members can offer a full labor unit (280 days) and the wife, due to other obligations, can offer up to 150 days. Students between 15 and 18 years old contribute 60 days and the elderly (over 65), 100 days.

In 1983, family labor availability in the four agroeconomic zones was 489–664 days per farm. Of the available labor, 28–46% was used on the farm, about 25% was employed off-farm, and the remaining 37–47% was

Table 7. Family labor availability and use on the average farm by zone, 1983.

	Zone			
	Coastal	Dryland	Vines	Mountain
Available family labor (days)	574	664	552	489
On-farm use of				
family labor (days)	209	172	203	175
hired labor (days)	55	13	26	6
Per cent of available				
family labor	46	28	42	37
Off-farm use of family				
labor (days)	120	159	128	134
Per cent of available				
family labor	21	24	23	27
Unused family labor (days)	245	313	221	180
Per cent of available				
family labor	43	47	40	37

Source: Papayiannis et al. (1982–84).

underutilized or completely unused (Table 7). Nevertheless, the average farm had to hire extra labor during peak periods (e.g., for harvesting, pruning, etc.).

Hired labor is recruited either within the village or from neighboring villages, but in certain periods it is hard to find so some farmers make labor exchange arrangements with other farmers. This is most common for wine grape harvesting where large quantities of produce, sometimes up to 10 t, must be harvested in a day to fill a truckload for the factory. In this case, the labor of six or more households is usually combined and the grapes from one farm are harvested each day. This saves money and ensures that labor is available when required. This arrangement is not feasible for town-based farmers or farmers with full-time, off-farm employment who employ hired labor instead.

Ansel et al. (1984) found that on-farm labor use also depends on the status of the farmer, with a clear tendency for less labor to be used on part-time rather than full-time holdings (Table 8). Not only does the operator employed off-farm spend less time on his farm, but also his family works fewer days on-farm than those of full-time farmers. In addition, part-time

Table 8. Labor inputs by farmer status (days per year).

Farmer status	Operator	Wife	Other family	Hired labor	Total labor
Full-time	145	101	21	301	568
Part-time off-farm	107	93	4	245	449
Full-time off-farm	53	69	1	152	275
Over 65 years	53	49	12	120	234

Source: Ansel et al. (1984); ARI (1981, unpublished data).

Table 9. Percentages of old (over 65) and female holders by zone, 1977 and 1985.

		Zone				All Cyprus
		Coastal	Dryland	Vines	Mountain	
Holders	1977	14	20	32	28	21
over 65 years	1985	12	19	36	35	21
Female	1975	6	8	12	13	9
holders	1985	5	7	12	14	8

Source: SRD (1979; 1986b).

farmers appear to employ less hired labor, possibly because they operate smaller holdings and tend to concentrate on one or two crops, mainly fruit trees and vines, so they have a low labor requirement except at pruning and harvest. Such seasonal activity may be integrated more conveniently with off-farm occupations than other types of farming.

The majority of farmers with part-time, off-farm occupations (62%) are self-employed while the remainder are seasonally employed casual workers. Farmers with full-time, off-farm occupations tend to be salaried (52%) or wage earning (15%) and only 24% are self-employed (Ansel et al. 1984). Most children have salaried or wage earning jobs and in general wives are casually employed.

Employment in the Agricultural Sector

Prior to 1974, agricultural employment decreased by an average of 0.1% per annum compared with an increase in the rural population of 0.5% per annum. The low productivity in this sector (less than half of the average for all sectors in 1973 as measured by value added per worker) suggests that a significant proportion of agricultural workers were underemployed. Demetriades (1982) estimated underemployment at 14% in 1966.

Although full employment has prevailed in the non-agricultural sectors there has been only a small decline in the labor force engaged in agriculture. As there has been little expansion in the rural population, the source of labor in the non-agricultural sectors is probably the growing urban labor force and not migrating agricultural workers. The stability of the agricultural labor force is probably due to its large proportion of elderly people and housewives who find it difficult to move away from rural areas for psychological, social, and economic reasons. In 1960, 26.6% of the persons employed in agriculture were aged 55 and over, of whom 54.3% were women (SRD 1963). In 1977, 35.5% of those employed in agriculture were aged 55 and over, and 50.7% were women (SRD 1979). There were more old and female operators in the poorer and more remote Vines and Mountain Zones (Table 9). Subsequent studies (ARI 1981, unpublished data; Neocleous 1982) confirm the steadily aging farm population in all zones, and suggest an increase in the number of farmers with full- and part-time off-farm occupations (Ansel

Table 10. Holders and family members by main occupation, 1985.

	Agriculture			Other sectors		
	Male (%)	Female (%)	Total no.	Male (%)	Female (%)	Total no.
Only occupation	34	66	38,327	62	38	25,227
Main occupation	73	27	1,859	83	17	25,317
Secondary occupation	83	17	25,317	73	27	1,859

Source: SRD (1986).

et al. 1984). Some elderly farmers may continue farming as a hobby rather than a full-time occupation.

During 1960–73 agricultural income more than doubled and although it was below the average for the whole economy, taking into account the higher cost of living in urban areas it was fairly attractive. Finally, a large part of the agricultural labor force did not have the skills required by the other expanding sectors.

After the 1974 war and the loss of much farmland, the number of people engaged in agriculture declined substantially, then changed little between 1975 and 1984 although as a percentage of the total labor force they have dropped from 30.5 to 19.9%. In 1984 43,100 people were gainfully employed in agriculture and their numbers are now exceeded by those in manufacturing and trade, restaurants, and hotels.

Since 1984, the methodology for computing employment in agriculture has changed. With the previous system, all holders and family members working less than 3 months on the farm were ignored, while all those working for more than 3 months were counted equally as gainfully employed, irrespective of actual time spent on the farm. This led to confusion when employment data for agriculture were compared with other sectors where employment was reported in full-time working equivalents.

In the agricultural census of 1985 all employment data were reported in 7–day weeks of actual work on the farm by all farmers and family members. By dividing the total work input on the farm by that corresponding to 9 working months (i.e., 38.7 weeks) the full-time equivalent number of persons employed in agriculture was calculated. For off-farm employment, a work year of 52 weeks was used.

The number of persons from farm households participating in the labor force in 1985 was 90,730. Of these, 38,327 were employed full-time in agriculture and 25,227 in other sectors. The rest combined farm and off-farm occupations (Table 10). It appears that there is considerable underemployment as 90,730 persons only worked 75,477 full-time equivalent labor units (Table 11). This underemployment may be due to casual farm work by students, housewives, and old people, who often have agriculture as their only occupation but do not work full-time. As expected, there is less underemployment in the non-agricultural sector.

Table 11. Employment of holders and family members in full-time equivalents[1] by zone and occupation, 1985.

Zone	Agriculture (On-farm)	Agriculture (Off-farm)	Other sectors	Total
Coastal	11,839	737	14,707	27,283
Dryland	8,895	523	21,584	31,002
Vines	5,231	213	2,602	8,046
Mountain	4,130	257	4,759	9,146
All zones	30,095	1,730	43,652	75,477

Source: SRD (1986b).
[1]Full-time equivalent for agriculture is 38.7 weeks and for other sectors is 52 weeks.

Based on the new approach described above, the data for employment in agriculture were revised for the years prior to 1984. This revision suggests that the actual labor input in agriculture for 1976–85 was more or less stable at around 36,500 persons (SRD 1986a).

Female Labor Force Participation

Female labor force participation rates were very low but started increasing after 1974, rising by 31% during 1976–80 compared with only 14% for men (SRD 1976; 1980). Women are now very important in the supply of labor to the agricultural sector. House (1981) estimated that 54% of the agricultural labor force are women (relatively old on average) and of these 86% are unpaid family workers. The rate of participation in the labor force for married women is 38.6% (Table 12).

Female participation is positively correlated with age in rural areas up to the age of 50, but negatively correlated in urban areas especially after the age of 40. The low participation rates of younger rural women are attributed by House to their dislike of farm work, due to higher average levels of education than their mothers and grandmothers have. If there is a generation effect, then the farm labor supply could shrink dramatically when the females currently over 65 grow older and move out of the labor force.

Since 1980, female participation in the general labor force has increased further and Demetriades (1984b) reported a rate of 48% for the years 1982–83. A number of factors underlie this change. To sustain economic growth, marginal groups, such as women, have been encouraged to work. As industrialization has proceeded, new jobs have been created and, as women represent a cheaper source of labor than men, employers are more willing to employ them in new vacancies (SRD 1980; House 1982b). Furthermore, alterations in the internal structure and distribution of roles in the family, coupled with the decline in cottage activities have induced women to join the labor force. The demand for higher standards of living has increased acceptance of a wage-earning role for both husband and wife. Finally, Cypriot

Table 12. Female labor force participation rates (%) by age (currently married women) and location, 1980.

	Age									
	15–19	20–24	25–29	30–34	35–39	40–44	45–49	50–59	60+	Mean
Urban	13.3	44.3	41.7	41.4	45.0	37.9	37.4	27.0	9.4	34.2
Rural	21.1	19.7	33.6	38.4	53.8	61.4	64.8	57.5	31.5	44.4
Total	17.7	34.2	38.6	40.3	48.3	46.9	48.1	40.0	21.8	38.6

Source: House (1981).

154

Table 13. Incidence of part-time farming by zone, 1977 (% of holdings).

	Zone				All Cyprus
	Coastal	Dryland	Vines	Mountain	
Full-time farmers	47	28	63	38	46
Farmers with part-time off-farm work	3	2	8	2	3
Farmers with full-time off-farm work	50	70	29	60	51
Total no. of holders	13,706	16,654	6,714	7,448	44,522

Source: Ansel et al. (1984).

women are nowadays better educated to take employment in almost any sector (Demetriades 1982).

Part-Time Farming

Only one third of the population lives in rural areas. The need to prevent depopulation of these areas, by maintaining farm incomes and enabling marginal farmers to survive, conflicts with the need to transfer labor from agriculture to more productive industry. However, distances from the remotest villages to the main urban centers are short enough to permit commuting and part-time farming may be the solution. Migration will be avoided if people continue to live in rural areas to tend their farms while also working in the non-agricultural sector.

According to the 1977 census of agriculture (SRD 1979), in all except the Vines Zone over half of all farmers have off-farm occupations (Table 13). However, this includes farmers with very small holdings, so the proportion of farmers with viable holdings who worked off-farm might be lower than this (Ansel et al. 1984).

In 1979, 59.5% of total holders and family members were employed in agriculture for 3 months or more ("gainfully employed"), ranging from 52% in the Dryland to 71% in the Vines Zone. Another 17–21% were underemployed and 12–30% were part-timers (Table 14).

The 1985 census shows that only 40% of the holders reported agriculture

Table 14. Employment of holders and family members in agriculture, 1979.

Zone	Gainfully employed[1] (%)	Underemployed (%)	Part-time (%)	Total no.
Coastal	65.8	18.7	15.5	19,710
Dryland	51.8	20.5	27.7	21,930
Vines	70.6	16.8	12.6	11,200
Mountains	52.8	16.9	30.3	11,960
All zones	59.5	18.6	21.8	64,800

Source: Phillippides and Papyiannis (1983).
[1]Employed in agriculture for 3 months or more.

Table 15. Main occupation of holders by zone, 1985.

Zone	Agriculture on-farm		Agriculture off-farm		Other sectors		Total
	No.	%	No.	%	No.	%	No.
Coastal	5,782	37	377	3	9,309	60	15,470
Dryland	6,932	34	300	2	13,133	64	20,365
Vines	3,512	64	125	2	1,867	34	5,504
Mountain	2,708	44	154	2	3,320	54	6,182
Total	18,934	40	956	2	27,629	58	47,521

Source: SRD (1986b).

as their main occupation, plus 2% reporting off-farm agriculture (usually tractor owners who work by contract on other farms or permanent agricultural workers), and 58% of holders had other main occupations (Table 15).

Part-time farming may be associated with a shortage of land or other resources which forces farmers to supplement their incomes. Alternatively, the farmer who adopts an off-farm occupation for other reasons may wish to limit the size of his holding. In either case, a negative association might be expected between the extent of off-farm activity and farm size.

Ansel et al. (1984) found a negative correlation (-0.4) between farm size and the incidence of off-farm employment. Part-time farmers operated a significantly smaller area than full-time farmers in the Coastal and Dryland Zones but there was no significant association between off-farm activity and holding size in the Vines and Mountain Zones. This suggests that off-farm work competes with farming for labor in the Coastal and Dryland Zones while in the poorer Vines and Mountain Zones there is some disguised unemployment.

Farm income per unit of labor, land, and capital is generally higher on full-time farms than on part-time farms, suggesting that the latter are less efficient than the former in using limited resources. However, part-time farmers procure additional income off-farm so their productivity as individuals may be greater than that of full-time farmers. The total income of full-time farmers is higher than others in the Coastal and Dryland Zones (Table 16). While the farm income of farmers with part-time jobs represents 70–75% of their total incomes, farmers with full-time, off-farm jobs get most income from their off-farm work.

The persistence of the traditional inheritance system suggests that land-ownership will continue to be widespread throughout Cyprus and since many of those who have inherited land have remained as part-time farmers, often with full-time, off-farm jobs (Ansel et al. 1984), it is likely that future generations will do the same. Similarly, continued agricultural instability will encourage some farmers to seek off-farm work for security but their attachment to land will deter them from leaving agriculture. Therefore, part-time farming will probably remain a feature of Cypriot agriculture. Because part-time farming maintains the links between farms and holders and family

Table 16. Income from on- and off-farm employment, 1981.

	Full-time farmers	Farmers with part-time off-farm work	Farmers with full-time off-farm work
Coastal zone			
farm	5,850	3,634	1,837
off-farm		1,334	2,142
total	5,850	4,968	3,979
Dryland zone			
farm	5,429	2,192	845
off-farm		728	1,509
total	5,429	2,920	2,354
Vines zone			
farm	1,869	1,742	978
off-farm		801	1,049
total	1,869	2,543	2,027

Source: Ansel et al. (1984).

members employed off-farm, labor availability for agriculture will be maintained. It is more likely that there will be labor shortages in the production of certain labor-intensive crops as part-time farmers tend to concentrate on the production of few crops, mainly permanent tree crops.

The Nature of the Farm Household

The average farm household consists of 3.7 persons (1.9 male and 1.8 female), usually a husband and wife and their unmarried children. Papayiannis et al. (1982) found that there were 1.0–1.8 children of different ages living with their parents and 1.8–2.8 children living off-farm, with more children living off-farm in the poorer areas of the Mountain, Dryland, and Vines Zones. Those off-farm were married children living with their own families or unmarried children who had emigrated to the towns or abroad. Also, farmers were older in the poorer areas (Table 17).

Table 17. Family structure of the average farm by zone, 1982.

	Zone			
	Coastal	Dryland	Vines	Mountain
Farmer's age (years)	48.4	50.9	53.9	59.4
Farmer's education (years)	6.5	6.1	6.0	6.8
Children living on-farm	1.8	1.6	1.1	1.0
0–10 years	0.7	0.2	0.3	0.1
11–16 years	0.6	0.5	0.4	0.4
>16 years	0.5	0.9	0.4	0.6
Children living off-farm	1.8	2.1	2.0	2.8
Family members on-farm	3.8	3.6	3.1	3.0

Source: Papayiannis et al. (1982).

Ansel et al. (1984) reported that in addition to a spouse and children, approximately 18% of households had elderly parents living with them permanently. These elderly people usually offer no assistance in farm work but the females help to raise the children and tend the house.

Data from 1982 show that operators received 6.0–6.8 years of formal schooling, so on average all went through primary school and some received secondary education (Table 17). Women spent fewer years in education than their husbands but children spent much more. In 1982, 95% of farm families had children with more than 3 years secondary education, 34% had children studying abroad, 7% had children with higher education in Cyprus, and 17% had children with vocational training (Ansel et al. 1984).

The head of the household usually maintains the family. Resources other than labor are pooled and used by the whole family so personal income from off-farm occupations and the farm income are used for food, clothing, child rearing, and education. The wife usually performs the household operations and helps on the farm. If she works off-farm, she contributes her income to the common fund. Male and female children working off-farm, although sharing household resources, may not and usually do not make any contribution to the household funds. However, brothers may agree to inherit less land than their sisters and sometimes they contribute to their sisters' dowry.

Until recently, rural families were patriarchal with all decisions being taken by the male head of the family. Although a recent study in Pitsilia showed that this system still exists in remote villages (Neocleous 1982), it is now widely accepted that other members of the family participate in decision making. Evidence from the survey for the Young Farmers Project (ARI 1984, unpublished data) and from other sources suggests that rural women today have an increasing role in decision making in agriculture.

Day-to-day decisions i.e., when to plant, apply fertilizer, weed, or harvest are taken by the operator in the case of the full-time farmer or his wife in the case of the part-time farmer, especially if the latter operates away from the village. In most cases, however, decisions are taken by the operator and his wife. More important decisions, such as choosing farming activities (annual crops, trees, or livestock), buying a piece of machinery or equipment, or buying or selling livestock, are again taken by the operator and his wife but the elder children, especially the boys, must be informed and consulted. On more serious matters, such as buying or selling land, giving a plot of land to a marrying child, or exchanging land for other property, the consensus of the whole family is usually necessary.

Where a son decides to stay in the village and engage in agriculture full-time, he takes all agricultural decisions in cooperation with his father. The take-over process is now much faster than in the past due to mechanization, education, and new methods of farming. Even so, the son is fully responsible only after he gets married and forms his own family. The land allocated to him will then be officially transfered to him.

If grown-up children are working full-time off-farm, there can be a conflict

Table 18. Capital investment on farms by farmer status and zone (CYP/ha).

Farmer status	Zone			Whole sample
	Coastal	Dryland	Vines	
Full-time	1912	412	990	1335
Part-time off-farm	1912	698	825	1372
Full-time off-farm	2288	1012	915	1445
Over 65 years	1432	584	750	952
All farmers	1928	645	900	1320

Source: Ansel et al. (1984).

for labor. This is especially true when the operator is also working full-time off-farm. The wife cannot undertake all farming operations and at peak periods extra labor must come from members of the family working off-farm or from hired labor.

The Process of Substitution between Labor and Capital

Mechanization of a number of operations has already been discussed. The only labor-intensive activity which is still to be substituted by capital is harvesting.

The literature (ADAS 1975; CEAS 1977) suggests that part-time farmers invest more heavily in labor-saving machinery and equipment than full-time farmers. This is supposed to result in a high capital-output ratio, i.e., a low average return on capital but high labor productivity on part-time farms. The Agricultural Research Institute Survey of 1981 (ARI, unpublished data) provides some evidence of higher investment per ha on part-time farms, although this varies between zones (Table 18). For full- or part-time farmers, total and per ha investment is considerably higher in the Coastal Zone than elsewhere.

A study on marginal productivity of the various inputs for wine grape production in the three regions of the Vines Zone (Panayiotou 1980) showed that the marginal return to opportunity cost for labor was below unity except for the Commandaria region, suggesting that too much labor was used in the other two regions (Table 19). The corresponding ratio for capital was above unity indicating that there was room for increased capital investment in wine grape production. The substitution of labor by capital seems profitable except in the Krasokhoria region where the productivity of capital is very low, probably because vineyards in that region are very old and planted randomly making mechanical cultivation difficult. However, with the establishment of new plantations under the Vines Replanting Scheme, substitution of labor by capital in this region should become more profitable.

As mechanization in vine cultivation started only recently and has grown very slowly due to sloping terrain, lack of access roads, and fragmentation

Table 19. Production elasticities, marginal value products, and opportunity costs for labor and capital in grape production in the three regions of the Vines zone.

	Paphos	Krasokhoria	Commandaria
Production elasticities			
labor	0.328	0.590	0.804
capital	0.456	0.443	0.898
Marginal value product			
labor (CYP/day)	1.65	2.03	2.26
capital (CYP/CYP)	1.27	0.89	1.76
Marginal return to opportunity cost ratio			
labor	0.66	0.92	1.08
capital	1.17	0.82	1.63
Marginal rate of substitution of labor by capital (CYP/CYP)	1.78	0.89	1.45

Source: Panayiotou (1980).

of fields, it can be assumed that this is the least mechanized and modernized sector of Cypriot agriculture. The upper limit is represented by irrigated crops such as vegetables under cover, citrus, and bananas, where modern techniques like fertigation (simultaneous application of water and fertilizers) are practiced. The degree of mechanization in the production of other crops fluctuates between these two extremes. Research is currently being done on irrigation, fertigation, weed control, and agricultural mechanization for field crops, with emphasis on mechanical harvesting.

Long-Term Perspectives

Actual labor use in the agricultural sector of Cyprus has decreased only slightly in the past 10 years, allowing sustained growth of agricultural production. However, for economic and social reasons, labor will become less available in the future and labor will become a constraint. On the other hand, expansion of part-time farming will lead to differentiation of the cropping pattern in favor of tree crops with a low labor requirement. Mechanization of farming practices, especially harvesting, will also play a key role in the change in cropping patterns. These factors, together with rapidly increasing labor costs, will result in intensification of production and a further reduction in cultivated land, as marginal land and scattered trees will be abandoned in favor of more fertile, irrigated land and compact plantations. As Cypriot agriculture is market oriented, the most important determinant for its future development and restructuring will be the marketing prospects in both the local and the export markets.

160

References

ADAS (Agricultural Development and Advisory Service). 1975. Part-time Farmers in the United Kingdom. Socio-economic Paper 3. London: Ministry of Agriculture, Fisheries and Food.

Andreou, P. 1972. An Empirical Investigation into the Main Factors that led to the Successful Operation of the Cooperative Marketing Societies in Cyprus. Pages 135–148 *in* Cooperative Institutions and Economic Development. Nairobi: East African Literature Bureau.

Angastiniotis, K.M. 1965. (The Cooperative Movement: Its Birth and Development in Cyprus.) In Greek. Nicosia.

Ansel, D.J., Bishop, C. and Upton, M. 1984. Part-time Farming in Cyprus. Development Study No. 26. Department of Agricultural Economics and Management, University of Reading, England.

Apostolides, C. 1980. The Role of Agriculture in the Economic Development and Trade of Cyprus from 1878–1971. Mimeo.

Camelaris, G. 1977. Land and Water Tenure: The Determinants of Agricultural Potential in Cyprus. MA thesis, Institute of Social Studies, The Hague.

CEAS (Centre for European Agricultural Studies). 1977. Part-time Farming: Its Nature and Implications. Seminar Paper 2. Wye College, Ashford, UK.

CGC (Cyprus Grain Commission). 1986. (Annual Report for 1983/84.) In Greek. Nicosia: CGC.

Chappa, I. 1982. Demographic Trends in Cyprus. Pages 124–152 *in* Proceedings of the National Workshop of the ILO/UNFPA Project on Population, Employment Planning and Labor Force Mobility in Cyprus, 22–23 February 1982. Nicosia: Ministry of Finance.

Christodoulou, D. 1959. The Evolution of the Rural Land Use Pattern in Cyprus. Geographical Publications: Bude, Cornwall, England.

Demetriades, E.I. 1982. The Statistical Base for Population and Employment Planning in Cyprus. Pages 49–67 *in* Proceedings of the National Workshop of the ILO/UNFPA Project on Population, Employment Planning and Labour Force Mobility in Cyprus, 22–23 February 1982. Nicosia: Ministry of Finance.

Demetriades, E.I. 1984a. Population Policy Issues in Cyprus. International Population Conference, Mexico City, 6–14 August 1984. Nicosia: Ministry of Finance.

Demetriades, E.I. 1984b. Tapping the Female Labour Resource in Cyprus and Government Perspectives on Utilization of Research for Women's and Population Issues in Cyprus. Nicosia: Ministry of Finance.

Economides, S. 1985. The Roughage Situation in Cyprus in Relation to Animal Health and Productivity. Miscellaneous Report 18. Nicosia: Ministry of Agriculture and Natural Resources.

Georgallides, G.S. 1979. A Political and Administrative History of Cyprus, 1918–1926. Nicosia: Cyprus Research Centre.

Hill, G. 1972. A History of Cyprus. Vol. IV. Cambridge: University Press.

House, J.W. 1981. Patterns and Determinants of Female Labour Force Participation in Cyprus. ILO/UNFPA, Population, Employment Planning and Labour Force Mobility in Cyprus. Working Paper No. 8. Nicosia: Ministry of Finance.

House, J.W. 1982a. Population, Employment Planning and Labour Force Mobility in Cyprus. Terminal Report. Nicosia: Ministry of Finance.

House, J.W. 1982b. Labor Market Segmentation: Evidence from Cyprus. ILO/UNFPA, Population, Employment Planning and Labor Force Mobility in Cyprus. Working Paper No. 11. Nicosia: Ministry of Finance.

Lanitis, N.C. 1944. Rural Indebtedness and Agricultural Co-operation in Cyprus. Limassol.

MANR (Ministry of Agriculture and Natural Resources). 1984. Agriculture in Cyprus: An Analysis and Evaluation. Nicosia: MANR.

Matsis, S. 1982. Problems of Population and Human Resources Planning in Cyprus. Pages 30–48 *in* Proceedings of the National Workshop of the ILO/UNFPA Project on Population, Employment Planning and Labour Force Mobility. Nicosia: Ministry of Finance.

Neocleous, G. 1982. Rural Pluri-activity in Pitsilia Area. A Case Study. Nicosia: Ministry of Agriculture and Natural Resources. Mimeo.

Panayiotou, G.S. 1980. Interregional Variation in Production and Productivity of Wine Grapes in Cyprus. Agricultural Economics Report 11. Nicosia: Ministry of Agriculture and Natural Resources.

Panayiotou, G.S. and Papachristodoulou, S. 1983. Agroeconomic Analysis of Farms in the North-West Region of Cyprus. Miscellaneous Report 8. Nicosia: Ministry of Agriculture and Natural Resources.

Panayiotou, G.S., Papayiannis, Chr. and Papachristodoulou, S. 1981. Results of the Farm Management Survey for the Southern Conveyor Project. Nicosia: Ministry of Agriculture and Natural Resources.

Papachristodoulou, S. 1979. Socioeconomic Aspects of Rainfed Agriculture in Cyprus. Nicosia: Ministry of Agriculture and Natural Resources. Mimeo.

Papachristodoulou, S., Papayiannis, Chr. and Panayiotou, G. 1987. Norm Input-Output Data of the Main Crop and Livestock Enterprises. Nicosia: Ministry of Agriculture and Natural Resources.

Papayiannis, Chr., Panayiotou, G. and Papachristodoulou, S. 1982–84. Agroeconomic Analysis Summary for Farms in Cyprus 1981–1983. Nicosia: Ministry of Agriculture and Natural Resources.

Papayiannis, Chr. and Papachristodoulou, S. 1982. Farm Types in Cyprus. Agricultural Economic Report 15. Nicosia: Ministry of Agriculture and Natural Resources.

Philippides, P. and Papayiannis, Chr. 1983. Agricultural Regions of Cyprus (A Comparative Statistical and Technoeconomic Analysis). Agricultural Studies Report No. 1. Nicosia: Ministry of Finance and Ministry of Agriculture and Natural Resources.

Planning Bureau. 1972. The Third Five-Year Plan (1972–76). Nicosia: Ministry of Finance.

Planning Bureau. 1982. The Fourth Emergency Economic Action Plan (1982–86). Nicosia: Ministry of Finance.

SRD (Statistics and Research Department). 1949. Census of Population and Agriculture 1949. Nicosia: Government of Cyprus.

SRD (Statistics and Research Department). 1963. Census of Population and Agriculture 1960. Vol. VI. Nicosia: Ministry of Finance.

SRD (Statistics and Research Department). 1961a–86a. Economic Report 1960–84. Nicosia: Ministry of Finance.

SRD (Statistics and Research Department). 1961b–86b. Statistical Abstracts 1960--1985. Nicosia: Ministry of Finance.

SRD (Statistics and Research Department). 1976. Registration of Establishments. Nicosia: Ministry of Finance.

SRD (Statistics and Research Department). 1979. Census of Agriculture 1977. Vol. I. Nicosia: Ministry of Finance.

SRD (Statistics and Research Department). 1980. Manpower Survey. Nicosia: Ministry of Finance.

SRD (Statistics and Research Department). 1985. Sales of Vine Products Manufactured in Cyprus 1984. Nicosia: Ministry of Finance.

SRD (Statistics and Research Department). 1986a. Agricultural Statistics 1985. Nicosia: Ministry of Finance.

SRD (Statistics and Research Department). 1986b. Census of Agriculture 1985. (Preliminary data.) Nicosia: Ministry of Finance.

SRD (Statistics and Research Department). 1986c. Household Income and Expenditure Survey, 1984–1985. Nicosia: Ministry of Finance.

8. Agricultural Labor and Technological Change in the Syrian Arab Republic

M. AL-ASHRAM

Department of Agricultural Economics, Aleppo University, Aleppo, Syria

Introduction

The Syrian Arab Republic is one of the major agricultural countries of the Arab World. The Euphrates and other rivers and oases support irrigated production, over 5 million ha are under rainfed crops, and extensive low-rainfall pastures support small ruminant production. The wet coastal region lies between the sea and the mountains which run from north to south parallel to the sea, with the agricultural plains of Damascus, Homs, Hama, Aleppo, Hassakeh, and Dara'a to the east. The desert region consists of the plains to the southeast on the Jordanian and Iraqi borders (Figure 1). The climate is typically Mediterranean with a mean annual rainfall of about 300 mm in the agricultural plains, and higher levels to the west of the mountains (Saki 1984).

Fifteen million ha, 81% of Syria's area, are considered productive, including 6.1 million ha of cultivable land, 8.3 million ha of pastures, and 0.5 million ha of forests. Of the cultivable area, approximately 4 million ha, 84% of which are rainfed, are under crops in a given year.

In 1960 Syria's population was 4.6 million and, with an annual growth rate of 3.7%, by 1985 the population was 10.6 million (CBS 1986a; 1982b; El-Zoobi 1983). The FAO estimates that by 2000 the population will be 18.7 million (FAO 1982b), while the Syrian Central Bureau of Statistics' estimate is 17.1 million (CBS 1981b). In 1985, 51% of the total population, or about 5 million, lived in rural areas. The rural population is projected to increase to 8.4 million by 2000, but to form a slightly smaller proportion of the total (48.9%) (CBS 1981b; Schmid 1980).

The total value of agricultural production, at producer's value, was 5.8 billion SYP in 1963 (1980 prices) which increased to 12 billion SYP in 1985. Of this, 67% was for plant production, mostly vegetables, fruits, and cereals, and 33% for animal production, in particular milk and livestock (Table 1) (CBS 1986a).

From 1970 to 1982 the total value of agricultural exports increased from 0.6 billion SYP to 1.3 billion SYP, an annual increase of 11%, although as a proportion of total exports they fell from 72% in 1970 to 16.4% in 1982. Cotton is the major export, accounting for over 50% of the total in 1984.

Dennis Tully (ed.), Labor and Rainfed Agriculture in West Asia and North Africa, 163–183.

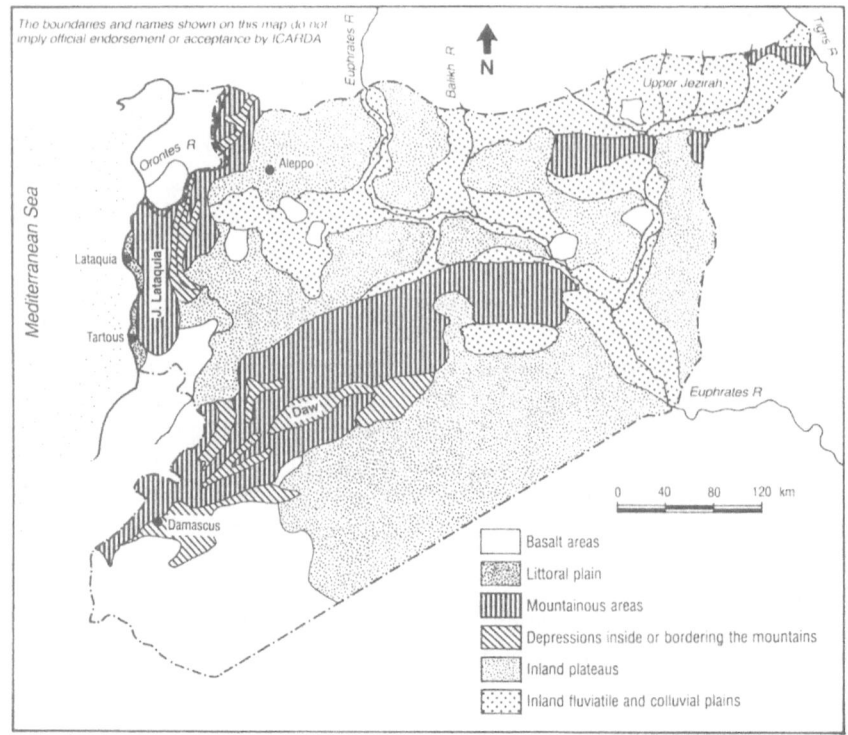

Fig. 1. Agricultural areas in Syria. *Source*: Syrian Central Bureau of Statistics (CBS 1986).

Other exports are tobacco, lentils, fruits, and vegetables (CBS 1986a; MAAR 1984).

Syria is only 62, 18, 0, and 25% self-sufficient in wheat, maize, rice, and sugar, respectively, and produces only 75 and 34% of its milk and fish requirements (Shafer and Blomo 1980) (Table 2). To meet requirements, therefore, imports increased at a rate of 30% per year between 1970 and 1983 when they were valued at nearly 2.6 billion SYP (650 million USD).

Agriculture continues to rank as one of the largest sectors of the Syrian economy although weather conditions cause much variability in agricultural output. In 1963 the agricultural sector accounted for 30% of gross domestic product (GDP), decreasing to 18% in 1985 (CBS 1986a). The two other major sectors, industry and mining and trade, contributed 14 and 24%, respectively, in 1985.

From 1963 to 1985, Syria's agricultural GDP rose from 4.7 billion SYP to 10.1 billion SYP (1980 prices). The average annual growth rate was 5.1% from 1963 to 1970, 9.3% from 1970 to 1978, and 7.2% from 1978 to 1985.

Per capita national income was 853 SYP at current prices in 1965, 1,050 SYP in 1970, 5,702 SYP in 1980, and 7,462 SYP in 1985. This means that

Table 1. Value of agricultural production (million SYP, 1980 prices), 1981–85.

Production	Year				
	1981	1982	1983	1984	1985
Cereals	2,831	1,906	2,202	1,222	2,139
Industrial crops	1,225	1,232	1,501	1,339	1,260
Fruits	2,045	2,452	1,929	2,213	2,159
Vegetables	2,834	2,736	2,689	2,692	2,533
Dry legumes	273	215	282	163	215
Forages	34	62	54	80	60
Others	141	152	159	163	180
Rural industries	89	96	100	138	138
Total from crops	9,371	8,850	8,915	8,010	8,684
Milk and products	1,647	1,723	1,759	1,576	1,709
Livestock	1,572	1,700	1,596	1,646	1,484
Eggs	536	584	599	626	530
Wool	127	140	151	138	133
Fisheries	31	22	34	53	46
Honey	18	22	25	25	19
Others	11	10	10	8	49
Total from animals	3,942	4,201	4,176	4,071	3,971
Total value of agricultural products	13,313	13,051	13,091	12,081	12,655
Customs	418	413	347	303	298
Gross output product in producer price	17,331	13,464	13,438	12,384	12,953

Source: CBS (1986a).

the per capita national income had increased about six-fold during 1963–85 (CBS 1986a). In 1983, the average income of a rural family reached 58% of the average income of an urban family (GUP 1984).

The main source of finance for agriculture is the Agricultural Cooperative Bank, which was established in 1884. It is presently owned and operated by the government and has 63 branches scattered throughout the country. During 1973–85, total credits for agriculture increased from 0.35 billion SYP to 1.5 billion SYP (CBS 1982b). The bank distributes its loans among the private, cooperative, and public sectors. By 1985 the private sector was receiving less than 48% of loans while the cooperative sector received over 51% and the public sector about 1% (CBS 1986a; GUP 1981). Most loans have gone to cotton production, followed by cereals and machinery.

In its program of 1965, the Ba'ath Party planned to form large agricultural units and apply modern technology to increase agricultural productivity, reorganize the agricultural sector into public, cooperative, and private subsectors, and nearly double the irrigated area from 522,000 ha to 1 million ha

Table 2. The food gap in Syria, 1983.

Commodity	Self sufficiency ratio (%)	Total requirement	Shortfall[1] Value (million USD)	Shortfall[1] Quantity (million ton)	Imports Value (million USD)	Imports Quantity (million ton)	Exports Value (million USD)	Exports Quantity (million ton)	Production Quantity (million ton)
Cereals	67.3	3,999.00	327.30	1,305.00	336.26	1,362.02	8.96	57.02	2,694.00
Wheat	61.84	2,606.61	225.05	994.61	225.05	994.61	n.a.	n.a.	1,612.00
Maize	17.80	150.55	18.52	123.75	18.52	123.75	n.a.	n.a.	26.80
Rice	0	243.31	92.66	243.31	92.66	243.31	n.a.	n.a.	n.a.
Barley	105.77	986.34	+8.93	+56.96	0.03	0.06	8.96	57.02	1,043.30
Potatoes	101.34	310.93	+0.38	+4.17	2.85	8.49	3.23	12.66	315.10
Pulses	115.38	184.60	+9.37	+28.40	0.73	1.01	10.10	29.41	213.00
Vegetables	99.57	3,415.75	+1.75	14.75	6.26	34.51	8.01	19.76	3,401.00
Fruits	92.35	1,049.18	20.39	80.28	25.75	88.89	5.36	8.61	968.90
Refined sugar	24.85	828.91	182.53	622.91	182.53	622.91	n.a.	n.a.	206.00
Edible oils and fats	72.28	131.49	32.62	36.45	32.84	36.52	0.22	0.07	95.04
Meat	97.57	235.33	12.32	5.73	27.09	8.18	14.77	2.45	229.60
Red	96.42	160.23	12.31	5.73	27.08	8.18	14.77	2.45	154.50
White	100.00	75.10	0.01	n.a.	0.01	n.a.	n.a.	n.a.	75.10
Fish	34.37	12.83	17.06	8.42	17.14	8.46	0.08	0.04	4.41
Eggs	100.05	86.31	+0.04	0.01	*	0.05	0.04	86.35	
Milk (liquid)	75.59	1,535.91	69.47	374.91	69.47	374.91	n.a.	n.a.	1,161.00
Total			650.15		700.93		50.78		

Source: AOAD (1985b).
[1]Plus sign indicates net export.
*negligible.
n.a. = not available.

by the year 2000 and so intensify agricultural production (Al-Ashram 1975; 1972). The Fourth and Fifth National Conferences of the Ba'ath Party (1968 and 1971) reemphasized the application of new technologies, especially mechanization, which has had a big influence on the agricultural labor force and on rural migration.

Farming Systems

Agricultural land in Syria has been divided into zones based on rainfall patterns, which are closely related to the crops grown, crop rotations, and land use. The Ministry of Agriculture and Agrarian Reform (MAAR) has developed a comprehensive inventory of agricultural land resources based on these zones.

The first zone has a mean annual rainfall of over 350 mm and is divided into two subzones. One is the area with annual rainfall over 600 mm where many rainfed crops can be successfully grown. The other subzone has mean annual rainfall of 350–600 mm, and no less than 300 mm falling during two thirds of the years recorded. Thus at least two harvests are expected every 3 years. The main crops are wheat, pulses, and summer crops. The second zone has an annual rainfall of 250–350 mm with no less than 250 mm falling during two thirds of the years, so at least two barley crops are possible every 3 years. Wheat, pulses, and melons are also grown. In the third zone, the annual rainfall is over 250 mm and no less than 250 mm fall during half the years. Thus one or two barley harvests are expected every 3 years. The annual rainfall in the fourth zone is 200–250 mm and not less than 220 mm fall during half the years. This zone is only useful for barley or as permanent grazing land. The fifth zone, the desert or steppe region, covers what remains and is not suitable for rainfed crops (CBS 1986a) (Figure 2).

The area under rainfed crops has ranged from approximately 2.8 million to 3.5 million ha in recent decades (CBS 1984b). Crop rotations in rainfed areas generally vary by soil type and zone. In better soils and in wetter zones, cropping is continuous or nearly so, in medium conditions fallowing every other year is practiced, and fallowing 2 years out of 3 may be found in Zone 4. The main crops are wheat, barley, lentils, chickpeas, watermelons, musk-melons, olives, and grapes (CBS 1984b), covering, in 1981, 99% of the total rainfed area planted. Zones 1 and 2 contain 62 and 65% of irrigated and rainfed cultivated land, while 87% of steppe and pasture land are in Zone 5 (Table 3).

Official estimates of crop production costs, estimated yields, and prices are shown in Table 4. The relative importance of labor costs for major crops is shown in Table 5, based on an AOAD study. Labor, primarily hired, makes up over 40% of total costs for sugar beet and cotton, 37% for lentils, and 32% for chickpeas. Rainfed cereals, on the other hand, depend less on human labor relative to other inputs.

168

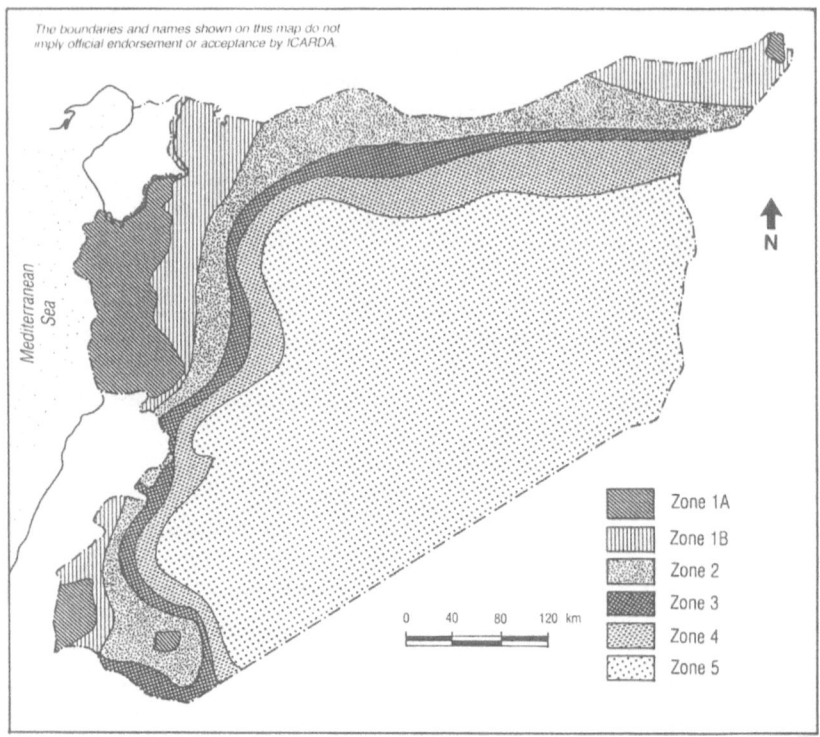

The boundaries and names shown on this map do not imply official endorsement or acceptance by ICARDA.

N

Zone 1A
Zone 1B
Zone 2
Zone 3
Zone 4
Zone 5

0   40   80   120 km

Mediterranean Sea

Fig. 2. Geographical features of Syria. *Sources*: Syrian Central Bureau of Statistics (CBS 1986), Arab Center for the Studies of the Arid Zones and Dry Lands.

There was no appreciable change in the area of irrigated land or in its share of the total during 1963–85, with a range of 450,000–670,000 ha (CBS 1986a; 1984b). In the fourth and fifth 5-year plans (1976–80 and 1981–85) priority was given to intensifying production in irrigated areas and to increasing the irrigated area to 31% of the total by the year 2000. This will require the reclamation and cultivation of 880,000 ha by 1995, an annual increase of 60,000 ha per year (Al-Ashram 1972; MOP 1981).

Total livestock production is valued at 4.0 billion SYP (1980 prices) compared to 8.7 billion SYP for crop production (CBS 1986a). In 1985 there were about 11.0 million sheep, 1.06 million goats, and 0.7 million cattle (CBS 1986a). In 1984, 97, 46, 30, and 68% of sheep, goats, cattle, and poultry, respectively, were owned by farmers, the remainder being managed by cooperatives (GUP 1984). Livestock, particularly sheep and goats, are largely dependent on natural pastures and cereal stubble which are mainly found in rainfed areas. Accordingly, the livestock population, especially sheep and goats, fluctuates with rainfall (Table 6). Cattle numbers are generally more stable than small stock because farmers' herds are small, most being owned by cooperatives.

Table 3. Land use by stability zones, 1985 (×1,000 ha).

Stability zone	Total area	Cultivable lands				Steppe and pasture	Forest	Uncultivable
		Under crops		Fallow	Uncultivated			
		Irrigated	Rainfed					
1	2,698	246	1,153	50	289	185	305	470
2	2,474	160	1,017	710	69	144	19	355
3	1,306	42	446	415	27	165	9	202
4	1,823	43	484	474	7	613	10	192
5	10,217	161	218	4	112	7,221	173	2,328
Total	18,518	652	3,318	1,653	504	8,328	516	3,547

Source: CBS (1985a).

Table 4. Prices and production costs of important crops in Syria, 1984 (SYP/ha).

Crop	Mechanical operations	Input costs	Land rent (15% of total income)	Interest at 5%	Other costs	Total costs	Yield (kg/ha)	Cost per kg	Official price/kg	Total income	Net profit
High yielding wheat (I)	1,671	1,585	664	123	163	4,205	3,600	1.17	1.23	4,428	223
Durum wheat (R)	610	422	228	39	52	1,351	1,100	1.23	1.38	1,518	167
Barley (R)	459	209	135	22	33	858	1,100	0.78	1.00	1,100	242
Chickpeas (R)	1,256	461	334	41	86	2,178	850	2.56	3.00	2,500	372
Lentils (R)	1,908	401	240	34	82	2,002	1,000	1.50	1.80	1,800	-202
Lentils (I)	1,208	1,555	750	118	173	4,504	2,000	1.75	2.60	5,200	696
Maize (I)	2,536	1,683	765	156	211	5,351	3,000	1.78	1.70	5,100	-251
Peanut (I)	5,052	2,547	1,575	310	380	9,864	2,000	4.78	5.50	11,000	1,136
Cotton (I)	7,440	8,010	3,188	556	773	10,366	28,000	3.70	4.00	112,000	101,634
Sugar beet (I)	7,410	3,640	2,574	379	553	14,556	50,000	0.29	0.31	15,500	944
Potato (spring) (I)	6,471	8,918	2,790	589	769	19,537	20,000	0.98	1.25	25,000	5,463
Tomato (I)	8,327	8,195	2,730	426	826	20,504	28,000	0.73	0.65	18,200	2,304
Red onion (I)	7,440	8,021	3,188	556	773	19,978	25,000	0.80	1.00	25,000	5,022

Source: MAAR (1984).
Note: I = irrigated; R = rainfed.

171

Table 5. Distribution of production costs of major crops (% of total cost).

Crop	Inputs	Mechanical operations	Animal labor	Family labor	Hired labor	Interest on capital	Land rent
High yielding wheat (I)	40	18	—	5	9	4	26
High yielding wheat (R)	38	24	—	3	2	3	31
Local wheat (R)	24	27	—	8	8	3	28
Barley (I)	24	25	9	7	17	4	23
Barley (R)	24	24	1	8	3	3	37
Lentils (R)	19	15	1	11	27	4	23
Chickpeas (R)	19	17	—	14	18	3	28
Sugar beet (I)	34	13	—	10	31	4	8
Cotton (I)	33	11	—	13	29	4	10

Source: AOAD (1985a).
Note: I = irrigated; R = rainfed: — = < 0.5%.

Table 6. Numbers of sheep, goats, and cattle (×1,000) in Syria, 1963–85.

Year	Sheep	Goats	Cattle
1963	4,297	581	449
1965	5,075	818	506
1970	6,046	774	528
1975	5,809	814	557
1980	9,301	1,025	768
1985	10,993	1,059	742

Source: CBS (1984a; 1986a).

Land Tenure

At the end of the French occupation in 1946, big landowners and Bedouin sheiks were encouraged to occupy the good land previously owned by the state. As a result, in the 1950s 87% of agricultural land was owned by the large and medium-sized owners, while small holdings made up 13% of total agricultural land. In 1958 the government issued the agricultural relations law which specified the types of contracts and the proportion of production to be exchanged between landowners and small farmers and workers. In addition, it outlined the rights, duties, and wages of laborers.

The disparity in holding size (Table 7) was the main reason for issuing the land reform laws in 1958 and 1963 and there are many publications on this issue (Shakra 1966; Khader 1984; 1975; 1974; Bakour 1978; Keilany 1980; Hosry 1981; Al-Ashram 1980). The laws also specified limits to the size of property owned by landlords depending on location and water source.

Table 7. The effect of the agrarian reform on the distribution of land in Syria as of 1975.

Holding size (ha)	% of area before redistribution	% of area after redistribution
Small		
<1	1	14.9
2–5	7	27.7
5–7	5	8.8
Total	13	50.7
Medium		
7–25	17	42.4
25–50	11	5.5
50–100	10	1.2
Total	38	49.1
Large		
100–500	24	0.1
500–1,000	9	0.0
>1,000	16	0.1
Total	49	0.2

Source: IBRD (1955); FAO (1982a); Wazzan (1967).

For rainfed areas the upper limits were 80 ha in areas with over 500 mm rainfall, 120 ha in areas with 350–500 mm, and 200 ha in areas with less than 350 mm rainfall, as well as Hassakeh and Deir Ezzor provinces.

The nationalization of big farms and redistribution of land among small or landless farmers, as outlined in the reform laws, has been very slow. Up to 1975, only 60,700 ha (13%) of irrigated and 405,400 ha of rainfed land had been distributed to 52,504 families in 1,386 villages, with a mean holding size of 8 ha in irrigated areas and 30 ha in rainfed areas (FAO 1982a; CBS 1986a; Al-Ashram 1981; Wazzan 1967). The laws also obliged those receiving land to participate in established cooperatives and to pay a quarter of the value of the distributed land to the cooperative over 20 years.

As a result of the reform laws, the area of holdings with less than 25 ha increased from 30% of the total before the laws were implemented to 93% afterwards (Table 7). The area of holdings over 100 ha decreased from 49% to 0.23% of the total.

In 1985, of all land under crops (and excluding fallow), state farms covered about 1.4%, 37.1% was being farmed by 4,156 cooperative farms, and 61.5% was on private farms (Al-Ashram 1987). Cooperative farming may involve the sharing of machinery between individual farmers, joint marketing, and joint purchasing of agricultural inputs and other goods. There are a few true production cooperatives where the basic agricultural operations are also done jointly. Cooperatives are organized at the village level and their members are part of a sub-province association. Representatives of these associations make up the Provincial Peasants' Union from which the General Peasants' Union is formed (CBS 1986a; GUP 1974; Kasswani 1969).

The state agricultural sector has nine state farms involved in crop production, with a total area of 54,458 ha in 1985, and 19 others involved in livestock production, with a total area of 3,455 ha and 6,810 head of cattle (CBS 1986a). Another 43,300 ha of cultivable land are farmed in the Euphrates Basin by the General Organization for Exploitation and Development of the Euphrates Basin, where the farm size ranges from 1,500 to 4,130 ha (MOI 1985). The majority of crop and livestock production is still in the private sector, which is presently considered a permanent element of the agricultural system. In 1985 private farms made up about 50 and 67% of the total irrigated and dryland areas, respectively, and produced 63, 74, 69, 50, 69, 63, and 74% of total wheat, barley, sugar beets, onions, eggs, fish, and meat chickens, respectively (Table 8).

Agricultural Labor Force

Of the total population in 1984, 6.5 million were over 10 years old and therefore considered to be part of the labor force. Of these, 2.8 million lived in rural areas and 30% or about 1 million of these formed the active rural labor force of 839,843 males and 187,223 females (CBS 1986a). In addition,

Table 8. Distribution of agricultural production according to farm type, 1985.

Commodity	Total production (×1,000t)	Private farms		Cooperative farms	
		Production (×1,000t)	% of Total	Production (×1,000t)	% of Total
Wheat	1,714	1,074	62.6	640	37.4
Wheat (Mexican)	1,130	723	64.0	407	36.0
Lentils	47	25	52.3	22	47.7
Barley	740	545	73.6	195	26.4
Cotton	487	299	59.4	198	40.6
Sugar beet	412	284	68.9	128	31.1
Tomatoes	778	308	39.6	470	60.4
Potatoes	280	120	42.9	160	57.1
Olive	185	97	52.5	88	47.5
Apple	125	48	38.4	77	61.6
Maize	60	25	43.3	34	56.7
Milk	1,006	253	25.0	753	75.0
Meat	235	98	102.0	137	98.0
Eggs (mill.)	1,619	1,120	69.0	499	31.0

Source: CBS (1986a); GUP (1986a).
Note: The production of state farms is insignificant and is not included.

Table 9. Distribution of rural labor in different sectors, 1984.

Economic activity	Male	Female	Total
Agriculture and forestry	385,488	137,631	523,119
Mining and quarrying	9,737	300	10,037
Electricity, gas, and water	6,725	–	6,725
Converting industry	50,294	10,339	60,633
Building and construction	146,465	1,606	148,071
Trade	36,541	1,806	38,347
Transportation and storage	35,135	1,003	36,138
Banking and insurance	1,405	300	1,605
Community and social services	135,824	14,455	150,279
Sub total	807,614	167,340	974,954
Unemployed	32,229	19,893	50,122
Total	839,843	187,233	1,025,076

Source: Al-Ashram (1986): CBS (1986a).

about 48,385 persons from urban areas worked in agriculture and forestry, 90% of whom were male. We can distribute the workers in rural areas in 1984 by employment as shown in Table 9. The records do not distinguish between permanent, temporary, and seasonal labor. The labor force in agriculture and forestry is primarily self-employed. Of the females, 78% are unpaid workers (Table 10), presumably working on family farms (Al-Ashram 1986).

As a proportion of the total, the agricultural labor force declined between the early 1970s and 1980 and by 2000 is expected to be 27.5% or less of the total labor force (CBS 1981a). In the same period the male farm workforce

Table 10. Employment status in agriculture and forestry.

Kind of employment	Male	Female	Total
Employer	41,962	1,607	43,569
Self-employed	222,760	14,757	237,517
Paid worker	44,271	13,953	58,224
Unpaid worker	76,495	107,314	183,809
Total	385,488	137,631	523,119

Source: CBS (1986a).

also declined, from 48.9% of the total workforce in 1970 to 19% in 1984, partly due to mechanization and partly to the low incomes in agriculture encouraging workers to seek urban employment. In 1984, 5% of the agricultural labor force was unemployed (32,335 males and 19,877 females) (CBS 1986a), 8,833 less than the figure recorded in 1971 (CBS 1974a).

The number of women employed in agriculture and forestry varies depending on the harvest. According to a WCARRD draft report on Syria in 1983, the female agricultural labor force was 309,000 in 1975 but fell to 62,500 in 1976 (FAO 1983). National experts estimated female farm labor during the 1970s to have been 250,000–300,000 (MSAL 1982). Women tend to do the lighter work such as planting, weeding, thinning, threshing, and harvesting, which are done manually or with hand tools. Women are also responsible for poultry raising, care of livestock, milking, and milk processing at the family level. A good deal of the agricultural work performed by women is done within family farms as unpaid labor (Rassam and Tully 1988).

A study in four villages in northwestern Syria found that female and male contributions to agricultural labor (in terms of hours of physical work and including both family and hired labor) are almost equal. Household labor provides 61% of total work hours in agricultural operations and 57% of this is provided by females, while hired labor is equally divided by gender. The kind of crop grown and the level of mechanization influence male and female labor. For cereal crops, where most of the operations are mechanized, males contribute 56% of total hours and 87% of hired hours. The opposite is found for legume crops, where most operations are not mechanized and females contribute 70 and 75% of total and hired labor hours, respectively (Rassam and Tully 1988).

The percentage of workers in agriculture under 15 years old is 26.5% for females and 7% for males, possibly because a larger proportion of girls drop out of formal schooling early (Dawood 1978).

An estimate of hired farm labor can be made from the number of holdings, farmers, and farm laborers. There were about 396,053 landholders in Syria in 1978 (FAO 1982b). Holdings of less than 10 ha constitute only 23% of total agricultural land, but they represent more than 75% of the farms. Holdings above 10 ha constitute 77% of agricultural land and are owned by 5% of the holders (97,800 holdings with an average area of 29 ha). These holders usually need hired labor. The non-farm rural labor force decreased

Table 11. Labor requirements for crops (10 most labor-using only).

Crop	Hectares	Man-days
Olives	228,263	14,486,909
Cotton	186,507	12,284,364
Wheat	1,527,718	11,076,714
Barley	1,021,492	5,770,180
Watermelons	87,728	4,057,235
Grapes	93,973	3,934,490
Lentils	178,346	3,274,774
Tobacco	15,331	3,184,186
Tomatoes	32,791	3,103,719
Vetches	82,544	1,711,787

Source: Schmid (1980).

from 552,438 in 1970 to 451,845 in 1984, but as a proportion of the rural labor force it increased from 42.9 to 46.3% in this period (CBS 1982b; 1986a; MAAR 1984). The higher figure for 1970, before full implementation of land reform, may have been due to the large number of landless people in rural areas who sought work.

Labor Requirements for Plant and Animal Production

Schmid estimated an overall labor requirement for crop production in 1976/77 of 78 million man-days, and 12 million for livestock (Table 11). Rainfed arable crops (cereals, lentils, chickpeas, vetches, and melons) account for 35% of crop labor and rainfed tree crops (olives, grapes, apples, and apricots) account for another 26%.

If evenly required over the year, labor force estimates suggest that this labor would be available. However, labor demand peaks in the May-July grain and legume harvest period. There is a lower demand peak in September-November for harvesting olives, cotton, and sugar beet, and as these are more restricted geographically, local shortages of labor may occur. Shortages appear to be met by working extra hours and by persons not normally working in agriculture (students and off-farm workers) taking part in peak season activities. On the other hand, there is substantial underemployment of the farm labor force during much of the rest of the year (Schmid 1980).

"Underemployment, from the viewpoint of hours worked, is due more to the biological nature of farming than to any particular organization of farming" (Schmid 1980). Thus, Schmid argues, productivity and not hours worked should be the criterion of underemployment in agriculture.

The productivity of agricultural labor is a function of productivity per land and animal unit and the number of land or animal units handled per worker. The area planted to crops for the country as a whole increased about 20% from 1970 to 1976 while the number of farm workers declined by about 30%

Table 12. Net migration change by Mohafaza in Syria.

Mohafaza	Population (millions)		Net migration	Change as % of pre-1970 population
	Pre-1970	1970		
Damascus City	0.69	0.79	0.10	14.7
Damascus	0.52	0.60	0.07	14.3
Aleppo	1.32	1.29	−0.03	−2.5
Hama	0.50	0.49	−0.01	−3.0
Lattakia	0.39	0.38	0.00	−1.9
Deir Ezzor	0.30	0.29	−0.01	−4.3
Idleb	0.41	0.38	−0.03	−7.8
Al Hassakeh	0.45	0.46	0.00	0.9
Sweida	0.13	0.13	−0.01	−4.9
Dara'a	0.23	0.22	0.00	−2.1
Tartous	0.30	0.29	0.00	−1.5
Quneitra	0.12	0.02	−0.11	−86.7
Homs	0.52	0.52	0.00	0.0
Al-Raqqa	0.20	0.24	0.04	22.0

Source: Williams (1980).

to 1976. As a result, the planted area per worker increased from 4.5 ha in 1970 to 7.3 ha in 1976 (Schmid 1980). By 1984 this figure had reached 7.6 ha/worker (CBS 1986a).

In spite of this improvement, productivity is limited by the low value of production per worker, which Schmid calculated to be 10,088 SYP in 1977, rising to 23,286 SYP in 1985 (GUP 1986a). The latter figure equals 77 and 30% of a worker's productivity in industry and commerce, respectively (Al-Ashram 1984). Thus migration out of agricultural areas may be expected.

Migration

During the period 1950–80 the population in the urban areas of developing countries increased from 275 million to 972 million (Alloush 1984). Syria has faced a similar increase in urban population and in the past 25 years there has been significant internal migration. For example, Mugharbel (1966) found very high levels of migration to Aleppo and other centers from 12 villages around the city. Data from 1970 on total migration flows show that 66% was rural to urban, 24% was inter-urban, and the remaining 10% was urban to rural (FAO 1983). The net annual migration of rural populations to urban areas was estimated to be 40,000 in 1967–77 (Dawood 1978; Yafi 1978).

In 1980, Williams presented a schematic diagram of flows between the major sectors using data from the 1970 census of population (Williams 1980). Two thirds of total migration was from rural areas to Mohafaza and other urban centers. Only the four Mohafazas of Damascus City, Damascus, Al Raqqa, and Al Hassakeh experienced net immigration (Table 12). Most rural-urban migrants are in the 15–39 year age bracket, the majority being male, and thus their departure makes a disproportionately large impact on the availability of rural labor (Dwaiger 1978; Williams 1980).

International migration is reported to involve only small parts of the population but is most significant in the male 15–49 year age group. It has been estimated that in 1975 70,415 Syrians were working in the Gulf countries and another 25,000 in Jordan (Meyer 1984; based on Birks and Sinclair 1980; Garguar 1980).

Growth of Mechanization and Use of Other Technology

After the Second World War, demand for food and prices of agricultural commodities were high. This encouraged landowners to buy new agricultural machinery and so extend the area of cultivated land and increase production (Binjarrow 1984). By 1951 there were significant numbers of tractors, pumps, modern plows, seeders, and combine harvesters (Table 13). Most machinery was used on big farms as it was uneconomical over small areas. Since 1965 policy has encouraged mechanization, and there have since been increases of two- to eight-fold in the numbers of agricultural machines of all types. Tractor numbers reached 7,675 in 1965 and 43,959 in 1985 (CBS 1986a; Singabeh 1984). According to the FAO, tractors will increase during 1975–90 at an average of 10.4% per annum, a moderate rate by international standards (FAO 1981). Allowing for replacement of 1,800 tractors annually, this would require an additional 2,700 tractors per year (GUP 1981).

In 1980, there was one tractor per 116 ha and by 1985 this figure was one tractor per 81 ha. This is still a large area per tractor considering that nearly 15% of the total cultivated area is irrigated. Production of all major crops has now been mechanized for at least one of the stages of production (Table 14). For example, land preparation, fertilizing, harvesting, and threshing of wheat and barley have been mechanized. For lentils and chickpeas, only land preparation is mechanized; mechanization of planting is variable and harvesting is still done by hand at great expense (20–30% of total production costs) (Binjarrow 1984). For cotton and maize, land reparation is mechanized while hoeing, weeding, irrigation, harvesting, and transportation are still done by hand. This is typical of irrigated summer crops, which employ many agricultural laborers. The General Organization for Agricultural Machinery is attempting to mechanize planting, harvesting, and threshing (Binjarrow and Malek 1984).

Table 15 shows the rapid increase in the use of chemical fertilizers from 1963 to 1985. The value of other chemicals used has also increased, from 15.4 million SYP in 1981 to 59.7 million SYP in 1985 (CBS 1986a).

The General Organization for Seed Multiplication started operations in 1976 with cotton and later expanded to wheat, potatoes, corn, and broad beans. In 1985 they distributed 88,000 t of improved wheat seeds (Arab Socialist Ba'ath Party 1965; GUP 1986a).

Table 13. Machines used in Syrian agriculture, 1951–85.

Year	Tractors		Pumps	Modern plows	Seeders	Combine harvester-threshers	Fixed threshers	Sprayers by motors	Sprinklers by motors
	>50 HP	<50 HP							
1951	760	n.a.	5,626	768	430	426	n.a.	n.a.	n.a.
1955	1,786	n.a.	13,267	1,940	627	693	n.a.	n.a.	n.a.
1960	4,754	n.a.	19,444	4,480	953	1,031	n.a.	n.a.	n.a.
1965	5,150	2,525	23,969	8,571	1,597	1,324	380	4,522	1,346
1970	6,102	2,929	29,042	11,797	1,838	1,455	474	7,267	1,163
1975	6,273	9,030	40,416	20,253	1,903	1,664	1,367	3,706	1,325
1980	6,399	21,145	47,206	35,245	3,483	2,329	2,301	14,806	1,081
1985	9,765	34,194	73,315	60,281	5,442	2,976	3,091	21,151	270

Source: CBS (1984a; 1986a).
n.a. = not available.

180

Table 14. Degree of mechanization of agricultural operations in Hassakeh Mohafaza.

Agricultural operation	Degree of mechanization (%)
Land preparation for all crops	
Tillage	100
Picking	100
Leveling	0
Seeding and planting	
Wheat, barley, and fodder crops	90
Chickpeas, lentils	85
Cotton	5–15
Maize	30
Agricultural crops services	
Chemical fertilizing for fodder crops	85
Organic fertilizing (transport)	10
Weed control	90
Ridging (after seeding for cotton and maize)	100
Leveling for irrigated crops	70
Transportation	85
Water pumping	100
Baling the fodder crops	70
Loading	5
Harvesting	
Wheat and barley (including threshing)	95
Threshing lentils and chickpeas	25
Harvesting lentils and chickpeas	5

Source: Binjarow (1984).

Table 15. Chemical fertilizer application (×1,000t) in Syria, 1963–85.

Year	Total	Compound fertilizer	Potassium fertilizer	Phosphatic fertilizer	Nitrogen fertilizer
1963	61,900	8,400	300	19,200	34,000
1965	75,065	13,000	60	13,000	49,000
1970	111,780	14,300	230	22,250	75,000
1975	164,642	6,788	1,825	35,861	120,168
1980	305,365	53,877	7,115	36,941	207,432
1985	497,075	0	11,278	161,363	324,444

Source: CBS (1968a–85a).

Conclusion and Recommendations

Syria is still primarily dependent on rainfed farming, which constitutes 86% of the cultivated area. As a result, production is unstable and fluctuates according to rainfall and climate. This has a direct impact on peasant incomes and the national economy. In spite of the general increase in the area of cultivable land during 1963–85, it has not increased as fast as population, so the per capita share fell from 0.7 ha in 1979 to 0.63 ha in 1983.

Increased migration, particularly in the 18–45 year old age group and the educated, has had a negative effect on the production and productivity of the agricultural sector. But there are insufficient data on internal migration and this warrants further study.

A policy for soil and water conservation must be established and implemented to help improve yields, reduce the area of fallow land, and minimize production instability. A policy of land consolidation would contribute to this by allowing more efficient use of machinery, irrigation, and drainage.

Efforts must be made to ensure that coordinated and cooperative research is carried out by specialized organizations, universities, international and local centers, and different government ministries. The planning of agricultural extension should be jointly undertaken by the Ministry of Higher Education, the Ministry of Agriculture and Agrarian Reform, and international organizations.

Mechanization studies should be given high priority, with in-depth studies on agricultural productivity and the scope for future improvement. Particular attention should be given to alternative mechanization strategies and their implications for agricultural labor. In-depth studies should also be carried out on the trends, potential, and problems of male and female farm labor, rural non-farm labor, and rural employment and unemployment under existing policies to decentralize employment opportunities.

A critical evaluation should be made of the role of rural women in agriculture and food production to establish a baseline for future activities to integrate women in rural development. Finally, the present structures and functions of agricultural cooperatives and state and private farms should be reviewed, and new policies up to the year 2000 be identified for the current 5-year economic and social development plan.

References

Al-Ashram, M. 1972. Zur Betriebswirtschaftlichen Gestaltung der Landwirtschaftlichen Produktions Prozess auf den Euphrates Projektes in der SAR. PhD thesis, Karl Marx Universitat, Leipzig.

Al-Ashram, M. 1975. Stand und Perspektiven der Mechanizietung der Agrarproduktion in der SAR. Beitrage zur Tropischen Landwirtschaft und Veterinarmedizin 2: 155–163.

Al-Ashram, M. 1980. The Agricultural Cooperative. Aleppo: Faculty of Agriculture, Aleppo University.

Al-Ashram, M. 1981. Agricultural Production, Before and After the Agrarian Reform in Syria. Paper presented at the Fifth International Summer Seminar on Agrarian Reform and Development in Asia, Africa and Latin America, 23 June–2 July 1980, Institute of Tropical Agriculture, Karl Marx University, Leipzig.

Al-Ashram, M. 1984. (The Negative Sides of Using Scientific Technologies Concepts in Arab Rural Society). In Arabic. Paper presented at the UNESCO conference on Technology, Science and Society, 24–29 January, Tunis.

Al-Ashram, M. 1986. Agricultural Extension's Role in Increasing Yields and Agricultural Production in SAR. Agricultural Development Week, 12–17 July, Aleppo. Aleppo: Aleppo University.

Al-Ashram, M. 1987. Farm Management. Aleppo: Faculty of Agriculture, Aleppo University.

Alloush, K. 1984. (Internal and External Migration, Its Estimation Methods). In Arabic. Damascus: Central Bureau of Statistics.

AOAD (Arab Organization for Agricultural Development). 1985a. Cost of Agricultural Production Improvement Projects in Arab Countries. Phase I. Khartoum: AOAD.

AOAD (Arab Organization for Agricultural Development). 1985b. Yearbook of Agricultural Statistics. Vol. 5. Khartoum: AOAD.

Arab Socialist Ba'ath Party. 1965. (Program of Arab Socialist Ba'ath Party). In Arabic. Damascus: National Leadership.

Bakour, Y. 1978. Supporting Policies and Services for Agrarian Reform Programme in Syria. Meeting Papers, WCARRO, Rome, July 1979. Rome: FAO.

Binjarow, H. 1984. (Agricultural Development and Agricultural Mechanization in Hassakeh Province). In Arabic. Hassakeh: General Organization for Agricultural Mechanization.

Binjarow, H. and Malek, G. 1984. (Report on Mechanization of Maize Crop in Al Manajeer Farm). In Arabic. Hassakeh: General Organization for Agricultural Mechanization.

Birks, J.S. and Sinclair, C.A. 1980. International Migration and Development in the Arab Region. Geneva: International Labour Office.

CBS (Central Bureau of Statistics). 1968a–86a (Annual). Statistical Abstract. Damascus: CBS.

CBS (Central Bureau of Statistics). 1981b. (Estimation of Syrian Population by Province, Region, and District Level During 1980–2000). In Arabic. Damascus: CBS.

CBS (Central Bureau of Statistics). 1982b. Agricultural Census Data in the SAR. Damascus: CBS.

CBS (Central Bureau of Statistics). 1984b. (Time Series Data of the Agricultural Sector, 1963–82 in Syria). In Arabic. Damascus: CBS.

Dawood, A.M. 1978. The Integration of Women in Rural Development in the Near East Region. FAO Regional Study. Cairo: WCARRD-RNEA.

Dwaiger, F. 1978. (International Migration in SAR, Its Results and Effectiveness). In Arabic. MSc thesis, Planning Institute for Socio-economical Development, Damascus.

El-Zoobi, M.A. 1983. Alleviation of Rural Poverty Through Agrarian Reform and Rural Development in the SAR. Damascus: FAO.

FAO (Food and Agriculture Organization). 1981. Agricultural and Rural Development up to the Year 2000. Rome: FAO. FAO (Food and Agriculture Organization). 1982a. Regional Study on Rainfed Agriculture and Agro-climatic Inventory of Eleven Countries in the Near East Region. Rome: FAO.

FAO (Food and Agriculture Organization). 1982b. Program Development Mission: Findings and Recommendations. Rome: FAO.

FAO (Food and Agriculture Organization). 1983. Draft report of WCARRD Follow-up High Level Mission to the SAR. Unpublished.

Garguar, T. 1980. Migration from Rural to Urban Areas in SAR. Damascus: Ministry of Culture.

GUP (General Union of Peasants). 1974. (Law number 21 for Peasant Organizing). In Arabic. Damascus: GUP.

GUP (General Union of Peasants). 1981. (Report of the Fifth General Conference of the General Union of Peasants). In Arabic. Damascus: GUP.

GUP (General Union of Peasants). 1984. Yearly Statistical Abstracts. Damascus: Office of Planning and Statistics.

GUP (General Union of Peasants). 1986a. (The Economic Report presented to the General Conference of the General Union of Peasants). In Arabic. Damascus: GUP.

GUP (General Union of Peasants). 1986b. (The Organizing Report presented to the General Conference of the General Union of Peasants). In Arabic. Damascus: GUP.

Hosry, M. 1981. (Socio-economic Effects of Agrarian Reform in Syria: Results of an Empirical

Investigation in Selected Villages). In German. Sozialokonomische Schriften zur Agrarentwicklung, No 43. Saarbrucken: Verlag Breitenbach.

IBRD (International Bank for Reconstruction and Development). 1955. The Economic Development of Syria. Baltimore: Johns Hopkins Press.

Kasswani, B. 1969. General Management of the Agricultural Sector and Its Development Possibilities. Diploma thesis, Planning Institute for Socio-economical Development, Damascus.

Keilany, Z. 1980. Land Reform in Syria. Middle Eastern Studies 16(3): 209–224.

Khader, Bichara. 1974. Agrarian Structures and Reforms in Syria. Maghreb 65: 45–55.

Khader, Bichara. 1975. Propriete Agricole et Reforme Agrarie en Syrie. Civilisations 25(1/2): 62–83.

Khader, Bichara. 1984. La Question Agraire dans les Pays Arabes: le Case de Syrie. Louvain-la-Neure, Belgium: CIACO Editeur.

Meyer, G. 1984. Landliche Lebens- und Wirtschaftsformen Syriens im Wandel. Erlanger Geographische Arbeiten No. 16. Erlangen: Frankischen Geographischen Gesellschaft.

MAAR (Ministry of Agriculture and Agrarian Reform). 1984. The Annual Agricultural Statistical Abstract. Damascus: Department of Planning and Statistics.

MOI (Ministry of Irrigation). 1985. Annual Statistical Abstracts of the General Organization for the Exploitation and Development of the Euphrates Basin. Raqqa: GOEDEB.

MOP (Ministry of Planning). 1981. The Fifth 5-year Plan for Socio-economic Development, 1981–85. Damascus: Ministry of Planning.

MSAL (Ministry of Social Affairs and Labor). 1982. Annual Statistical Bulletin of the MSAL for 1980. Damascus: MSAL.

Mugharbel, S. 1966. Investigation of the Causes of Internal Migration around Aleppo City. Aleppo: Faculty of Agriculture, Aleppo University.

Rassam, A. and Tully, D. 1988. Gender Related Aspects of Agricultural Labor in Northwestern Syria. Pages 287–301 in Gender Issues in Farming Systems Research and Extension (Poats, S.V., Schmink, M., and Spring, A., eds). Boulder, Colorado: Westview Press.

Saki, C. 1984. Management and Organization of State Farming in the SAR. Draft Report. Damascus: FAO.

Schmid, L. 1980. Agricultural Manpower. Chapter 2 in Syria: Agricultural Sector Assessment. Vol. 5, Human Resources Annex. USDA in cooperation with USAID and SPC.

Shafer, C.E. and Blomo, V.J. 1980. Agricultural Trends, Demand and Pricing Policy. Chapter 1 in Syria: Agricultural Sector Assessment. Vol. 4, Agricultural Marketing Annex. USDA in cooperation with USAID and SPC.

Shakra, A. 1966. Land Reform in Syria. PhD dissertation, University of Oklahoma.

Singabeh, A. 1984. Agricultural Mechanization in Syria. Pages 242–248 in Development of the Agricultural Machinery Industry in Developing Countries. Proceedings of the Second International Conference, 23–26 January 1984, Amsterdam. Compiled by A. Moens and A.H.J. Siepmann. Wageningen: Centre for Agricultural Publishing and Documentation.

Wazzan, S. 1967. (From Underdevelopment to Socialist Development). In Arabic. Damascus: Ministry of Culture.

Williams, J.D. 1980. Population and Migration. Chapter 1 in Syria: Agricultural Sector Assessment. Vol. 5, Human Resources Annex. USDA in cooperation with USAID and SPC.

Yafi, A.K. 1978. Rural Migration and Labour Force. Paper presented at the Symposium on Population and Working Power in SAR, Damascus.

9. Agricultural Labor and Technological Change in Jordan

ZULKUF AYDIN

Institute of Archaeology and Anthropology, Yarmouk University, Irbid, Jordan

Introduction

Only 9% of Jordan receives more than 200 mm of rainfall (Table 1; Figure 1) and is therefore potential agricultural land (NPC 1976). But these areas are generally mountainous and rocky. In the south especially, rainfall and the resulting productivity are highly variable (Qasem and Mitchell 1986; Barham 1979). Of the 364,000 ha of total cultivable land only 10% is irrigated (DOS 1985).

Agriculture's contribution to GDP is low (8.2% in 1985) as are the contributions of other commodity-producing sectors, e.g., mining 4.2%, manufacturing 12.3%, water and electricity 2.6%, and construction 9.1%. The bulk of the GDP is created in the service sector, which contributed 63.6% in 1985 (MOP 1986). Nevertheless, a considerable number of people derive their livelihood or supplement their incomes from agriculture.

Although rainfed areas constitute 93% of cultivable land they produce 60% of total agricultural production, while irrigated areas (7% of total land) produce nearly 40% of total agricultural production and 85% of agricultural exports (Duwayri 1985). Wheat occupies most of the rainfed area, but Jordan still produces only a fraction of its wheat requirements, importing the rest. Lentils, rice, and other products are also imported. Irrigation projects have helped to increase production of vegetables and citrus fruits but rainfed agriculture is still characterized by low productivity, fragmentation, risk, lack of well tested and adapted technological improvements, and lack of credit and inputs (Aresvik 1976; Kanaan and Attieh 1972).

Land Tenure and Reform

At independence in 1946, Jordan had an agrarian economy based on traditional farming systems with a low level of technology. Until the 1950s the Ottoman land system prevailed in Jordan and other Arab countries. The two basic categories of land were *mulk* and *miri* (Warriner 1962; Baer 1957; Khalidi 1984). Holders of *mulk* land had the right of absolute ownership (*raqaba*) and usufructuary rights (*tasarruf*). *Mulk* land was mostly confined to village settlements, and since 1953 it has excluded agricultural land (Ares-

Dennis Tully (ed.), Labor and Rainfed Agriculture in West Asia and North Africa, 185–208.
© 1990 *ICARDA*.

Table 1. Agroclimatic zones in Jordan and their areas (ha).

Governorate	Arid zone (<200 mm)	Marginal zone (200–350 mm)	Semi-arid zone (350–500 mm)	Semi-humid zone (500–800 mm)	Total
Irbid	2,094,060	137,696	83,908	55,360	2,371,025
Amman	1,650,389	151,574	26,725	20,492	1,849,180
Balqa	36,928	27,606	19,219	23,048	106,801
Karak	257,657	204,419	6,048	–	468,124
Maan	4,417,866	42,104	–	–	4,459,970
Total	8,456,900	563,400	135,900	98,900	9,225,100

Source: MOA (1974).

vik 1976). *Miri* land, owned by the state, was more widespread. The occupiers of *miri* land had usufruct rights and were *de facto* owners; they could keep it indefinitely, pass it to their children, or sell it.

Waqf land was given for charitable purposes and was similar to *mulk* land, while *mewat* or dead land (e.g., desert or waste land), belonged to the state. *Metruka* land, allocated for public use, covered an insignificant area.

1   The most widespread form of land tenure was the *musha'a* system. *Musha'a* land was *miri* land that was held communally by villages and rotated among individuals (Luke 1932). Lack of continuity of occupation discouraged farmers from improving the land and the system caused disputes and friction. This situation was resolved gradually through cadastral surveys and allocation of land among the peasants in the 1930s and 1950s (Abu Shaika 1971). However, this encouraged fragmentation under the traditional inheritance system. Also, much land was registered in the names of tribal leaders rather than members. Under the land settlement law, the tribal leaders became owners of large pieces of land, and privatization created a market in which merchants and wealthy individuals purchased land (Hezelton 1974).

There has not been any significant land reform in Jordan except in the East Jordan Valley area. In 1959 it was stipulated that all areas which had benefited from the East Ghor Canal Project should become privately owned with a minimum holding size of 3 ha, based on recommendations of a World Bank Mission (IBRD 1957; Baker and Harza 1955). However, later amendments reduced the scale of lands to be expropriated and "allowed many larger landholders to avoid the more harsh effects of expropriation by registering some of their land in the names of other family members as separate farm units but continuing to operate them as one" (Anbar 1984).

This has had a direct effect on the labor market in the Jordan Valley. Instead of the intended farming system based on owner-operators, the prevailing pattern is absentee landownership coupled with sharecropping and wage labor, including foreign labor. But the land distribution laws have prevented fragmentation through the traditional inheritance system in areas with irrigation projects and this has made the application of some technologies easier.

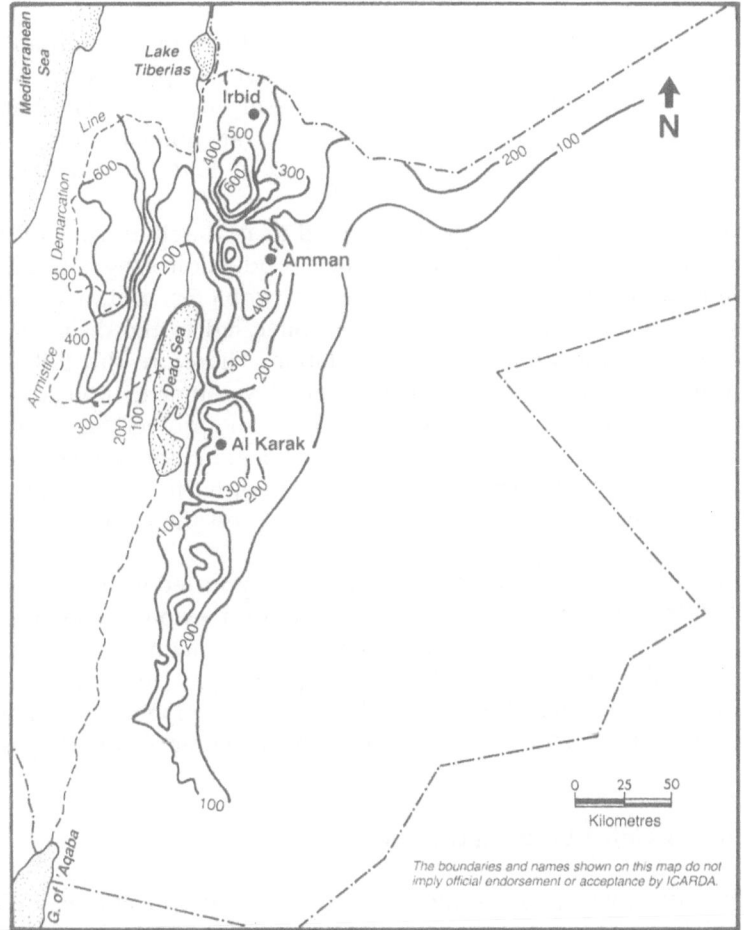

Fig. 1. Rainfall distribution in Jordan. *Source*: MOA (1973); United Nations Map No. 1569 Rev. 2 (1986).

Agricultural Policies

As early as 1950, the government attempted to improve agricultural production, mostly by introducing irrigation, and until 1969 the annual growth of the agricultural sector averaged around 8%, although there was no increase in grain production. Development efforts slowed with the 1967 war, and then increased from 1972 onwards.

Some of the aims were to increase agricultural production both to meet local requirements and to increase exports, reduce unemployment by absorbing more labor in irrigated agriculture, and support the growth of other sectors (services and industry) by transferring resources from the growing agricultural sector. However, between 1950 and 1970 these aims were not fulfilled. Since 1973, plans have emphasized the development of the agricultural sector, giving priority to irrigated agriculture in the Jordan Valley. Nevertheless, as stated in the 1986–90 5-year plan, despite consistent growth in production the sector cannot meet increasing demand. The balance of trade deficit in food products increased from an annual average of 32.8 million JOD in 1973–75 to 74.8 million JOD in 1976–80 and 41.2 million JOD in 1981–85.

Public investment has been concentrated in irrigated areas with unfortunate consequences for cereal and livestock production. Vegetables are now overproduced which has created marketing problems. Better employment opportunities in other sectors and abroad have led to the disintegration of family farms and the import of foreign labor. Therefore the 1986–90 plan aims to implement comprehensive rural development policies, reduce surpluses, conserve basic agricultural resources, and protect the environment. The plan also aims to increase the return on agricultural investment and improve the incomes of farmers and laborers in order to promote investment in agriculture and encourage farmers to remain on their farms and in their villages.

A number of projects will be introduced or supported including the development of farming in the highlands, agricultural extension in the Jordan Ghors, soil surveys and land classification, development of the Zarqa river catchment, soil conservation and fruit tree planting, seed improvement and multiplication, mechanized agricultural services, development of rangelands and fodder, and introduction of forage crops in the agricultural cycle.

Overview of Rainfed Farming Systems

In rainfed zones wheat is the most widely grown crop followed by barley, lentils, chickpeas, olives, and grapes (Figure 2). In marginal zones a 2-year rotation is common in which either wheat or barley is alternated with fallow. Where rainfall is 300–400 mm, for example in Irbid and Amman, there is generally a 2-year rotation, mainly of wheat with legumes or summer vegetables, or a 3-year rotation of wheat, legumes, and summer vegetables (Duwayri 1985).

Seedbed preparation for wheat starts in October. In most cases plowing is done by tractors and large three-disc harrow plows. Sowing, usually by hand broadcasting, starts around the middle of November but in drier areas it may be as late as February or March (Arabiat et al. 1983). Only a few farmers use chemical fertilizers. Harvesting is in late June and July by combine harvester or manually with sickles (Duwayri 1985; Arabiat et al. 1983).

For lentils, seedbeds are prepared between July and November, and sowing is done by hand in November or December, with weeding being done mostly by family members. Lentils are normally harvested manually in May by hired and family labor but there is often a shortage of labor during harvest so foreign, mainly Egyptian, laborers may be employed. Manual harvesting

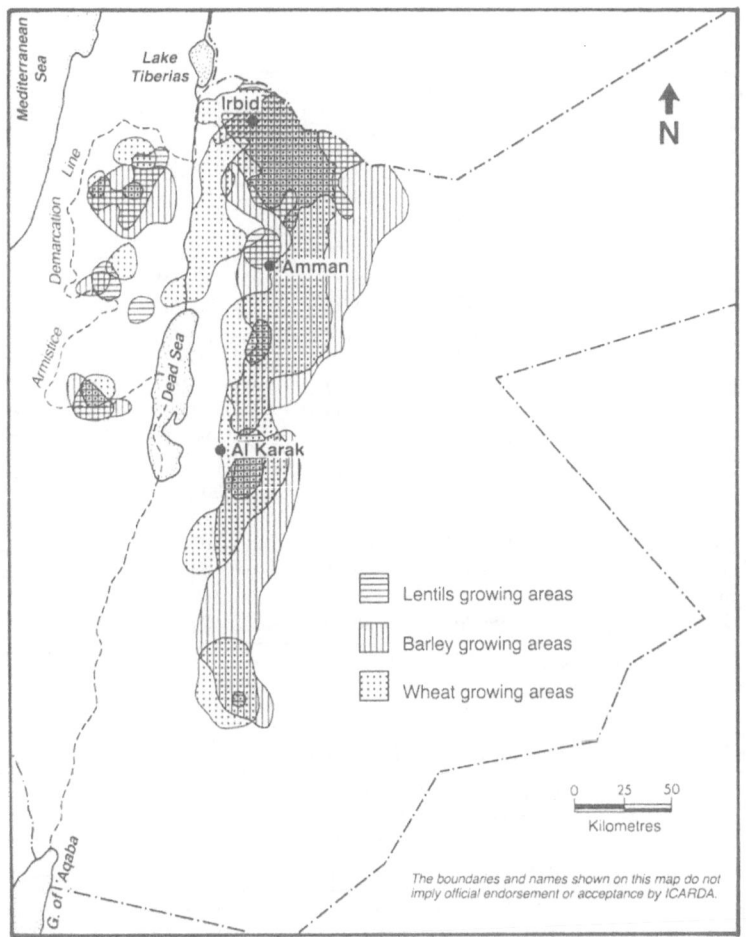

Fig. 2. Main rainfed crop production areas of Jordan. *Source*: MOA (1973); United Nations Map No. 1569 Rev. 2 (1986).

accounts for over 50% of total production costs (Haddad and Arabiat 1985; Table 2) and unless harvesting is mechanized labor shortages will lead to dwindling lentil production (Duwayri 1985).

In rainfed areas, the cultivation of field crops decreased between 1975 and 1984 due to the unreliable rainfall, low returns, and small holdings. Of all rainfed land, 34% is uncultivated mainly because of absenteeism and increasing land prices since 1973 which have rendered rainfed crops economically unattractive. Increasing off-farm job opportunities have reduced the farm labor supply, which in turn has reduced farmers' interest in planting field crops. Also, towns have encroached onto agricultural land and fields have been planted with trees (Table 3). Despite government efforts to encourage the planting of trees on hilly land, farmers convert good flat land

Table 2. Production costs and net profit (JOD/ha) for lentils, 1979/80.

Operation	Governorate			
	Irbid	Amman	Balqa	Karak
Plowing	5.00	5.00	5.00	5.00
Cost of seed	20.00	15.00	15.00	12.05
Broadcasting seed	2.60	2.50	2.00	2.30
Covering seed	2.00	2.00	2.00	2.00
Weeding	3.35	3.30	3.35	3.55
Harvesting	56.60	46.00	53.70	52.60
Transport	2.80	3.10	2.05	2.65
Threshing	3.05	3.40	2.40	2.65
Winnowing	3.00	2.80	2.75	2.70
Others	2.60	2.60	2.60	2.60
Total cost	101.00	85.70	90.85	88.55
Average yield (kg/ha)				
grain	800	557	606	707
straw	1,227	743	912	1,552
Value of grain	125.70	81.15	101.20	112.10
Value of straw	70.00	40.85	64.35	45.60
Total income	195.70	155.00	165.55	157.70
Gross margin	94.70	36.30	74.70	69.15

Source: El-Hurani and Duwayri (1986).

Table 3. Changes in cropping pattern, 1975–79 to 1980–84.

Crop	Average area utilized (×1,000 ha)		Change in cropped areas	
	1975–79	1980–84	(×1,000 ha)	%
Field crops	200.4	165.4	−35.0	−18
wheat	123.5	99.0	−24.5	−20
barley	49.8	47.5	−2.3	−5
legumes	16.1	11.4	−4.7	−29
others	11.0	5.5	−5.5	−50
Vegetables	24.7	32.2	+12.5	+50
tomatoes	8.4	12.3	+3.9	+46
others	13.3	20.9	+7.6	
Fruit trees	33.7	38.4	+4.7	+14
olives	26.8	30.0	+3.2	+12
citrus	2.2	3.3	+1.1	+50
bananas	0.4	0.5	+0.1	+25
others	3.6	4.0	+0.4	+11
Total cropped area	258.8	236.0	−22.8	−9
Fallow or uncropped	131.6	154.4[1]	+22.8	+9
Grand total	390.4	390.4[1]		

Source: Qasem (1986a).
[1]The total agricultural land was reported to be 360,000 ha in the agricultural census of 1983. The decrease may be due to urban expansion and other uses.

suitable for field crops into olive or fruit orchards, which have much better returns (El-Hurani and Duwayri 1986; Qasem 1986a).

In 1983, there were 980,000 sheep, 441,000 goats, and 34,000 cattle in Jordan. Sheep and goats are generally reared under nomadic and semi-nomadic conditions, grazing the grasses and brush of the eastern and southern steppe in spring, the uplands in summer, then the steppes again in the autumn and early winter. From November to April, sheep graze during the day and are given supplementary feed containing grain in the evening (Aresvik 1976; Duwayri 1985). In rainfed areas, sheep are also kept on small farms. Most goats are reared in marginal areas by the Bedouin while cattle are kept mainly in the northwestern farm area. About 75% of the cattle are of the local Baladi type and the remainder are Friesians. Baladi cows are generally kept in pairs by small cultivators and used as draft animals while Friesians are kept for milking as their productivity is superior to that of the Baladi.

In dry years the lack of grazing is a problem. Also, pastures are becoming over-grazed and, as long as they remain communal, their deterioration is unavoidable. Controlled access to grazing land is extremely difficult to impose. An alternative is to increase fodder production in dryland areas by incorporating fodder crops, such as vetches, clover, and sorghum, into crop rotations (Mazur 1979).

The mechanization of livestock production is fairly limited except on large dairy farms, which also employ some wage laborers. Family labor is utilized in sheep and goat husbandry, especially on small farms. The level of technology is very low and traditional methods are used for milking and the production of dairy products.

Farm Economics

In Jordan, land fragmentation is considered to be one of the main causes of low productivity (El-Hurani and Duwayri 1986; Qasem 1985; 1986a) and while the various laws passed to limit fragmentation have been successful in the irrigated Jordan Valley, they have not been practicable in rainfed areas (Table 4).

Qasem (1986b) has shown that during 1982–84 an average family of six should generate 3,600 JOD per year to be on a par with national average per capita income. He finds that, "holding sizes in the Jordan Valley are sufficient to generate incomes around national averages in the majority of cases". But in rainfed areas up to 100 ha may be required to generate an income of even 2,000 JOD depending on the crop, agroclimatic zone, and method of cultivation (Table 5). Even with high rainfall and improved practices, 28 ha would be needed to reach the average national income by growing wheat. Therefore, most holdings in rainfed areas are too small to generate

Table 4. Number and area of holdings by size, including the Jordan Valley.

Holding size (ha)	Number of holdings in 1975	1983	Percent change 1975–83	Total area held (×1,000 ha) 1975	1983	Percent of 1983 total No.	Area
<2	19,269	24,156	25	13	19	42	5
2–5	12,971	15,899	23	40	49	28	13
5–10	8,634	8,981	4	57	59	16	16
10–20	5,479	4,947	−10	70	63	9	17
20–100	4,078	3,179	−22	139	108	6	30
>100	356	276	−22	71	66	0	18
Total	50,787	57,438	–	391	364	100	100

Source: Qasem (1986b).

Table 5. Net income generated by different crops with different water resources and area needed to generate 2,000, 3,000, and 4,000 JOD.

	Net income (JOD/ha)	Area (ha) needed to generate JOD/year 2,000	3,000	4,000
Irrigated land cultivated with				
Tomatoes or cucumbers in plastic houses with high inputs	8,270	0.2	0.4	0.5
Tomatoes or cucumbers in plastic houses with normal inputs	5,070	0.4	0.6	0.8
Vegetables with drip and mulch in open field	1,970	1.0	1.5	2.0
Vegetables in open field with traditional methods	260	7.7	11.5	15.4
Citrus	2,090	1.0	1.4	1.9
Stone fruits, pomegranates, or grapes	2,080	1.0	1.4	1.9
Wheat	360	5.6	8.3	11.1
Average		2.4	3.4	5.0
Rainfed land cultivated with				
Wheat at ≥400 mm rainfall with improved practices	130	15.4	23.1	30.8
Wheat at ≥300 mm rainfall with improved practices	100	20.0	30.0	40.0
Wheat traditional	50	40.0	60.0	80.0
Barley traditional, 200–300 mm rainfall	20	100.0	150.0	200.0
Legumes traditional	80	25.0	37.5	50.0
Olives ≥400 mm	310	6.5	9.7	12.9
Grapes ≥400 mm	370	5.4	8.1	10.8
Stone fruits ≥400 mm	950	2.1	3.2	4.2
Average for ≥400 mm		7.2	11.0	14.5
Average for <400 mm		46.3	69.4	92.5

Source: Qasem (1986b).

sufficient income. Qasem suggests that the minimum holding size in rainfed areas should be increased (Qasem 1986c).

Duwayri (1985) calculates that in wetter areas, assuming a wheat yield increase from 1.1 to 1.6 t/ha with modern technologies, net income would increase from 49 to 93 JOD/ha (cf. Goetze 1977). However, fewer than 30% of farmers in rainfed areas use available modern technology, such as improved seed, weed control, and fertilizers (Qasem 1985).

In Jordan, irrigated agriculture is more market oriented than rainfed agriculture. Arabiat et al. (1983) found that 50% of the wheat produced in rainfed areas is consumed at home, 6% is stored, and 44% is sold. Of the farmers interviewed, 95% used wheat for home consumption.

Animal production presents a different type of economic problem. There was no significant growth in animal production between 1974 and 1984. Imports of red meat increased by 48% during 1973–80 and by 74% during 1981–83, with local production only meeting 30% of demand. During the same period poultry production increased considerably. Abdulrahim and Arabiat (1986) explain this in terms of relative profitability. While investments in broilers and layer hens give a return on capital of 25 and 17%, respectively, Awassi sheep give less than 12%, and lamb fattening gives only 4%. For livestock production to be more profitable it must be carried out together with crop production so that crop by-products can be used as feed. Government policies have had both negative and positive effects. Subsidizing feed ingredients such as corn, barley, and wheat bran reduces costs and so may increase meat production. On the other hand, fixing the price of cows' milk has a negative effect on dairy production and has discouraged investors. The import of fresh red meat at less than half the price of local meat has also deterred investors in livestock production (Abdulrahim and Arabiat 1986). Significant changes in the production system, import policies, and the pricing system will be necessary to avoid a deficit in livestock products (Aresvik 1976; Abdulrahim and Arabiat 1986).

The Social Organization of Agriculture

Statistics on the distribution of land in Jordan are mostly presented in terms of holdings. A holding is defined as "all the land which is used wholly or partly for agricultural production and is operated as one technical unit by one person alone or with others, without regard to title, legal form, size, or location". The term holder denotes "the person who has responsibility for the operation of agricultural holding". He could be the owner or the tenant who "exercises the technical initiative and responsibility for the operation of the holding" (DOS 1985).

These statistics offer valuable information on the efficiency of production, but the distribution of landownership would be a better indicator of wealth distribution (Mazur 1979). Data are available on the distribution of landownership for the early 1970s. In 1971 on the East Bank, excluding the Ghor, 9.3% of landowners owned 44.4% of the land. In 1974 in the Ghor, ownership was even more unequal but in parts of the Ghor where land had been distributed, landownership was much more equitable. For example, in the East Ghor Canal Project area, owners with 5 ha or less constituted 94.9% of all landowners (Dajani et al. 1980). In rainfed areas land distribution is very unequal, as 75% of owners hold 30–40% of the land, and the largest landowners (10% of all owners) own 35–45% of the land area (UNDP/FAO 1970; Abu Shaika 1971; El-Zoobi 1973).

Of the 47,438 agricultural holdings, 80% are owner-operated, 13% are rented, and the remainder are under other forms of tenure (DOS 1985). Few farms under 1 ha (9.5%) are rented, compared to 17.6% of farms of 1–5 ha. Possibly an area of less than 1 ha is not worth renting except in irrigated agriculture with intensive use of inputs. In large holdings rental is also uncommon (8.5% of farms over 10 ha). Owners in this category may operate their own farms because of the good returns. However, these statistics should be treated with caution as they do not differentiate between irrigated and rainfed land.

One of the few studies of dry farming areas (FAO 1970) shows that in the Abu-Neseir and Mubis villages in the Baq'a Valley absenteeism is widespread: there were 73 owner-operators and 112 absentee landowners, who included 34 merchants, 30 white-collar workers, 21 retired people, 11 army members, 9 workers, 4 farmers, and 3 students. Of the 73 owner-operators, 56 (77%) were full-time farmers, 13 were part-time laborers, and 4 were merchants. The majority of the 37 sharecroppers (59%) were full-time farmers; the rest included 10 part-time laborers, 4 merchants, and a white-collar worker.

In rainfed areas, sharecropping arrangements are for one season. The sharecropper provides half of the seed, all labor and draft animals, and gets from half to two thirds of the grain and straw from a rocky field; from a cleared field he only receives one third of the produce (Aresvik 1976). There is an unequal relationship between the tenant and the landowner as the latter can terminate the arrangement. Citing an FAO report (1967), Aresvik (1976) maintains that, "The insecurity of tenure and the absence of protective legislation have serious implications for agricultural production. The relations of the tenant with the landlord are usually such that the tenant is not likely to invest in the land. Indeed, his attitude to the land logically becomes that of the miner who tries to get as much out of it as he possibly can during his stay, not mindful of the future since he has no assurance that his association with the land will be continued even for the next season". For additional information on forms of sharecropping arrangements see El-Hurani (1985b), Steitieh and Musa (1983), Sharab (1975), and Hezelton (1978).

Several categorizations of agricultural holders in the Jordan Valley have been produced (UNRWA 1957; DOS 1961; Anbar 1980) but they concentrate on the form of tenure and do not consider the social categories involved in agriculture. Lanzendorfer (1985) attempted to do this using a sample of 308 farmers from irrigated and rainfed areas. His classification recognizes the following 12 groups.

Commercial farmer, investors (2.3% of sample) cultivate crops intensively on 40–50 ha of irrigated land obtained on a fixed rent or sharecropping basis. The farm is run by a manager who employs permanent or seasonal laborers and production is completely profit-oriented. The investors visit the farm periodically but are not involved in cultivation.

Commercial farmer, supervisors (2.6%), located mostly in the middle Ghor, are medium or large landowners who manage their holdings personally but use permanent and casual laborers, not family labor. Permanent laborers may be sharecroppers. Renting and leasing are widespread and the same person may rent land from others and lease it to a sharecropper. Farmers in this category are mostly well educated, tend to use advanced technology (such as plastic houses for vegetable production), and have occupations outside agriculture.

Landlord, supervisors (3.3%) operate in good rainfed areas and in the south of the Jordan Valley where there has been no land reform. Each of the 10 cases studied owned more than 200 ha, mostly irrigated. Sharecropping is predominant. In rainfed areas livestock are kept and wheat, barley, and lentils are grown as well as olives, fruit trees, and vines. In the southern Jordan Valley vegetables, such as tomatoes, squash, and eggplants are predominant. Although farming is their main occupation, neither the landlords nor their families are actively involved in cultivation, only in management. They have a patron-client relationship with the sharecroppers and they do not invest in high technology.

The number of *owner, operators* (28.9%) is well below the Jordanian average, due to Lanzendorfer's definition of this category as "landownership without additionally rented land". These farmers are found throughout Jordan and use family labor, usually full-time but in some cases family members may have another occupation. Casual laborers are employed at certain times of the year and full-time laborers are employed only on irrigated holdings that cultivate vegetables.

Owner-cum-renter, operators (4.9%) are found in all except arid areas. They own medium-sized farms which are run by the family. Land is also rented on a fixed rent basis, either to increase the holding size or for lease to sharecroppers. They are more commercial in attitute than owner, operators. *Owner-cum-sharecropper, operators* (11.4%) are generally found in rainfed areas. They lease additional land on a sharecropping basis and cultivate it with full-time family labor. They do not rent land and may employ casual but not permanent laborers. *Renter, operators* (2.6%) are found in irrigated areas. They rent their holdings on a fixed basis, cultivating vegetables with family and paid labor.

Sharecropper, managers (14.9%), mainly Palestinian, are found in irrigated areas and differ from the traditional sharecropper in that one "acts partly as a manager on behalf of, and together with, the landlord, and is mainly responsible for the organisation of the labor input". As the complex production requires considerable management, these sharecroppers are in a strong bargaining position vis-a-vis the landowner. The sharecropper's family actively participates in cultivation, and casual or even permanent laborers are employed. *Sharecropper, laborers* (8.4%) predominantly cultivate wheat and lentils in rainfed areas. Some may also own a piece of land too small to make a living. They rely mainly on family labor but may occasionally employ casual laborers. Most holdings in this category are small, and yield a much lower income than the sharecropper-manager receives. Livestock are kept for household needs.

Part-time holder, supervisors (5.8%) are found in the best rainfed areas. They have occupations other than agriculture and run their holdings with minimum labor input, choosing appropriate crops such as cereals. Production activities are carried out by custom operators and casual laborers. This type of farming is the result of increasing job opportunities in urban areas, coupled with the mechanization of agriculture and the emergence of custom operators.

Part-time holder, operators (7.8%) are found in all rainfed areas and rarely in the Jordan Valley. Their main source of income is outside agriculture but they continue to cultivate their own small holdings. The family provides little labor, and temporary laborers are occasionally employed.

Semi-nomads (7.1%) depend upon their livestock and inhabit the borders between settled areas and the desert. Although they have permanent houses, they move seasonally with their flocks. They mostly own land and cultivate it with family labor, but some are sharecroppers. They cultivate cereals for animal feed and domestic consumption.

Despite the comprehensiveness of Lanzendorfer's classification it does not consider the recent emergence of cooperatives using modern technology in irrigated areas of the Jordan Valley, which operate on rented or purchased land by hiring a manager and Egyptian wage laborers (Hezelton 1978). Furthermore, Lanzendorfer does not consider the legal status of holders. Of the 62,162 holdings in Jordan (including 4,724 holdings without land, not shown in Table 4), 3,001 are registered companies, 100 are government holdings, and 5 are tribal holdings (DOS 1985). An analysis of these would give additional insights into the relations of production in Jordanian agriculture.

Technological Change in Agriculture

In recent years, Jordanian agriculture has become mechanized to such an extent that Lanzendorfer (1985) has argued that "an extension of already mechanised production techniques to areas in Jordan where they are not yet in use is only possible on a very limited scale . . . . Agricultural mechanisation

Table 6. Users of agricultural machinery (owned or hired) according to socioeconomic classification (percentages of respondents in each socioeconomic group).

Socioeconomic group	Land preparation	Transport	Spraying	Threshing cereal grain	Harvesting cereal grain
Commercial farmer, investor	100	86	71	100[1]	0[1]
Commercial farmer, supervisor	100	100	88	100[1]	100[1]
Landlord, supervisor	100	100	60	100	100
Owner, operator	93	74	21	67	39
Owner-cum-renter, operator	93	80	53	75	38
Owner-cum-share- cropper operator	89	74	26	81	31
Renter, operator	100	100	88	50[1]	0[1]
Sharecropper, manager	100	93	57	92	33
Sharecropper, laborer	100	65	19	74	43
Part-time holder, supervisor	100	78	22	82	76
Part-time holder, operator	96	63	25	60	53
Semi-nomads	77	86	9	91	27
Total	95	79	34	76	43

Source: Lanzendorfer (1985).
[1]Very few cases of cereal cultivators in these groups.

in Jordan has reached limits beyond which it can no longer grow continuously but only transgress in a leap". However, mechanization has not taken place at the same pace for every operation. Land preparation has become almost completely mechanized as a result of tractorization, whereas mechanization of threshing and spraying varies according to crop and holding type (Table 6). Furthermore, topographic conditions limit mechanization of harvesting in some areas (Lanzendorfer 1985).

The extensive use of machinery, encouraged by the government, is due not to any dramatic increase in the number of machines but to the expansion of custom services. The parastatal Jordan Cooperative Organisation (JCO) opened machine-hiring stations in Irbid and Madaba in 1973/74 and 1981, respectively. The JCO and other organizations offer centralized custom services, and most farmers who own machinery also offer their machines to neighbors on a custom service basis. However, the efficiency of custom operations could be improved (Peckover 1983).

Tractors were first imported in 1930 (Lanzendorfer 1985) and the number of tractors rose from 883 in 1960 to 2,856 in 1971 (Kanaan and Atieh 1972) and to 4,343 in 1979 (DOS 1980). This increase was accompanied by a decline in the use of traditional implements. By 1983 farmers owned 3,103 four-wheel tractors, which does not include machinery owned by organizations like the JCO or by non-agricultural holders (DOS 1985). The number of combines also increased, from 15 in 1960 to 157 in 1971 (Kanaan and Atieh 1972). In 1983 the agricultural census recorded 226 harvesters, 4,367 harvesters and threshers, and 3,652 threshers (DOS 1985).

It is commonly believed that mechanization will reduce the demand for labor, which is true in rainfed farming where mechanization has reduced labor input without much increase in output (Mazur 1979). However, this has not caused an unemployment problem because work is available in the Gulf states and in fact the exodus from rural areas has recently reached such proportions that it has created a net labor shortage in commercial agriculture which has been met by imported foreign labor. Between 1961 and 1979, as the East Bank population grew from 0.9 to 2.2 million, rural share in the total population decreased from 49.2 to 39.6%. In irrigated areas the commercial production of vegetables using new technology has actually created jobs.

Mechanization alone is insufficient to increase productivity. Policies to increase production through the introduction of improved seed and varieties, insecticides, and other new inputs have not been very successful in Jordan, especially in rainfed farming (Aresvik 1976; Mazur 1979; Mitchel 1986; El-Hurani and Duwayri 1986; Qasem 1985; 1986c; Arabiat et al. 1983). Adoption of new technologies has been much faster in irrigated agriculture, which has received more support from the government, parastatal organizations, and private firms (Qasem 1985; El-Hurani 1985a). Many reasons have been suggested for the slow adoption of new technology, including the reluctance of farmers to invest in risky rainfed farming (Arabiat et al. 1983; Mitchel 1986), ignorance (El-Sherbini 1979; Duwayri 1985; El-Hurani 1985a; Mazur 1979), land fragmentation, and small farm size (El-Hurani and Duwayri 1986; Mazur 1979).

Lack of appropriate technology may also be important. Most green revolution plant varieties available so far are not suitable for Jordan's rainfed farming as they require too much moisture. Also, Dorling and Mutlu (1985) argue that small farms favor the use of small tractors, which do not exist in Jordan, and there are insufficient credit facilities to enable farmers to buy the large tractors which are available. Nevertheless, dryland farmers are encouraged, without much success, by private suppliers to buy unsuitable machinery (El-Hurani and Duwayri 1986).

New technologies from demonstration stations have not been adopted by farmers as a complete package. For example, in the Cooperative Winter Cereal Project, Arabiat et al. (1983) maintain that the package was not applied *in toto* because of the risk and because the infrastructure necessary for the increased supply of inputs was not sufficiently well developed. This suggests that packages are too complex to be adopted by dryland farmers and their adoption may only be gradual and long-term.

Certain technologies are not adopted because their return does not justify the cost (Mazur 1979). The government's price policy for wheat exacerbates poor returns by controlling wheat but not input prices, which reduces profitability (Qasem 1985; Mazur 1979; Gotsch 1976). Under these circumstances, households which satisfy most of their wheat needs are unlikely to adopt new technologies to increase production, especially if they can buy subsidized wheat or wheat products.

The limited extension services are cited as a secondary factor in the slow adoption of new technologies. Mazur (1979) suggests that inadequate adaptive research is the most important single factor in slow adoption, so top priority should be given to adaptive research strategies. The extension services may then become important.

Information and Decision Making

Many factors may constrain decisions taken by farmers, such as the physical features of the land, agroclimatic conditions, expected profitability of crops, the risk of drought, diseases, and prices. The availability of information and technical services or inputs and conditions imposed by state agencies also have an effect (Qasem 1986b). A typical farmer makes his own decisions concerning what to grow and how to grow it but he may be influenced by extension agents and parastatal organizations.

The extension division of the Ministry of Agriculture aims to educate and guide farmers in modern methods and techniques by disseminating information from various research stations, distributing new technology such as improved seeds and fertilizers and other inputs to farmers, controlling crop and livestock diseases (Kanaan and Attieh 1972). Kirsch (1975), Kanaan and Attieh (1972), Aresvik (1976), Abu Howayej (1985), and Qasem (1985) suggest that the formal extension services cannot attain the desired objectives because of insufficient personnel and training, inadequate means of transportation to villages, poor management, and the low status of and lack of incentives for extension agents. Also, the personnel are unevenly distributed across regions: of 180 agricultural engineers employed as extension agents, many would not live outside Amman or Irbid (Qasem 1985). Therefore, the role of extension agents in promoting information or material inputs appears to be very weak: Qasem (1985) states that only 10% of the farm labor force have contact with extension services. Dryland farmers benefit from extension services even less than the farmers in irrigated areas and so they are extremely reluctant to take risks with unknown inputs.

Parastatal organizations like the JCO and the Jordan Farmers' Association also offer some technical help to farmers but only to their own members. The 60 commercial companies involved in extension work have been much more successful than the formal extension institutions (Qasem 1985). They employ highly qualified personnel and are quite popular with farmers. They provide services and inputs on credit, deliver to the farm, and give up-to-date technical assistance. However, because rainfed production does not use many new inputs, private companies show little interest in dryland farming.

To minimize risk, farmers in rainfed areas in most cases opt for conditional decision making: waiting for the autumn rains limits investment in tillage and seedbed preparation, but undermines the economic viability of fertilizer use (Arabiat et al. 1983). With new technology, poor farmers may not choose to take the risk of trying it. Wealthy and educated farmers are usually first to increase their productivity using modern technology (Qasem 1986c). In addition, farmers often do not use an optimal combination of inputs (Qasem 1985). Input use may be limited by lack of working capital and inadequate facilities (Taqieddin 1985; 1986; Haldorson 1975; Aresvik 1976; Mazur 1979; Sharab 1975).

A farmer's decisions may be affected by his relations with the landowner. Traditionally, the sharecropper and the landowner would decide together on the crop to be grown, but the deployment of labor would be decided by the tenant. With the distribution of land and development projects in the Jordan Valley, relations between landowners and tenants have been transformed. Important decisions on the choice of crop, land preparation, quantity and type of fertilizer, and the sale of produce were made mutually in most cases, but decisions concerning the date of planting, harvesting, and the method of harvesting were generally made by the sharecropper (Sharab 1975).

The introduction of new technology has also brought about new configurations of landlord-tenant relations (Steitieh and Musa 1980). For instance, in vegetable production in open fields with drip irrigation the landlord pays for the drip system but the tenant pays 200 JOD/ha to the landlord. The tenant supplies all the labor and half the cost of all other inputs, while the landlord provides the land and water. Similar changes might be expected as new technology is adopted in rainfed areas.

Labor Availability

While the labor force in the 1950s was predominantly agricultural, in 1961 the agricultural labor force had decreased to 33.5% of the total and the rate of labor participation remained quite low. The share of agricultural labor in the total labor force declined dramatically in subsequent years (Table 7) and the rate of participation in the agricultural labor force declined from 24% in 1961 to 22% in 1979 and 21% in 1985 (MOP 1986).

The decline in the agricultural labor force is quite often explained by national and international emigration of labor. El-Hurani (1985b) suggests that prior to 1970 the government actually encouraged migration in order to reduce unemployment, estimated at 12–15% in the 1960s (Clawson et al. 1971).

There are no accurate figures for the numbers working abroad, and in most cases the data that are available do not differentiate between Jordanians

Table 7. Jordanian manpower according to economic activity.

Economic activity	1961		1979		1985[1]	
	No.	%	No.	%	No.	%
Agriculture	72,977	33.5	46,728	11.5	39,237	7.8
Mining and manufacturing	22,276	10.2	34,935	8.6	53,053	10.6
Electricity and water	925	0.4	2,472	0.6	5,520	1.1
Construction	22,187	10.2	52,645	13.0	55,263	11.0
Trade	17,452	8.0	41,541	10.3	50,239	10.0
Communications	7,624	3.3	28,977	7.2	47,225	9.4
Financial services[2]			8,673	2.1	17,132	3.4
Social services, public administration and defence	74,398	34.2	189,303	46.7	234,718	46.7
Total	217,840	100	405,274	100	502,343	100

Source: MOP (1986).
[1]Data for 1985 are estimates.
[2]Included in social services in 1961.

and Palestinians (Kirman 1982; Seccombe 1981). The figures for those working abroad in 1980 vary from 101,000 (Kirman 1980) or 250,000 (Birks and Sinclair 1980) to 400,000 (MEED 1980), while one source puts the figure in 1985 at 339,000 (MOP 1986).

Meanwhile, rural-urban migration was caused by the poor incomes of small farmers, tenants, or herders (Dajani and Murdock 1978). The army provides employment and many urban development efforts attract rural people to wage employment. Dajani's study (1979) of the Southern Ghor and Wadi Araba gives an example of this rural-urban trend. Thus an internal demand for labor was created and the loss of labor to the oil-rich states, especially from 1973 onwards, exacerbated the labor shortage in rural areas. Wages in agriculture were forced up, and the scarcity and high cost of hired labor had a discouraging effect on farmers (Snobar 1984).

The shortage had different consequences for rainfed and irrigated agriculture. It reduced the scale of rainfed farming, but most farmers in irrigated areas merely relied more on imported labor. Egyptian immigrant workers became a significant part of the Jordan Valley labor force. One estimate suggests that in 1985 the number of guest workers reached 143,000 or 22% of the total 645,000 labor force (MOP 1986). Steitieh and Musa's survey (1980) showed that more than 90% of permanent workers employed in the production of fruits and vegetables in the Jordan Valley were Egyptian guest workers. Thus Jordan has become both an importer and an exporter of labor. While people from rural and urban areas are leaving for countries in the Gulf, the shortage of labor is being met by foreign laborers from Egypt, Pakistan, and Sri Lanka.

Workers' remittances have helped the balance of payments and continue

Table 8. Labor utilization in Jordanian agriculture, 1983.

Type of labor[1]	Under 15 years	15–34 years	35–64 years	Over 64	Total
Household (unpaid)					
Male	4,638	25,312	39,723	8,003	77,676
Female	3,866	19,499	19,990	888	44,243
Total	8,504	44,811	59,713	8,891	121,919
Permanent (paid)					
Male	125	5,326	4,944	106	10,501
Female	115	528	345	5	993
Total	240	5,854	5,289	111	11,494
Temporary (paid)					
Male	99	5,367	4,096	22	9,584
Female	62	546	304	0	912
Total	161	5,913	4,400	22	10,496
Occasional (paid)					
Male	325	25,865	21,177	96	47,643
Female	292	4,193	1,519	4	6,008
Total	617	30,058	22,696	100	53,651
Total					
Male	5,187	61,870	69,940	8,227	145,404
Female	4,335	14,766	22,158	897	52,156
Total	9,522	76,636	92,098	9,124	197,560

Source: Computed from DOS (1985).
[1]Permanent laborers work more than two thirds of the agricultural year, temporary laborers work between one third and two thirds, and occasional laborers work less than one third.

to be a major source of GNP in Jordan (Saket 1985; RSS 1978; Kirman 1982; Birks and Sinclair 1978; Seccombe 1980; 1981; McClelland 1979; Kirman 1980; RSS 1983), although those remittances which are spent on conspicuous consumption and family maintenance (food, clothing, and education) have an inflationary effect on the economy. On the other hand, the labor shortage created by international migration has increased wages and the cost of training a skilled worker who migrates is considered to be a loss for his country (RSS 1978).

Remittances do not seem to be invested in agricultural land or in modernizing agriculture. In rainfed areas this may be due to the unreliable and unprofitable nature of dryland farming. Kirman (1982) argues that remittances may have increased land prices, especially around urban areas, since large sums have been invested in construction.

The most recent and comprehensive data on the composition of the labor force are in the agricultural census of 1983 (DOS 1986). The predominant form of agricultural labor is family labor, with unpaid family labor contributing 61.7% of the total (Table 8). Children below 15 years contributed a minor part of farm labor and only 1,018 of the total 9,522 child laborers were paid, with the majority of these only being employed as occasional laborers. Many children miss school to work but the introduction of compulsory education up to the sixth grade has forced farmers to employ wage laborers (Snobar and Arabiat 1983).

Sociocultural values hinder large-scale participation of women in agricultural activities other than family farming. In 1983 females contributed 26.4% of total agricultural labor, but 36.3% of family labor. Of the total paid laborers, 89.5% were males and 10.5% females. Tractor driving and other jobs are considered men's specialities. Raising children, looking after household animals, and doing household chores also deter women from taking paid jobs. Women's participation in agricultural wage labor is correlated with their socioeconomic status, with paid female workers being mostly widows, single women, or from poor families (Dajani et al. 1980).

The poor wages and the stigma attached to agricultural wage labor also keep many women away from paid agricultural work and participation has not grown enough to meet the demand for labor. However, the demand for labor has resulted in increasing participation rates of women in the labor force, from 3.1% in 1961 to 7.7% in 1979 and about 12.5% in 1985 (MOP 1986).

As a result of the rural exodus, the family farm is declining (MOP 1986) and agriculture has become a part-time occupation for a considerable number of farmers who now employ wage laborers. In 1983, 25,547 holders were engaged in agriculture as a secondary occupation while 33,403 holders had farming as their main occupation (DOS 1985). Paid laborers constituted 38.2% of the economically active population in agriculture; of these, 15, 14, and 71%, respectively, were permanent, temporary, and casual workers (DOS 1985). Snobar and Arabiat (1983) confirm that farming is becoming a part-time occupation and that agriculture is losing farmers and farm workers.

Various surveys in the Jordan Valley indicate a significant increase in the number of paid workers (DOS 1973; 1981). Family participation in farming is highest in the south (Kerak and Wadi Musa), followed by the Jordan Valley where sharecropper, managers rely mostly on family labor. In Lanzendorfer's survey, the highest family participation was among semi-nomads and lowest among the part-time holder, supervisors, tractor-owning commercial farmers, and landlord, supervisors. Most permanently employed laborers were unskilled and non-Jordanian, and received very low wages (Lanzendorfer 1985).

Substitution Between Labor and Capital

In Jordan, capital is invested in agriculture in the form of machinery and advanced technology. But neither technology nor capital displaces labor. Whether or not the introduction of machinery and technology will increase, decrease, or reschedule labor depends on the type and nature of the technology and the agricultural activity. Mechanization has had different consequences for rainfed and irrigated agriculture although tractorization of land preparation has reduced labor requirements in both types of farming. Gadi et al. (1975) have shown that, for wheat production, man-hours and costs were much higher in manual agriculture using draft animals than in mechanized agriculture. Also, the net return on mechanized wheat production was much higher than on partially mechanized operations using manual and animal power traditionally. But Dorling and Mutlu (1985) argue that there is a need to develop indigenous technology in rainfed crop production because imported technological inputs like machinery and agrochemicals have not been adopted to the extent necessary for their translation into economic gains. They say that machinery has often replaced labor without much increase in the productivity of other cooperant resources.

However, both groups agree that labor is substituted by machinery in dryland farming. This view is also shared by Lanzendorfer (1985), who nevertheless suggests that mechanization has not had a dramatic effect on employment in dryland areas. His view is that, as underemployment was widespread before 1970, the mechanization of dryland agriculture merely converted underemployment into full unemployment and these people have emigrated to the Gulf states since the early 1970s.

Lanzendorfer also found that mechanization has reduced the participation of family labor in the cereal-producing Irbid area, where farming is now secondary to urban employment. Cereal-producing farmers may now supervise custom operators. Thus hardly any permanent paid laborers are required, and a few casual laborers are only occasionally employed for certain operations like seeding, harvesting, and threshing. So, the mechanization of plowing, harvesting, and threshing has reduced the demand for labor, but has not created many jobs; those for tractor drivers, combine operators, and repair workers are too few to significantly reduce the number of unemployed. In areas like Kerak and Wadi Musa where mechanization has not been extensive, the participation of family labor remains greater.

Contrary to dryland farming, in irrigated farming the demand for labor has continuously increased and will probably continue to do so (Salt and Keeley 1976) through the expansion of area, crop intensity, and labor-intensive vegetable production (Seccombe 1981).

Conclusion

Jordan's dependence on imported basic foodstuffs has necessitated optimum utilization of agricultural resources. Rainfed areas are particularly important as they constitute about 90% of all cultivable land. However, the realization of the potential of rainfed areas is difficult because production and related decisions are limited by general economic conditions, availability of land, capital, and labor, government policies, the level of infrastructural development, and agroclimatic conditions.

Despite government efforts, improvements in rainfed agriculture have been far from satisfactory and new technologies are being adopted slowly even in irrigated agriculture where returns are much higher. Mechanization and the increasing availability of off-farm employment have had different consequences for rainfed and irrigated agriculture: labor from rainfed areas has moved to urban areas and neighboring oil-rich countries, while the demand for labor has increased in irrigated areas resulting in greater use of foreign labor. It is difficult to assess the importance of mechanization as a labor-saving factor, and therefore as a push factor, in dryland farming relative to the importance of off-farm employment as a pull factor. It is certain, however, that Jordan has become both a labor-exporting and a labor-importing country.

Social, political, and cultural factors may be very influential in determining the impact of new technology, and information about these factors needs to be made available to planners. Unfortunately, in Jordan such qualitative data are virtually non-existent. There is a great demand for detailed socio-economic and sociocultural studies of rural communities in Jordan that would give a deeper insight into the working of such communities.

References

Abdulrahim, S. and Arabiat, S. 1986. Trends and Policies in the Animal Sector. Pages 73–84 in Agricultural Policy in Jordan. (Burrel, A., ed.). London: Ithaca Press.

Abu Howayej. B. 1985. Farm Systems in Irrigated Areas. Pages 89–125 in Agricultural Sector of Jordan: Policy and Systems Studies (Zahlan, A.B., ed.). London: Ithaca Press.

Abu Shaika, A. 1971. Land Tenure in Jordan: A Case Study of the Bani-Hassan Area. MA thesis, American University of Beirut.

Anbar, A.H. 1980. Socio-Economic Aspects of the East Ghor Canal Project. PhD thesis, University of Southampton.

Anbar, A.H. 1984. Changing Land Tenure Patterns in the East Jordan Valley. Paper presented at the workshop on Some Aspects of Agricultural Modernization and their Socio-Economic Impacts, 18–23 May, Amman.

Arabiat, S., Nygaard, D. and Somel, K. 1983. Factors Affecting Wheat Production in Jordan. Aleppo: ICARDA.

Aresvik, O. 1976. The Agricultural Development of Jordan. New York: Praeger.

Baer, G. 1957. Land Tenure in the Hashemite Kingdom of Jordan. Land Economics 23(3): 167–183.

Baker, M. Jr. and Harza Engineering Company. 1955. Yarmouk Jordan Valley Project: Master Plan Report. Chicago, Illinois.

Barham, N. 1979. Geographical Problems of Dry Farming in Jordan. PhD dissertation, Technische Universitat, Hanover.

Birks, J.S. and Sinclair, C.A. 1978. Country Case Study: The Hashemite Kingdom of Jordan, International Migration Project. Durham: University of Durham.

Birks, J.S. and Sinclair, C.A. 1980. The Socio-economic Determinants of Intra-regional Migration. Paper presented to the ECWA Conference on International Migration in the Arab World, 11–16 May, Nicosia, Cyprus.

Clawson, M., Landsberg, H.H., and Alexander, L.T. 1971. The Agricultural Potential of the Middle East. New York: Elsevier.

Dajani, J.S. 1979. A Baseline Socio-Economic Study of Southern Ghors and Wadi Araba. USAID: Amman.

Dajani, J.S., Hazelton, J., Rhoda, R. and Sharry, D. 1980. An Interim Evaluation of the Jordan Valley Development Effort: 1973–1980. USAID. Mimeo.

Dajani, J.S. and Murdock, M.S. 1978. Assessing Basic Human Needs in Rural Jordan. USAID. Mimeo.

DOS (Department of Statistics). 1961. The East Jordan Valley: A Social and Economic Survey. Amman: DOS.

DOS (Department of Statistics). 1973. Social and Economic Survey of the East Jordan Valley. Amman: DOS.

DOS (Department of Statistics). 1980. Statistical Yearbook 1979. Amman: DOS.

DOS (Department of Statistics). 1981. Results of the Agricultural Census in the Jordan Valley 1978. Amman: DOS.

DOS (Department of Statistics). 1985. General Results of the Agricultural Census 1983. Amman: DOS.

DOS (Department of Statistics). 1986. Five Year Plan for Economic and Social Development 1986–1990. Amman: DOS.

Dorling, M.J. and Mutlu, S. 1985. Commercialization of New Indigenous Technology in Jordanian Agriculture. Geneva: International Labor Office.

Duwayri, M. 1985. Farm Systems in Rainfed Areas. Pages 126–158 in Agricultural Sector of Jordan: Policy and Systems Studies (Zahlan, A.B., ed.). London: Ithaca Press.

El-Hurani, M.H. 1985a. The Supply of Purchased Agricultural Inputs. Pages 274–287 in Agricultural Sector of Jordan: Policy and Systems Studies (Zahlan, A.B., ed.). London: Ithaca Press.

El-Hurani, M.H. 1985b. The Supply of Agricultural Labor. Pages 68–87 in Agricultural Sector of Jordan: Policy and Systems Studies (Zahlan, A.B., ed.). London: Ithaca Press.

El-Hurani, M.H. and Duwayri, M. 1986. Policies Affecting Field Crop Production in the Rainfed Sector. Pages 55–72 in Agricultural Policy in Jordan (Burrel, A., ed.). London: Ithaca Press.

El-Sherbini, A.A. 1979. Food Security Issues in the Arab Near East. Report for the United Nations Economic Commission for Western Asia. Oxford: Pergamon Press.

El-Zoobi, A.M. 1973. Socio-Economic Survey of the Operator Farmers in the Three Pilot Areas of the Project. Dryland Farming Project Jor/18 Karak. FAO/UNDP. Socio-Economic Series No. 8. Karak: Ministry of Agriculture.

FAO (Food and Agriculture Organization). 1967. Mediterranean Development Project. Jordan Country Report. Rome: FAO.

FAO (Food and Agriculture Organization). 1970. Dryland Farming Jordan: A Socioeconomic Study with Special Reference to Land Tenure Problems in Abu-naseir and Mubis Villages, Baq'a Valley. Technical Report 1. Rome: FAO.

Gadi, A.R. et al. 1975. (Cost of Field Crops in Jordan.) In Arabic. Amman: Ministry of Agriculture.

Goetze, N. 1977. Wheat Improvement Project on Jordan. AID Staff Papers. Amman: USAID.

Gotsch, S.H. 1976. Wheat Price Policy and the Demand for Improved Technology in Jordan's Rainfed Agriculture. Discussion Paper No. 2. Amman: The Ford Foundation.

Haddad, N. and Arabiat, S. 1985. (Problems and Methods of Growing Lentils in Jordan.) In Arabic. Dirasat (Agricultural Sciences) 12.

Haldorson, L. 1975. Availability of Goods and Services. Pages 162–142 in Jordan Wheat Research and Production (Goetze, N., ed.). Final Report to USAID. Oregon State University.

Hezelton, J.E. 1974. The Impact of the East Ghor Canal Project and Land Consolidation, Distribution and Tenure. Amman: The Royal Scientific Society, Economic Research Department.

Hezelton, J.E. 1978. Social Soundness Analysis of the Jordan Valley Farmers Association Credit Program. Amman: USAID.

IBRD (International Bank for Reconstruction and Development). 1957. The Economic Development of Jordan. Baltimore: Johns Hopkins Press.

Kanaan, W. and Attieh, Y. 1972. Jordan: Agricultural Development. Amman: Ministry of Culture and Information.

Khalidi, T. (ed.). 1984. Land Tenure and Social Transformation in the Middle East. Beirut: American University of Beirut.

Kirman, F.X. 1980. The Impact of Labor Migration on the Jordanian Economy. Unpublished manuscript.

Kirman, F. 1982. Labor Exporting in the Middle East; The Jordanian Case. Development and Change 13: 63–89.

Kirsch, J.E. 1975. Report to the Government of Jordan on Dryland Farming Extension Programs in the Irbid Project Area. UNDP/FAO Projects, Amman. Mimeo.

Lanzendorfer, M. 1985. Agricultural Mechanization in Jordan. A Study of its Processes in a Socioeconomic Context. Socioeconomic Studies on Rural Development, Volume 62. Gottingen: Edition Heredot.

Luke, H. 1932. The Handbook of Palestine and Transjordan. London: MacMillan.

Mazur, M.P. 1979. Economic Growth and Development in Jordan. London: Croom Helm.

McClelland, D.H. 1979. Worker Migration and Worker Remittances. Amman: USAID.

MEED (Middle East Economic Digest). 1980. 24(28): 16–18.

Mitchell, M. 1986. Cereal Production and Risk in the Rainfed Sector. Pages 41–54 in Agricultural Policy in Jordan (Burrel, A., ed.). London: Ithaca Press.

MOA (Ministry of Agriculture). 1973. Agricultural Atlas of Jordan. (Prepared by B. Abu Howayej). Amman: MOA.

MOA (Ministry of Agriculture). 1974. Report on Agricultural Zoning by the Department of Agricultural Research and Extension. Mimeo.

MOP (Ministry of Planning). 1986. Five-Year Plan for Economic and Social Development 1986–1990. Amman: MOP.

NPC (National Planning Council). 1976. Five-Year Plan for Social and Economic Development 1976–1980. Amman: NPC.

Peckover, T.E.F. 1983. Comparisons of Machinery Utilization in the Rainfed Agricultural Areas of Jordan and Australia. The Jordan-Australia Dryland Farming Project, Madaba. Mimeo.

Qasem, S. 1985. Agricultural Research and Extension. Pages 343–411 in the Agricultural Sector of Jordan: Policy and Systems Studies (Zahlan, A.B., ed.). London: Ithaca Press.

Qasem, S. 1986a. Agricultural Land Use Policies. Pages 11–20 in Agricultural Policy in Jordan (Burrell, A., ed.). London: Ithaca Press.

Qasem, S. 1986b. The Size of Agricultural Holding. Pages 21–29 in Agricultural Policy in Jordan (Burrell, A., ed.). London: Ithaca Press.

Qasem, S. 1986c. The Productivity of Irrigated Agriculture. Pages 89–100 in Agricultural Policy in Jordan (Burrell, A., ed.). London: Ithaca Press.

Qasem, S. and Mitchell, M. 1986. The Problems of Rainfed Agriculture. Pages 30–40 in Agricultural Policy in Jordan (Burrell, A., ed.). London: Ithaca Press.

RSS (Royal Scientific Society). 1978. Report on the Study Group on Worker Migration Abroad. Amman: RSS.

RSS (Royal Scientific Society). 1983. Workers Migration Abroad: Socioeconomic Implications for Households in Jordan. Amman: RSS, Economic Research Department.

Saket, B. 1985. The Jordanian Economy. Pages 9–39 in Agricultural Sector of Jordan: Policy and Systems Studies (Zahlan, A.B., ed.). London: Ithaca Press.

Salt, A. and Keeley, W. 1976. Manpower Developments in the Hashemite Kingdom of Jordan with Special Reference to East Jordan Valley. Washington, D.C.: US Department of Labor.

Seccombe, I.J. 1980. Jordanian Labor Migration: The Impact on Domestic Development. MA thesis, University of Durham.

208

Seccombe, I.J. 1981. Manpower and Migration: The Effects of International Migration on Agricultural Development in the East Jordan Valley, 1973–1980. Durham: University of Durham.

Sharab, H. 1975. Agro-economic Aspects of Tenancy in the East Jordan Valley 1975. Amman: Royal Scientific Society, Agricultural Economic Research Division.

Snobar, B.A. 1984. Some Special Aspects of Farm Mechanization in Jordan. Paper presented at the workshop on Some Aspects of Agricultural Modernization and their Socio-Economic Impacts, 18–23 May, Amman.

Snobar, B.A. and Arabiat, S.M. 1983. Mechanization of Agriculture and Socioeconomic Development in Jordan. Faculty of Agriculture, University of Jordan. Mimeo.

Steitieh, A.M. and Musa, A.H. 1980. Vegetables Grown under Plastic Covers and Drip Irrigation Systems in the East Jordan Valley. Amman: University of Jordan.

Steitieh, A.M. and Musa, A.H. 1983. Some Aspects of Technological Development in the Farming Systems of the East Jordan Valley 1977–1980. Dirasat (Agricultural Sciences) 10(1): 67–89.

Taqieddin, N. 1985. Agricultural Credit and Finance. Pages 288–324 in Agricultural Sector of Jordan: Policy and Systems Studies (Zahlan, A.B., ed.). London: Ithaca Press.

Taqieddin, N. 1986. Agricultural Credit. Pages 101–107 in Agricultural Policy in Jordan (Burrell, A., ed.). London: Ithaca Press.

UNDP/FAO (United Nations Development Project/Food and Agriculture Organization). 1970. Dry-Land Farming: A Socioeconomic Study with Special Reference to Land Tenure Problems in Abu-Nasiev and Mublis Villages, Baq'a Valley. Report prepared for the Government of Jordan. Rome: FAO.

UNRWA (United Nations Relief and Works Agency). 1957. Jordan Valley Land Tenure Survey. Amman: UNRWA.

Warriner, D. 1962. Land Reform and Development in the Middle East 2nd edition. London: Oxford University Press.

10. Agricultural Labor and Technological Change in Iraq

Department of Economics, Yarmouk University, Irbid, Jordan

Introduction

Iraq's economic and social history has seen cycles of tribalization and settlement (Haidar 1942) caused by instability of cultivation in the irrigated Mesopotamian plain and in rainfed areas. In the 200 years following the mid-17th century, nomadic and semi-nomadic pastoralism prevailed, and in the center and south, permanent cultivation was limited to small areas around a handful of towns. Under the Ottoman system, cultivation and settlement were more widespread in the rainfed foothills and upland plains, especially along main transport routes east of the Tigris which had government protection. In some mountain valleys farther north and east, agricultural settlements were protected by autonomous Kurdish principalities which survived until about the mid-19th century. Other mountain areas were inhabited by semi-nomadic tribes, some of whom only settled during the 1920s and 1930s (Khisbak 1972). Crop production and sedentary agriculture began to expand and the population grew rapidly from about the 1860s.

In settled areas of the northern rainfed zone, peasant land rights were generally better defined than in the south although land title was usually bestowed on village notables (Haidar 1942). Among those peasants who acquired title many, particularly in the plain areas, lost it to creditors following successive poor harvests (Haidar 1942; Warriner 1957). Some peasant proprietorship survived in the Kurdish mountain valleys with more reliable rainfall (Warriner 1948). In the mountain region, where land suitable for cultivation is limited, trade expanded mainly in livestock products and tobacco. Apart from tobacco and other irrigated summer crops, the landlord's share in dry farming was relatively small with little potential for a marketable surplus.

By World War I, the total cultivated area was only about 400,000 ha (Mahdee 1982), reflecting the prevalence of pastoral nomadism. During the inter-war period, agricultural settlement and the expansion of cultivation continued, particularly in the irrigated zone where land was mostly owned by ex-tribal chiefs and other rich absentees who provided almost no service in exchange for their large crop share. In the words of a prominent Iraqi economist (Hasan 1965): "The growth of the export trade and the expansion of agricultural production proceeded with the mobilisation of uncultivated

Dennis Tully (ed.), Labor and Rainfed Agriculture in West Asia and North Africa, 209–227.
© 1990 *ICARDA*.

agricultural land and with employment of unemployed labor. Traditional agricultural methods and primitive means of production were retained, and no increase in either land or labor productivity resulted from this process." This growth was wasteful of scarce autumn water, ecologically unbalanced, and depended on large supplies of formerly nomadic labor.

Around 1950, agricultural growth shifted to the rainfed zone and settlement was accompanied by the granting of land title to ex-tribal chiefs and urban tractor owners. Mechanization facilitated rapid expansion of dryland grain farming in the Jazira and elsewhere. Between 1948/49 and 1956/57 the area under wheat and barley in the rainfed zone more than doubled (CSO 1949a-86a) but there was apparently no radical change in agricultural practices or increase in productivity.

Extensive crop/fallow cereal monoculture with largely mechanized tillage and harvesting appears to have been more or less fully developed by the late 1950s, covering large areas of marginal land (Mahdee 1982). Coinciding with the introduction of land reform in 1958, a prolonged drought began which drastically reduced yields, and the cultivated area receded for several years only to recover its 1956/57 level by 1966/67. The total area under wheat and barley peaked around 1968, declining in the following decade and becoming more variable from one season to the next. Figures for the first half of the 1980s suggest that this trend is reversing, particularly for barley cultivation (Table 1).

Despite the availability of improved seeds and machinery for more timely operations, grain production in the rainfed zone was stagnant for almost 3 decades, and yields did not improve until the early 1980s. Nevertheless, low production costs and labor requirements have ensured that rainfed cereal cultivation has, subject to weather-induced variations, continued to be widely practised. Recent expansion has mainly been in vegetable, fruit, and poultry production, especially in the irrigated zone, while the potential for fruit, nut, and forestry products in the Kurdish mountain region has not been exploited.

Until recently, there has been little research into dry farming in Iraq. The Kurdish provinces which are part of the rainfed zone have been involved in civil war since 1961 and the economy and society of Iraq as a whole are changing due to war and fluctuations in the oil market. Thus sources for this paper are few and possibly out of date.

Rainfed Agriculture

Iraq's rainfed zone lies north and east of the 200 mm isohyet which is the official demarcation line between grazing and cultivable rainfed lands (Al-Ani 1977). Agricultural holdings in the five northern governorates which cover most of the rainfed area total 3.3 million ha. The area to the south and west covers over 70% of Iraq's 438,000 km^2 and is mostly uncultivable and poor grazing land, with as little as 50 mm mean annual rainfall. The 5.5

Table 1. Total area of wheat and barley in Iraq and in the rainfed region[1], and average wheat yields on rainfed land 1966/67–1984/85.

Year	Gross cultivated area[2]			Net cropped area[3]			Gross cultivated area[3]			
	Rainfed wheat & barley	Total wheat & barley	Average rainfed wheat yield	Rainfed wheat & barley	Total wheat & barley	Average rainfed wheat yield	Rainfed wheat & barley	Total wheat & barley	Average rainfed wheat yield	Rainfed output of wheat & barley
	(×1,000 ha)	(×1,000 ha)	(kg/ha)	(×1,000 ha)	(×1,000 ha)	(kg/ha)	(×1,000 ha)	(×1,000 ha)	(kg/ha)	(×1,000 t)
1966/67	1,658	2,927	308	1,386	2,253	460				658
1967/68	1,835	3,228	668	1,669	2,588	784				1,347
1968/69	1,927	3,307	572	1,626	2,538	516				1,051
1969/70		3,098		1,604	2,432	496				816
1970/71		2,132		604	1,344	572				345
1971/72		3,703		1,778	2,641	1,208				2,147
1972/73		2,114		884	1,621	472				439
1973/74		3,152		1,445	2,152	564				956
1974/75		2,311		1,308	1,975	432				596
1975/76				1,337	2,051	740				1,015
1976/77					1,394					
1977/78				1,567	2,210	502		2,366		840
1978/79					1,838			2,105		
1979/80					2,204			2,329		
1980/81					2,176			2,261		
1981/82					2,279		1,745	2,348	774	1,305
1982/83					2,408			2,673		
1983/84					1,018			2,754		
1984/85					2,897		2,083	3,016	914	1,965

Source: CSO (1949a–86a various issues); MAAR (1976).
[1]The rainfed region is defined here as the five northern Muhafadhas. Boundary changes implemented in 1976 have a minor effect on the delineation of the region and have been ignored.
[2]Ministry of Agriculture.
[3]CSO.

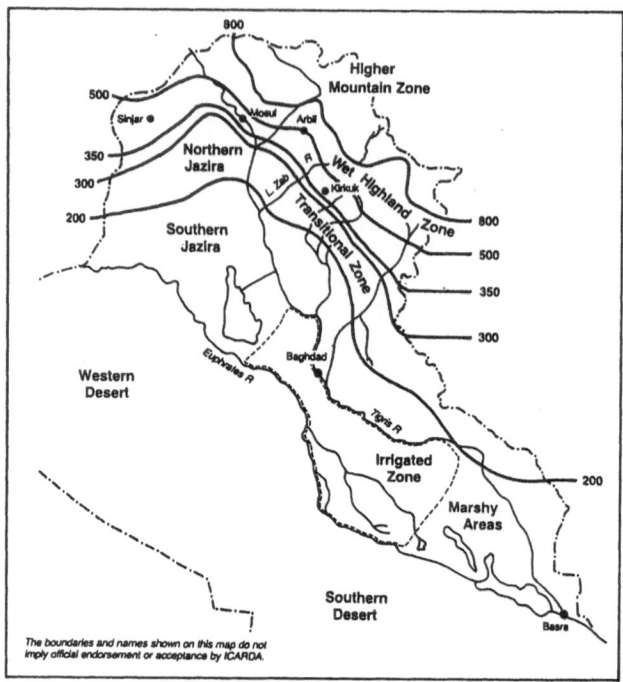

Fig. 1. Agroclimatic zones in Iraq. *Source*: AOAD (1983a).

million ha between the Tigris and Euphrates rivers are cultivable under irrigation (Figure 1), but most of this land is of low productivity in spite of heavy state investment in physical infrastructure, production, extension, and marketing. However, new, large-scale irrigation projects near Mosul and Kirkuk are extending irrigation into traditional rainfed lands.

The rainfed zone can be divided into three subregions. The wettest area lies in the northeast and consists of a series of plains interspersed with mountains and foothills. Rainfall averages above 350 mm but in the valleys of the higher mountains there are about 100,000 ha with rainfall of 900–1,000 mm (Clawson et al. 1971)

South of this zone is an area of plains and gently sloping lands with 200–350 mm annual rainfall, forming a transition zone between secure rainfed cultivation and the irrigated zone. Much of it is hilly and uncultivable, parts have been severely eroded, and the grazing potential is under-utilized with the existing open range system.

The third region is the northern half of the Jazira lying to the west of the other two zones. Average annual rainfall reaches about 500 mm in the north, but most of the plateau receives 200–350 mm. Since the 1950s, this area and the transitional zone to the east have been settled, and nomadic grazing has been replaced by mechanized cereal cultivation.

Rainfall occurs only from autumn until spring and is highly variable, especially in the drier areas. There are about 1 million ha with relatively reliable rainfall at an annual mean above 350 mm (AOAD 1983a; Clawson et al. 1971 give a higher figure), 1.38 million ha with uncertain rainfall (mean 250–350 mm), and 1.32 million ha with 200–250 mm (calculated from estimates in AOAD 1983a; 1984) where the crop is extremely precarious and average yields are low. After successive harvest failures, AOAD (1984) suggests that cultivation has receded somewhat from the driest area.

Low-input cereal monoculture is predominant in Iraq's rainfed agriculture with wheat and barley accounting for over two thirds of cultivated land. Farmers depend on tractor owners for mechanical operations in exchange for a share of the crop. In 1985, legumes (particularly chickpeas and lentils) covered a total of only 36,000 ha and melons covered 95,000 ha although there is no information on what proportion was irrigated. The main irrigated crops grown in the rainfed zone in 1985 were tobacco (16,000 ha) and sunflower (13,000 ha). Orchards covered 29,000 ha in 1978 (CSO 1985a) while an unknown but probably small percentage of the 195,000 ha of vegetable crops was in the rainfed region in 1985.

Sheep and goats may be kept as an alternative to cultivation, to reduce the risk of crop failure, or for domestic consumption (Al-Jabiri 1981; Muhammad 1977), but production is generally not compatible with cultivation as it requires mobility, and the best grazing period coincides with the spring harvest (AOAD 1983b). Cattle are generally only kept by farmers in high rainfall and more settled areas.

Free grazing of the common range is supplemented with cereal stubbles and fallow. In the summer and for part of the winter, animals must be fed supplementary feed, the cost of which forces many pastoralists to sell their lambs early (AOAD 1983b). Farmers grow their own barley for feed, although it is not widely or regularly used as green fodder, and tend to have smaller flocks than the semi-nomadic pastoralists, who raise their flocks on the open range as do farmers with large flocks (Khalaf 1984). The expansion of mechanized cereal farming has reduced the available natural forage. Livestock numbers have increased, so uncultivable shallow and steep soils and areas of sub-marginal rainfall are grazed more intensively and feed has become the main limitation to livestock production (AOAD 1983a).

Since the 1950s, policies have favored wheat production but more recently barley cultivation has increased, particularly in marginal areas of the northern rainfed zone, to meet the need for livestock feed caused by increased demand for meat and dairy products (AOAD 1984). This increased cultivation has not affected the pattern of land use or the structure, organization, and intensity of production, and extensive cultivation techniques will continue to limit output in the rainfed zone.

An important feature of rainfed agriculture in the plains is the lack of adaptation of the patterns of land use to different areas. For example, a grain/fallow rotation is widely practised in both wet and dry areas. A very small proportion of the wetter areas under winter grain is utilized in a rotation that includes legumes such as chickpeas and lentils. Even smaller proportions of the wetter areas are utilized in rotation with oilseed and green fodder crops (AOAD 1983a) or with summer crops. As farmers appear to be aware of alternatives to cereal/fallow rotations, their limited areas may be due to socioeconomic factors.

There is a dichotomy between the mountains and the plains in cropping patterns, crop-livestock interactions, degree of mechanization, and land/labor ratios. The least land per capita is found in the mountainous governorates, where crop-livestock interaction is greater and there are more labor-intensive activities than in the rest of the rainfed zone. Level land in mountain valleys is cultivated in summer whereas the gentler slopes are cultivated in winter. Where land is limited, it is terraced; sharper inclines are utilized for fruit and vines and land at high altitudes is utilized for pasture, with small plots being cultivated by livestock breeders during summer (Khisbak 1972). Valley land is cultivated under irrigation during the summer. Some farmers migrate to the higher altitudes after the spring harvest where they cultivate small plots and tend their animals and poorer farmers practice woodcutting in the more rugged and remote areas where trees still survive. Summer crops, particularly tobacco, are the most commercially valuable, although there has also been recent expansion of sunflower cultivation. Horticulture and orchard farming in the valleys require summer irrigation and products are usually sold at the small local markets. Periods of oversupply and a depressed market, particularly for grapes, have led farmers to abandon orchard farming (Khisbak 1972; Khammo 1977). However, whenever water has been available and markets accessible, commercial production has prevailed and farmers have concentrated on summer crop production (Khisbak 1972).

In the rainfed plains, powerful external economic and technological factors have affected the stability of patterns of land use, resulting in rapid soil erosion, even in higher rainfall areas. Overgrazing and cutting trees and shrubs for firewood have led to rapid deterioration of the range in recent decades. Also, the numbers of sheep have increased, and the conditions and intensity of grazing have changed with modern transport, mechanical provision of water, population pressure, and the settlement of nomads (AOAD 1984; Thalen 1977). Soil erosion also increases construction and maintenance costs for flood control, water storage, and irrigation structures.

Mechanization is usually limited to tillage and harvesting, and chemical fertilizers are not used on grain crops. Cultural practices are traditional and no guidance is given to farmers on the use of machines and implements for various operations or techniques of dry farming (AOAD 1984).

There is a dearth of research on machinery and implements suitable for Iraqi conditions (Khammo 1977) and very little attention has been given to soil conservation in the rainfed zone. But research has recently been started into the feasibility of instituting a more stable pattern of land use in the 200–300 mm rainfall belt, and mixed farming in parts of Nineveh governorate with average annual rainfall above 250 mm (AOAD 1984).

Land Tenure Reform: Fragmentation and Capital Dependence

Prior to the 1958 land reform, land was highly unequally distributed; there were 162,000 holdings of less than 7.5 ha with a total area under 300,000 ha. Under the reform and subsequent revisions the very large, semi-feudal estates were broken up and ceilings set on landownership. In the rainfed region 1.71 million ha, or 51% of land in agricultural holdings, were redistributed to eligible beneficiaries by 1979 (CSO 1985a) but former large landowners retained about 850,000 ha or 41% of the total landholdings, probably including the more productive land, in holdings of over 250 ha (AOAD 1983b). In Iraq as a whole, 3.5 million ha of land not affected by the reform remain unequally distributed. Approximately 1.9 million ha are held by about 33,500 holders each with 30 ha or more, while the other 1.6 million ha are held by nearly 300,000 holders.

The growth in the numbers and area of small holdings was not accompanied by an appreciable change in the pattern of land use. With production increasingly reliant on mechanization, the loss of economies of scale may have made fragmentation a more serious problem than the figures suggest.

Since 1959, fragmentation has occurred as a result of inheritance, sale, and the implementation of the reform. In a village in a marginal rainfed zone in the southern part of Nineveh governorate, the unit of land distribution was 7.75 ha including 0.5 ha of flow irrigated land on the Tigris and Zab riverbanks (Khalaf 1984). In the more settled and densely populated areas of higher rainfall, including the mountain valleys, fragmentation probably resulted from the inheritance of established land rights, large-scale population transfers, and land redistribution measures in 1975 in the Kurdish region. In the less densely populated Jazira, land was distributed in units of approximately 25 ha. These extensively cultivated lands are heavily reliant on machinery which is usually privately owned and operated over a large area. The farmer is therefore dependent on machinery hire, while labor requirements are seasonal (e.g., sowing and roughage collection) and very small with existing practices. The reform laws set limits on shares in sharecropping arrangements, but emphasized farm labor while neglecting the growing trend towards mechanization. Contracts between farmers and machinery owners appear to have become largely monetized, thereby reducing the risks for machinery owners and increasing the risks for farmers (Khammo 1977).

Recently, the restrictions on large-scale private farming have been relaxed and land which had been leased in small plots and land repossessed from agrarian reform beneficiaries is being leased in plots of over 250 ha (Springborg 1986) to individuals and companies at below market rents for up to 20 years. No figures are available so the recent expansion in cereal cultivation cannot be taken as proof of the re-emergence of large estates (see Springborg 1986).

Farmers' inputs seem to have increased, but incomes fluctuate from season to season, especially in the areas of lower and more variable rainfall. This fluctuation is increased by Iraq's oil-based economy, the government's fixed price policy, and the urban bias to its statutory monopoly of grain marketing (see Mahdee 1982).

Farmers now seek additional off-farm employment, leaving their families to operate the farm. Some have left farming altogether and either rent their plots or have returned them to the agrarian reform authorities. The 1977 census shows that only 52% of economically active rural males were engaged in agriculture, and that the proportion of non-agricultural income in total rural incomes was increasing (Al-Qaddu 1986). Khalaf (1984) noted that about half of the agrarian reform beneficiaries in the village he studied had moved to alternative occupations.

As Jaubert (1985) has noted for Syria, off-farm employment turns farming into a process of mining the land which is more effective in large-scale mechanized operations. The authorities have established institutions to combine small individual holdings and carry out larger scale operations. Cooperatives and government machinery rental stations (GMRS) served as channels for agricultural extension and other services, and for subsidized inputs.

Cooperative membership, which is compulsory for agrarian reform beneficiaries and open to others, grew most rapidly in the 1970s in spite of a reputation for inefficiency in previous state-sponsored cooperatives (Singh 1970). They were encouraged to engage in directly productive activities funded by government investment and received institutional agricultural credit. But after 1979 cooperatives began to be liquidated and machinery was sold (Springborg 1986; cf. Ibrahim 1979 on arguments preceding the formal policy change).

Population and Labor Force Structure and Trends

Iraq's population has grown continuously for more than a century and has changed from predominantly nomadic or semi-nomadic, to a rural agricultural society, to mainly urban (see Hasan 1965).

Between 1947–57 and 1965–77 the annual population growth rate increased from 2.7 to 3.4%. Rural population growth during the same periods fell from 1.9 to 0.8%.

In the 1977 census Iraq's population is given as 12 million, 3.2 million or 26.6% of whom were in Baghdad governorate, mostly in the capital city itself. The rural population accounted for 36% of the total but, excluding Baghdad governorate, this proportion was 46%. The population of the

Table 2. Population indicators 1973–75.

Rates per thousand	1973/74			1974/75		
	Urban	Rural	Overall	Urban	Rural	Overall
Crude birth rate	41.0	47.5	43.6	38.2	46.5	41.6
Crude death rate	9.4	13.3	11.1	8.8	12.2	10.1
Rate of natural increase	31.6	34.2	32.5	29.4	34.3	31.5
Net migration rate	11.4	−20.2	−	10.9	25.1	−
Growth rate	43.0	14.0	32.5	40.3	9.2	31.5

Source: CSO (1976b).

rainfed zone, estimated from the population of the five northern governorates, was 3.1 million, 46% of whom were in rural settlements. But the proportion of the country's total population resident in these governorates did not change from 1947 to 1977, despite an increase of over 10% in the share of Baghdad governorate.

A sample survey conducted during 1973–75 showed that the natural population growth rate was around 3.2% per annum. While rural population growth was higher than the annual rate derived from the 1965 and 1977 censuses, migration rates may have increased rapidly in the immediate aftermath of the 1970s oil boom. A summary of the survey's results is presented in Table 2.

As the proportion of the population in rural areas declines, the structure of that population and its labor force also changes. In 1977, 31% of the total urban male population were aged 20–44 compared to only 25% in rural areas, but in 1965 these figures were 29 and 26%, respectively. Of all agricultural workers in 1977, 42% were aged 20–44 with 71% in other occupations, although these figures are partly affected by the classification of farmers doing compulsory military service as employed in the service sector. Of those in agriculture, 41% were over 44 compared with 17% outside agriculture.

Agricultural employment, particularly of males, has declined (Table 3). The 1977 population census reports that agriculture accounted for only 52% of total male employment in rural areas, construction accounted for 11%, manufacturing and transport for 4% each, and trade for 2%. The service sector employed 20% of all economically active males, a large proportion of whom were probably engaged in compulsory military service.

There is a high proportion of females in the agricultural labor force; 41% in 1971 (CSO 1973b) and 37% (compared to 9% of the non-agricultural labor force) in 1977 (CSO 1987b). Since 1977 female participation rates have probably increased, possibly increasing the supply of agricultural labor, although the greatest scope for greater female participation is in urban occupations, which may also increase rural-urban migration. A long-term consequence of higher female participation may be a decline in fertility but the government is encouraging high fertility with financial rewards. In

Table 3. Employment in agriculture, 1971–77 (×1,000).

	1971			1977		
	Male	Female	Total	Male	Female	Total
	Whole country					
Permanent[1]	999	783	1,781	588	57	645
Seasonal[2]	255	74	329	–	296	296
Total	1,254	857	2,110	588	353	941
	Rainfed zone					
Permanent[1]	344	232	576			
Seasonal[2]	140	33	173			
Total	484	265	749			

Source: CSO (1973b; 1978b).
[1] Working not less than half the normal full-time work on the holding.
[2] Part-time work except by family not counted in 1977; procedures may differ for the two censuses.

agriculture, a rigid sexual division of labor may limit possibilities for females to replace male emigrants.

Detailed breakdowns of population and labor statistics for the rainfed zone or other areas are not readily available but we can infer, from available statistics and field observations (Al-Jabiri 1981; Al-Jeborey 1981; Salman 1980; Al-Deywachi et al. 1979; 1981; Al-Qaddu 1986; Khalaf 1984), that the rainfed areas have followed the general trends. However, the pace of urbanization in the northern governorates has been slower than elsewhere and rural-urban migration has been to urban centers within the region rather than to Baghdad.

Within the northern region, a large proportion of the population of the mountain provinces was in rural settlements rather than the foothills and marginal areas, despite the military conflict. This probably reflects a relatively weak urban pull, less severe inequities of the old land tenure system, slower and less mechanization, less commercialization, more stable yields with traditional farm practices, and greater possibilities for intensification within the traditional system. Additionally, political, national, and ethnic factors may have contributed to this pattern.

Since the mid-1970s, the counter-insurgency policy has resulted in the loss of cultivable and pasture land, increased rural population pressure in the remaining areas, and restricted access to farmland and pasture. In parts of northern Iraq, some of the administrative centers have become largely non-agricultural (Al-Deywachi et al. 1981). Thus, the labor supply may differ widely between the Kurdish mountain areas and the plains. Seasonal peaks in labor demand in mountain areas could be less pronounced than in the drier plains but output and employment prospects cannot be improved in this area due to a lack of political stability; investment in water development, terracing, afforestation, and pasture improvement, extension of infrastructure and services; and rehabilitation of village communities.

In the rest of the country, including the rainfed zone, the oil boom led to an acute urban labor shortage, which probably accelerated rural-urban migration and attracted foreign labor (Birks 1986). Guest workers are employed in all economic sectors, including agriculture, and represent a heavy drain on the balance of payments. One report estimated 200,000 guest workers engaged in Iraqi agriculture in 1982 (Bergmann 1984) although it is unlikely that there was a large proportion in the rainfed zone. In the early 1980s there were up to 1.5 million Egyptians in Iraq (El-Solh 1985) and they formed the majority of non-national labor. The number of foreign workers is probably still very high due to the heavy military demand on national manpower.

The Structure of Rural Employment and the Movement of Labor Out of Agriculture

A number of villages have been studied in the rainfed zone, two of which are near Mosul, just outside the marginal zone.

Al-Mahallabiyya lies in the Jazira, 40 km west of Mosul and, at the time of the study, had a population of 3,000. It is an administrative center for 37 other satellite settlements (Al-Deywachi et al. 1981). Agricultural activity in this village is almost entirely dependent on rainfall and is highly seasonal. Only one third of household heads were employed in agriculture, and two thirds of those in non-agricultural occupations had originally been employed in agriculture. Approximately half the household heads worked in Mosul and commuted to the city at least once a week. Urban employment was mainly in services and trade. The most common reason given for leaving agriculture was low income but many non-farmers continue to own land and a considerable number wish to return to farming, presumably if income or other conditions improve.

Many city employees continue to live in the village, maybe because city rents are too high for new migrants. Alternatively, many villagers may not have abandoned agriculture altogether or family members have remained to work the land or claim their right to it. Some may have shared risks with contractors and probably accepted lower incomes or rented the land, illegally if it was agrarian reform land.

The village of Al-Fadhiliyya, 12 km east of Mosul, with a population of 1,000–1,200, had more irrigation water than Al-Mahallabiyya, and was not a center of local administration (Al-Deywachi et al. 1979). Despite its proximity to the city, this village had a proportionately higher level of agricultural employment (54% of household heads) than Al-Mahallabiyya and a greater division between rural and urban employment.

In both villages urban employment, at least over short distances, does not involve permanent migration to the city and a complete break with the village as in the past (see Mahdee 1982), possibly due to more widespread landownership after the 1958 reform, residence requirements on reform beneficiaries, or high urban rents. Evidence from the south of Iraq confirms this new pattern (Salman 1980). Also, in villages which have more varied agriculture or are close to major towns, land may appreciate, encouraging residents to maintain their village residences.

In the village of Ash-Shak, landownership does not seem to prevent migration and farmers readily abandon land newly acquired under the reform laws in favor of urban employment (see Khalaf 1984). This village lies at the confluence of the Lesser Zab and Tigris rivers in the marginal zone in the south of Niniveh governorate and has limited rainfed and irrigable land. Few of those living in Ash-Shak work outside the village. Thus, the move out of agriculture usually involves a move out of the village. Possibly, the more inaccessible a village, the more likely is the peasantry to abandon the land in search of urban opportunities.

Large villages tend to be atypical. Both Al-Mahallabiyya and Ash-Shak are subadministrative centers and, as such, are strictly classified as urban. They are growing centers, attracting migrants from outlying rural areas while losing city-bound migrants.

Another factor affecting migration is education. It opens up urban employment opportunities which tend to have higher incomes and social status than agricultural employment. Education may have a greater effect close to major towns where educational services are well developed. Also, education is stratum-, gender-, and age-biased and tends to draw the most active and best potential recipients of improved cultural practices out of agriculture.

In Al-Fadhiliyya, the movement of younger, more educated people out of agriculture may account for the observation that 20% of household heads below 40 years old were farmers compared with 81% of those 40 years old and above. The illiteracy rate was higher among farming household heads and females than among non-farmers and males, and only about 1 in 400 rural adults had graduated from secondary schools (CS0 1987b).

Since the 1970s enrollment at secondary schools, universities, and vocational and teacher training schools, and students studying abroad have increased rapidly. But education and training are oriented towards urban, non-agricultural skills and the growth of vocational training in agriculture is comparatively slow (CSO 1985a).

The Impact of Urban Employment and Rural-Urban Disparities on Agriculture

After 1974 state consumption and investment grew rapidly for almost a decade. State and private employment and incomes increased but agricultural incomes were constrained by price controls, large-scale imports, resource constraints, and low productivity (Mahdee 1982). The average number of workers in the construction sector trebled from 64,000 to 186,000 between

1973 and 1978 with similar trends in the trade, transport, and service sectors and to some extent in industry. Immediately after the increases in oil revenues, state employment, including the armed forces, accounted for approximately 40% of all non-agricultural employment and increased annually at over 9% during 1973–76. In 1976 employment in government departments and state enterprises, excluding the armed forces, totalled 527,000 with 46,000 employed in the agricultural sector (CSO 1976a).

In addition to attractive salaries and benefits, government employment is more secure than farming and wages in the state sector have also generally been higher than in other paid employment (Mahdi 1977), although this may have changed with general urban labor shortages. In the private sector, wages of unskilled and semi-skilled workers employed in construction doubled between 1973 and 1975 while those of skilled workers quadrupled (CSO 1976a).

Urban wages continued to increase despite the influx of guest workers at the end of the 1970s. After 1981, however, nominal wage increases slowed and real wages declined. From 1981 to 1984, average wages in construction for the public sector increased nominally by 11%, while the (probably underestimated) consumer price index increased by 48%.

After 1973, agricultural incomes did not keep up with the rise in urban wages. Wholesale prices of meat, grains and pulses, and vegetables increased by 52, 21, and 58%, respectively, between 1973 and 1975, but output remained static (CSO 1976a; 1982a).

The disparity between agricultural and urban non-agricultural incomes was considerable by the late 1970s. A random sample study in the rainfed zone in 1978 showed that average family income from all sources totalled 552 IQD in a 200–350 mm rainfall zone, 647 IQD in an intermediate 350–500 mm zone, and 604 IQD in a zone with 500 mm and above (Al-Najafi 1979). During the same year the average wage of employees in the construction of public sector buildings and utilities was almost double the average farming family's income (CSO 1978a). The survey reported that 40–61% of average farm family income in the different zones was derived from non-agricultural sources.

Urban pull probably affected mostly skilled, young and male workers, and may have generated a new staggered pattern of rural-urban migration rather than the traditional pattern of a permanent family move preceded by deteriorating rural conditions. In other underdeveloped countries urban pull migration appears to attract the wealthier, more dynamic and successful farmers (Lipton 1982), resulting in endemic low productivity and rural poverty with no labor market adjustment.

Rural-urban disparities largely explain the massive decline in agricultural employment from 1971 to 1977. An AOAD team reports a decline in agricultural manpower of 42.9% between 1975 and 1977 alone (AOAD 1983b). The inflow of guest workers towards the end of the 1970s may only have slowed the outflow of labor from agriculture, given the extent of the growth of urban employment and income initially and the extremely heavy military demand on manpower since 1980.

Increased urban, non-agricultural employment has resulted in very serious agricultural labor shortages and the growing dependence on non-agricultural income makes a return to full-time farming less likely.

Seasonal migration supports the prevalent traditional system of land use with its rigidities, low factor productivities, and highly seasonal labor demands. Substitution of capital for labor can reduce peak labor demand (Khammo 1977) because wages are high and capital costs are low.

In northern parts of the rainfed zone, a labor shortage has severely restricted the cultivation of legumes normally produced in plains with assured rainfall. While wheat and barley are harvested mechanically in the northern plains, local legume varieties are unsuitable for mechanical harvesting. Labor shortages at harvest prevent the expansion of legume cultivation, despite the relatively higher output value per hectare (AOAD 1982).

A more general off-peak labor shortage may have been partly responsible for the deterioration of orchards. Springborg (1986) showed a very rapid growth in fruit production, but the increase in index numbers he cites can be attributed largely to melons, whose output increased rapidly during the 1980s; increases in orchard farming were modest. Expansion was greater in capital-intensive poultry farms and plasticulture and between 1980 and 1985 loans from the Agricultural Cooperative Bank for poultry enterprises were nearly 3.5 times those for orchards (CSO 1982a; 1985a). A proportion of the finance for orchards was probably for purchasing land, and some must have been used to develop new luxury products at very high capital costs (Springborg 1986). A large proportion of the foreign workers in agriculture are probably employed in poultry and plasticulture rather than low productivity dry farming. Apart from managerial and technical positions, these ventures rely on mobile wage labor close to towns. The capital-intensive ventures are commercial, have a high import content, and rely on heavy government subsidy and easy credit. Nevertheless, commercial success appears to depend on high prices and the long-term viability of some of these ventures is questionable.

Any movement of labor to the modern, high productivity, capital-intensive agricultural enterprises may draw labor away from traditional farming if it involves relocation away from the family farm. The continued expansion of capital-intensive agriculture may well accelerate the marginalization and decline of the traditional sector which will, nevertheless, continue to cover large areas and employ most of the agricultural labor force, including almost all family labor.

Agricultural Mechanization

Since 1973, the number of tractors in use has increased rapidly (Table 4).

Table 4. Numbers of tractors and combine harvesters in Iraqi agriculture.

Sector	Tractors					Combines				
	1970	1973	1979	1982	1984	1970	1973	1979	1982	1984
Private	n.a.	9,975	14,566	27,507	34,721	n.a.	1,706	946	1,908	2,053
Public	n.a.	1,483	5,492	2,449	1,284	n.a.	781	2,497	865	134
Total	10,400	11,458	20,058	29,956	36,005	2,280	2,487	3,443	2,773	2,187

Source: CSO (1982a; 1985a); Khammo (1977); Al-Rawi and Al-Khaffaf (1971).
n.a.: not available.

Private ownership of tractors and combines has also increased with an increase of over 40–fold in Agricultural Cooperative Bank credit for agricultural machinery and implements between 1978 and 1980 (CSO 1985a).

In the 1970s tractors and combines were utilized three times more intensively in the private than in the public sector (Khammo 1977). The rapid increase in available tractor power after 1979 was associated with an increase in the cultivated area. There was increased use of tractors with implements other than the plow, deeper and more timely plowing, plowing of fallow fields, and increased use of tractors for transport.

Small farmers frequently procured machinery services through the cooperative societies (Al-Deywachi et al. 1979). During the 1970s, machinery services became more widely available from the public and cooperative sectors at lower, effectively subsidized rents. In some areas, public sector machinery was more widely used than private services (Singh et al. 1978).

With rural migration, withdrawal of public machinery services, and the recent rapid growth in private ownership, custom services also appear to have expanded very rapidly. Custom operators of tractors were originally from small farming families while larger combines were operated by urban-based merchants or mechanics. Operators traveled widely to provide services to large numbers of small landholders, mostly beneficiaries of the land reforms. They charged for their services in cash or in kind at rates varying according to availability and field conditions (Khammo 1977).

In a recent sample survey (Al-Qaddu 1986) custom hire of machinery accounted for 26% of total rural family income compared with only 19% from crop production, 13% from livestock products, and 40% from the hiring out of labor, presumably mainly outside the agricultural sector. This suggests that mechanization and custom hire services are almost entirely labor-saving and labor-replacing and rather than encouraging modern intensive farming methods, mechanization appears to be associated with a decline in the relative shares of labor and land in agricultural production.

The closure of government machinery rental stations after 1979 would have resulted in a cost increase to many farmers. While total machinery costs, including depreciation, must have increased very rapidly, the marginal costs of tractor use in remote areas may have declined. The capital cost of tractors and combines depends on the repayment requirements of the

Agricultural Cooperative Bank, the rate of inflation, and the rate of depreci-ation of the goods. The relaxation of lending policy suggests official accept-ance of a lower rate of recovery of outstanding credit and that the subsidy for the use of machinery has shifted to the private sector instead of the disbanded government machinery rental stations. Policies may change as a result of wartime austerity measures or in response to a perceived need to increase domestic production.

The existing machinery provision policy has probably led to greater differ-entiation within the rainfed rural areas. On grain-producing land, access to harvesting machines by small farmers may have become more costly and difficult, although access to and cost of plowing are probably easier and cheaper. On more intensive farms machinery services have probably im-proved. With higher capital intensity, unequal distribution of the more pro-ductive farmland is likely to become an increasingly important source of differentiation within agriculture.

The Emergence and Stability of a New Farm Structure

Rural-urban migration, particularly by small landowners, has been associated with a new kind of absentee landownership and sharecropping in which small ex-farmers rather than large landowners rent their land. Tenancy nowadays "becomes more and more an instrument for wealthy farmers to balance and adjust farm size to economic requirements ... For the weaker cultivator, leasing out his land is the first step away from farming as a full-time occupa-tion" (Bergmann 1984). The impact of economies of scale associated with mechanization and of differential access to capital resources has been com-pounded by powerful urban pull factors and by the change in government policy in favor of large private landholdings. General observations point to an increasing land concentration (Springborg 1986). Decision-making has probably shifted from absentee small landowners to machinery owners, es-pecially with the increasing reliance on privately owned combines.

The tendency towards a new farm structure dominated by large holdings rests, among other things, upon unstable factors which are affected by current abnormal conditions and a volatile foreign exchange position. Changes in institutions, policies, and ecology may limit the trend towards large holdings, at least in certain regions and conditions. Therefore, policies and techno-logical research should continue to assume a composite farm structure.

With large-scale migration altering household decision-making patterns, the withdrawal of resources from the state and cooperative sectors, and the merging of cooperatives into larger units (see CSO 1985a), it is likely that cultivators' ability to rely on joint, cooperatively-procured combine harvest-ing services has also been reduced.

225

Conclusion

The oil boom appears to have resulted in a rapid and selective movement of labor away from agriculture but probably not all of this movement is permanent. The future size and structure of the labor force will depend on macroeconomic, policy, and international factors. Projections for labor requirements and supplies must be made separately for the different sub-regions, particularly the Kurdish mountain subregion which suffers from a politically-induced drain on its agricultural manpower more than the rest of the rainfed zone.

Apart from factors exogenous to agriculture, the labor market will depend upon the interaction of numerous factors at the village and household levels under a variety of conditions. But there is only limited information on the behavior of farming households. It is unclear whether and how large farms differ from their smaller counterparts in land utilization, or whether the growing role of custom contractors is leading to any change in risk sharing and management roles. It would be useful to have some precise information on the structure of the machinery hire service and on the distribution of capital ownership in agriculture.

It is not evident why farming families in the rainfed zone do not keep large numbers of livestock in a mixed enterprise. If high meat prices continue, a pattern of mixed farming may evolve and encourage a rotation of forage and grain crops. Unfortunately, potential rural innovators have probably been attracted by higher urban incomes and discouraged by the urban-biased fixed grain price policy. Research in the rainfed zone on the technical feasibility of a grain/green fodder rotation must take account of socioeconomic changes required for the adoption by small farmers of activities and patterns radically different from those existing at present.

Higher productivity may be technically achievable under a system of large-scale, specialized farming. However, this would require intensive use of wage labor for which there is no tradition in Iraqi agriculture. Wage labor is unlikely to become more available except at a cost too high for the uncertain conditions of Iraq's rainfed zone.

Evidence from community studies suggests that migrant labor is not replaced by family labor either through higher participation rates or through longer work hours. This needs to be supported by representative village surveys before any broad generalization can be made. Reliable and specific information is required on year-round and seasonal labor requirements, including skilled and managerial, under the existing patterns of land use. Such information should include details of how, in what institutional framework, and by what categories of labor these requirements are currently being met.

Because of low income and high risk, rainfed cultivation only continues to be widely practiced where farm income is supplemented by other sources and where rental and profit income prevail. Elsewhere, agriculture has declined as labor has become more mobile. In the better areas, the development of supplementary resources could reduce risk and increase income from dry farming. In the poor marginal areas, mechanized cereal cultivation may best be replaced by re-establishment and managed grazing of perennial natural cover. Traditional landholding and social organization have tended to break down under population and commercial pressures, and new appropriate institutions and policies have yet to emerge.

Trends in tenure and contractural arrangements need further study, in particular their relationship to farming patterns in each of the different subregions of the rainfed zone. It is also increasingly important to discern any changes in cultural practices, land productivity, and the use of inputs over different scale operations and tenure patterns.

National aggregate data cannot be used to relate land tenure, farm size, and availability of labor to the patterns of cultivation. Access to the now dated records of the 1971 Agricultural Census and to records of local Agricultural and Agrarian Reform offices would be useful. Detailed information from population censuses would also be useful in exploring changes in migration and rural settlement patterns and in studying labor effects in small rural settlements and further exploring non-agricultural employment in larger settlements.

References

Al-Ani, K.A.M. 1977. (Modern Iraq Encyclopedia.) In Arabic. Baghdad.

Al-Deywachi, Abi S. et al. 1979. (The Economic Future of Villages in the Vicinity of the City of Mosul: A Field Study of the Village of Al-Fadhiliyya.) In Arabic. Tanmiyat Al-Rafidain 1: 17–33.

Al-Deywachi, Abi S. et al. 1981. (Economic and Social Conditions in Rural Iraq: A Field Study of Mahallabiyya Nahiya.) In Arabic. Tanmiyat Al-Rafidain 5: 73–94.

Al-Jabiri, K.F. 1981. Stability and Social Change in Yezidi Society. PhD thesis, University of Oxford.

Al-Jeborey, Diham A.M. 1981. Agrarian Reform with Special Reference to Ishan, a Village in the Middle Euphrates Region of Iraq. PhD thesis, University of Newcastle-upon-Tyne.

Al-Najafi, Salem T. 1979. (The Economics of Rainfed Agricultural Production in Iraq.) In Arabic. Tanmiyat Al-Rafidain 1: 5–15.

Al-Qaddu, Badie J. 1986. (The Condition of Rural Women and its Effect on their Economic Decision-Making, A Field Study.) In Arabic. Tanmiyat Al-Rafidain 17: 145–164.

Al-Rawi, H. and Al-Khaffaf, A. 1971. (Views on Agricultural Mechanization.) In Arabic. Najaf, Iraq: Al Qadha Press.

AOAD (Arab Organization for Agricultural Development). 1982. (A Study of Mechanization of Harvesting Low Crops in the Iraqi Republic – First Stage.) In Arabic. Khartoum: AOAD.

AOAD (Arab Organization for Agricultural Development). 1983a. (Development of Production and Marketing of Animal Products in the Rangelands of the Iraqi Republic.) In Arabic. Khartoum: AOAD.

AOAD (Arab Organization for Agricultural Development). 1983b. (Arab Agricultural Policies. Vol. XI. Agricultural Policies in the Iraqi Republic.) In Arabic. Khartoum: AOAD.

AOAD (Arab Organization for Agricultural Development). 1984. (A Study of Technical and

Economic Feasibility of Cultivating Grain Crops and Grazing Shrubs in Marginal Lands in the Iraqi Republic.) In Arabic. Khartoum: AOAD.

Bergmann, Theodor. 1984. Mechanisation and Agricultural Development, 1. General Report. Gottingen: Edition Herodot.

Birks, J.S. 1986. The Demographic Challenge in the Arab Gulf. Arab Affairs 1 (1): 72–86.

CSO (Central Statistical Organization). 1949a-86a. Annual Abstracts of Statistics. Baghdad: CSO.

CSO (Central Statistical Organization). 1973b. (Results of 1971 Census of Agriculture, Parts 1 and 2.) In Arabic. Baghdad: CSO.

CSO (Central Statistical Organization). 1976b. (Survey Results of Vital Events in Iraq, 1974–75.) In Arabic. Baghdad: CSO.

CSO (Central Statistical Organization). 1978b. (Results of the General Population Census of 1977). In Arabic. Baghdad: CSO.

Clawson, Marion, Landsberg, Hans H., and Alexander, Lyle T. 1971. The Agricultural Potential of the Middle East. New York: Elsevier.

El-Solh, Camillia Fawzi. 1985. Migration and the Selectivity of Change: Egyptian Peasant Women in Iraq. Peuples Mediterraneens 31–32: 243–258.

Haidar, Salih. 1942. Land Problems of Iraq. PhD thesis, University of London.

Hasan, M.S. 1965. (The Economic Development of Iraq 1864–1958.) In Arabic. Vol. 1. Sidon: Al-Maktaba Al-Asriya.

Ibrahim, I. 1979. (The Cooperative Movement.) In Arabic. Talk given at the Eighth Agricultural Conference, Al-Jumhouriyya, Iraq. Baghdad Daily, 27 December 1979, page 3.

Jaubert, Ronald. 1985. Sedentary Farming in the Dry Areas of Syria: Problems of Development and the Role of Agricultural Research. Beirut: Dirasat Arabiyya 21 (9): 42–69.

Khalaf, S.H. 1984. Social Change and Rural Development in a North Iraqi Village: A Study of the Role of the Government and Popular Organisations, 1958–1981. PhD thesis, University of Keele.

Khammo, Awshalim L. 1977. The Role of Mechanisation in the Development of Agriculture in Iraq. PhD thesis, University of Leeds.

Khisbak, Shakir. 1972. (The Kurds.) In Arabic. Baghdad: Shafiq Press.

Lipton, Michael. 1982. Rural Development and the Retention of the Rural Population in the Countryside of Developing Countries. Canadian Journal of Development Studies (1): 11–37.

Mahdee, K.A. 1982. Agriculture and Agrarian Reform in Iraq with Special Reference to the Period 1952–1976: A Study of Aspects of a Rentier State and of its Impact on the Economy. PhD thesis, University of Birmingham.

Mahdi, F.A. 1977. (Economic Development and Planning in Iraq, 1960–1970.) In Arabic. Beirut: Dar al-Talia'a.

MAAR (Ministry of Agriculture and Agrarian Reform). 1976. (Agriculture in Iraq 1963–1973: A Statistical Pocket Book.) In Arabic. Baghdad: MAAR.

Muhammad, Khalil Ismael. 1977. (The Qadha of Khaniqin: A Study in Population Geography.) In Arabic. Baghdad: Al-Ani Press.

Salman, Abed Ali. 1980. (Rural Society in Iraq.) In Arabic. Baghdad: Dar Al-Rashid Publishing.

Singh, G.D. 1970. Study of Agricultural Administration and Important Agricultural Programmes in Babylon. UN Assistance in Development Planning and Execution, Baghdad. Mimeo.

Singh, Ranjit, Mohammed, Z.H. and Ghanem, M. 1978. Agricultural Cooperative Societies Contribution in Promoting Adoption of Improved Farming Practices. Mesopotamia Journal of Agriculture 13: 85–20.

Springborg, Robert. 1986. Infitah, Agrarian Transformation and Elite Consolidation in Contemporary Iraq. Middle East Journal 40 (1).

Thalen, D.C.P. 1977. Contribution to the Discussion of Experience in the Middle East. Paper presented at a Royal Society meeting, 17–18 March 1976. in Resource Development in Semi-Arid Lands. London: Royal Society.

Warriner, Doreen. 1948. Land and Poverty in the Middle East. London: RITA.

Warriner, Doreen. 1957. Land Reform and Development in the Middle East. London: RITA.

11. Agricultural Labor and Technological Change in the Yemen Arab Republic

RICHARD TUTWILER

The International Center for Agricultural Research in the Dry Areas (ICARDA), P.O. Box 5466, Aleppo, Syria

Introduction

In the early 1970s, rising petroleum prices brought new wealth to the major oil-exporting Arab states. This funded ambitious development projects, but these were impeded by an acute shortage of manpower. The Yemen Arab Republic (YAR), with a predominantly agricultural population estimated at 6–7.5 million, soon became their principal labor supplier and by 1975, 24–42% of Yemen's male labor force was working abroad.

Migration has had major implications for the Yemeni economy, which has received hundreds of millions of dollars in migrant remittances and repatriated savings. Rural disposable incomes have risen dramatically and money is now widely used. Yemen has become dependent upon imported commodities, chiefly foodstuffs. Remittances have fueled expansion of commerce and construction. These economic changes have challenged established social relations of production and led to differentiation among social groups. If labor export were to end, Yemen would remain among the least developed countries in the world, with a distorted economic structure that has evolved due to migration (Fergany 1980).

Prior to the rise in oil prices, nine out of ten Yemeni households depended on the land for their subsistence. Production was organized by individual households producing for direct consumption. In this paper, I have focused on changes in the system of agricultural production, the agricultural labor force, and the allocation of resources in the production process. Current aggregate agricultural statistics are scarce and often unreliable and sources are frequently contradictory. In contrast, diverse localized studies are generally of good quality. Therefore, the ethnographic literature has been used to texture the discussion of labor allocation and investment decisions at the household level.

Environment and Cultigens

Yemen is the most fertile area of the Arabian peninsula and its complex agriculture is based upon intensive cultivation of cereals, vegetables, and

Dennis Tully (ed.), Labor and Rainfed Agriculture in West Asia and North Africa, 229–251.
© 1990 *ICARDA*.

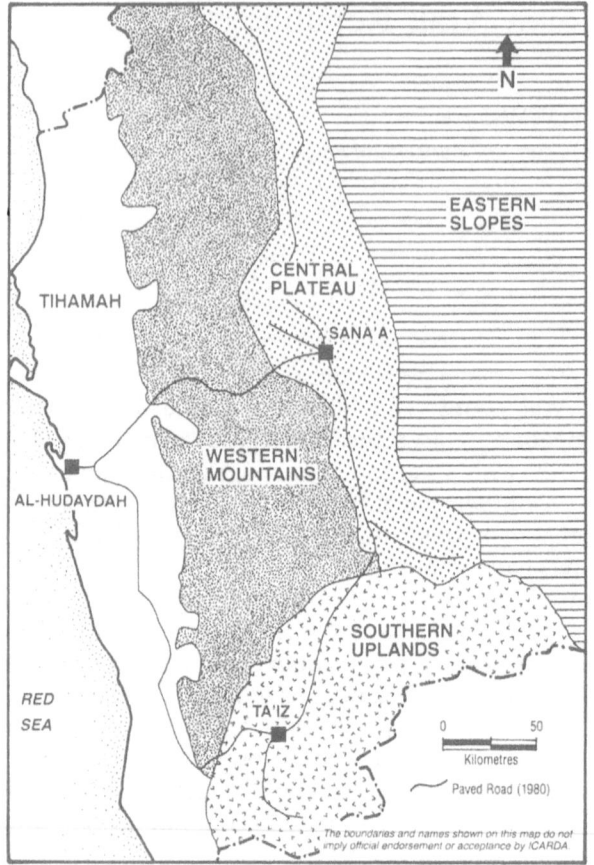

Fig. 1. Agricultural regions in Yemen AR. *Source*: Tutwiler (1987).

horticultural crops. Yemen has always been a predominantly village society and until the revolution in 1962 agrarian technology and socioeconomic processes were subject to solely endogenous forces.

Yemen has five distinct agricultural regions (Figure 1): the Tihamah coastal plain, the Western Mountains of the al-Sirat range, the Central Plateau in the geographic center, the Eastern Slopes which descend into the deserts of interior Arabia, and the Southern Uplands at the southern extension of the mountain and plateau zones. The annual rainfall and topography of each region are shown in Table 1.

The Western Mountains and Southern Uplands are difficult to cultivate because of the steep slopes and relatively few open areas, but they have a favorable climate for crop production. Tropical thermal conditions predominate, with elevation and winds determining local thermal conditions. Light frosts can occur in winter above 2,200 m but sheltered wadis in the mountains can be warmer (Schneider et al. 1981). In the Tihamah and the Eastern Slopes near the desert, daily highs of 36°C can occur throughout the year.

Table 1. Agricultural regions in the Yemen Arab Republic.

Ecological zones	Annual rainfall (mm)	Topography
Tihamah (Coastal Plain) (elevations: 0–400 m)	200–300	Sandy plains with gravel and seasonal water courses
Western Mountains (elevations: 400–3000 m)	400–900	Steep mountains and narrow valleys
Central Plateau (elevations: 1,800–3,000 m)	300–500	Plains, barren hills, and mountains
Eastern Slopes (elevations: 200–1,800 m)	100–200	Alluvial fans and barren ridges
Southern Uplands (elevations: 1,000–3,000 m)	500–1,100	Steep mountains and broad valleys

Source: CPO (1982b); Steffan (1978); Tutwiler et al. (1976).

Thus cultivation of some crops is restricted to certain locations, but thermal variations allow production of different cultigens within relatively short distances.

Rainfall is the most important variable determining the agricultural potential of the different regions. Maxima are in the spring and late summer, and there is a distinctive dry period in the early summer. Rainfall variability is high and is aggravated by topography. Successful harvests occur every year only in the mountains of the west and south with 500–1,100 mm of rain. Below 1,000 m in the mountains and in the Central Plateau, the monsoons are capricious, and there are frequently no spring rains. Therefore, farmers try to maintain sufficient stored grain to feed themselves during periods of dearth. Rainfed cultivation can be supported only intermittently in the Tihamah and almost never in the Eastern Slopes, where annual rainfall rarely exceeds 300 mm.

Spate floods following heavy rainstorms are the primary source of surface water and irrigation water for lowland farmers. Perennial streams in the highlands fluctuate according to rainfall in the catchment areas. Where the water table is sufficiently close to the surface, particularly in the Central Plateau and wadi floors, wells can be dug to irrigate small farms and gardens, but there is a real danger of over-exploiting this fragile resource. In the Tihamah there is the additional problem of saline water intrusion from the Red Sea.

Yemeni farmers have developed a surprising variety of cultigens and cultivation techniques adapted to their ecological circumstances (Bartelink 1974). The most prevalent cereals are native varieties of sorghum, millet, wheat, and barley. Other crops include lentils, peas, onions, radishes, garlic, fenugreek, and sesame. These crops probably have the longest history as Yemeni cultigens (Serjeant 1974). Other vegetables, fruits, and spices were introduced in classical and late pre-Islamic times while tobacco and coffee were probably introduced in the late medieval period, followed shortly after by New World cultigens and the cultivation of *qat* (*Catha edulis*). Other crops, such as maize, cotton, banana, and papaya have been introduced in the past 2 centuries.

Qat is a perennial shrub of contemporary economic importance. It produces mild euphoria when its freshly picked, tender leaves are chewed. It has long been a desirable consumption item in Yemen, but until recently only the wealthy could afford to enjoy it habitually (Weir 1985).

Rainfed Farming Systems

With existing technology, only 6.8% of the YAR's total land area is estimated to be cultivable, with moisture the main limiting factor (FAO 1986a). Of the 1.35 million ha of arable land, 83% (1.12 million ha) is rainfed and 17% (230,000 ha) is irrigable. Farmers have created cultivable surfaces in the mountains which receive most of the rainfall using terraces with stone walls or earthen embankments. It is estimated that 95% of cultivated land is terraced or leveled in some fashion (Kopp 1984). Terraces can be as large as a third of a hectare depending on the natural gradient and contour. In the Western Mountains and Southern Uplands most terraced fields are much smaller, sometimes only a few meters wide, and on some only a single row of grain is grown.

Terraces curb erosion and promote moisture retention, and also effectively use surface water harvested from adjacent steep or rocky land. In estimating labor inputs into cultivation the days spent each year constructing and maintaining terraces and runoff diversions, the principal fixed assets of traditional agriculture, must be included.

Sorghum constitutes perhaps 70% of the cultivated land in these regions, with the remainder under other grains, vegetables, and perennials (especially qat and coffee). Ideally, fields are fallowed about every fourth year, with sheep and goats browsing the stubble. In wetter areas farmers might plant alfalfa for at least one season to increase soil nitrogen and for fodder. In drier areas legumes are usually interplanted with sorghum.

The universal tool is a short-handled pick which is used primarily by men and boys to thin rows of sorghum or millet, loosen and mound soil at the base of plants, and remove weeds. A hand-held sickle is used principally by women to prune sorghum leaves, harvest grain heads, cut the stalks after harvest, and collect fodder for livestock.

These two tools are practically all that is needed for hand cultivation where the use of animal power is precluded by steep slopes and narrow terraces. On broader fields a single-bladed scratch-plow is used, as in dryland farming in other Middle Eastern countries (Harley 1942). One or two cattle, a camel, or a combination of ox and camel are customarily used for traction and a funnel and tube at the rear of the plow is used to apply seed. Draft animals are also used to pull flat boards, sometimes set with iron spikes, for soil scraping and field leveling.

```
Month          March April  May   June  July  Aug.  Sept. Oct.  Nov.  Dec.  Jan.  Feb.

Prevailing     WARMING            HOT          COOLING + HUMID         COLD + DRY
Weather          ≡RAINS≡        CLEAR      ≡≡≡≡RAINS≡≡ ≡≡≡MISTS≡≡≡≡    CLEAR

Seasons        --SHITA'--><----SAYF----><----JAHR----><----KHARIF----><----SURAB----><------SHITA'------

Sarabi crop
(sorghum)      PLOW -SOW---- KAHIF --- KHILFAH --- HAFRAH ---- HARVEST --- UPROOT ---- FERTILIZE
                          FAQUH        PRUNING            CUT, DRY, ETC.

Intercrop
(pulses)                  SOW  -------------  HARVEST

Qiyad crop
(wheat or barley)                              PLOW- SOW ----PLOW---    HARVEST

Jahari crop
(wheat or barley)  >---------------------HARVEST                          PLOW-SOW>

Mawashi
(sheep and goats)  MATING-----  ----------------------BIRTH------------------------  -----WEANING

Labor Demand
(sarabi crop)      <-----------MODERATE------------>  <----HIGH---->   <---------LOW--------->
```

Fig. 2. Annual cycle for grain production in the Highlands. *Source*: Tutwiler (1987); Tutwiler and Carapico (1981). *Note*: *Kahif*, plowing to raise young plants above troughs; *faquh*, removing weak plants; *khilfah*, plowing across *kahif* furrows; *hafrah*, mounding earth around plant bases.

Preparations for planting commence almost immediately after harvest. The field may be plowed to remove stover butts for fuel, or animals may be allowed to graze on stubble or a ratoon, or second growth, to improve soil fertility with their droppings. Good farmers try to plow several times before the next monsoon to conserve soil moisture, discourage weed growth, and mix in animal droppings. Just before the monsoon rains, manure is spread, fields are plowed and leveled, terrace walls are raised to hold rain water, and runoff surfaces and channels are prepared. After each of the early spring rains, terraces should again be plowed to help retain moisture and aerate the soil. When there is sufficient soil moisture and more rain is expected, the farmer will plant.

The majority of highland farmers plant their terraces with sorghum during *al-sayf*, which begins with the spring rains and ends with the period of dry, hot weather between the two monsoons (Figure 2). This is termed the *sarabi* crop because it is harvested during the *al-surab* season. If the spring monsoon is late, farmers who have carefully conserved soil moisture may still sow, hoping for rains in time to sustain maturation of their *sarabi* crop. Otherwise, they plant a lower yielding crop of wheat or barley in October or January, depending upon the *kharif* rains. These secondary crops are called *qiyad* and *jahari*, respectively.

Sorghum and millet (grown in drier and hotter areas than sorghum) mature in 5–6 months whereas wheat and barley can be harvested in 3–4 months. Farmers prefer sorghum because it produces more fresh fodder and stover than millet. *Sarabi* grains are commonly interplanted with pulses which improve soil nitrogen, provide green fodder, and are a nutritious addition to the family diet.

Livestock husbandry is timed to ensure adequate fodder throughout the year. Sheep and goats are mated when natural grazing is available following the spring rains and give birth about the time when sorghum is being pruned and thinned and the second monsoon rains have renewed natural pasture. Lambs and kids are generally weaned after harvest, but possibly earlier if the household wants milk, or later if offspring are being fattened. Cows are serviced in the fall to give birth in spring. Adult cattle are hand-fed dry sorghum stalks wrapped in green fodder for most of the year because natural pasture is inadequate (Tutwiler and Carapico 1981).

Field operations vary according to weather conditions, labor availability, and production objectives. Maximum productivity may depend upon six or eight mid-season plowings or hoeings as well as other laborious practices. High seed rates are commonly used with hand thinning at about 8 weeks after germination to select plants with preferred characteristics and provide green fodder for livestock. Plants are pruned to improve seed head growth during maturation.

The production system in the lowland Tihamah and Eastern Slopes is different due to the low rainfall. Most production and settlement surround the large interior wadi deltas where the spring and late summer floods are diverted for irrigation. Millet is the principal grain in the Tihamah supplemented by white sorghum, while wheat and barley predominate in the Eastern Slopes. The annual production cycle in the lowland wadis is about 2 or 3 months behind that in the highlands due to the nature of the water source. The major planting time for grains in Tihamah and the Eastern Slopes is August/September, with harvests in February (Eh-Basher 1986b).

The arable areas of the lowlands may be divided into land immediately adjacent to watercourses, land which receives regular annual spate floods, and land which is watered only in years with very large spates (Mundy 1984; Tutwiler 1984). The first category covers an extremely limited area but can be intensively cultivated. The second is larger and produces at least one harvest per year, while the third category is very large but can only be cropped about once every 4 years.

Animal husbandry is generally more important in the lowlands than in the highlands. People frequently live in fixed abodes only during 3 or 4 months of spate flooding, resuming a pastoral existence for the rest of the year.

In 1975 population distribution closely reflected rainfall distribution. Between 75 and 85% of the rural population lived in the Western Mountain and Southern Upland regions, with local densities often over 300 persons per km^2 (Steffan 1978). Towns of any consequence are few and far between, indicating a history of subsistence production with few market influences. In the past, each cluster of villages and hamlets constituted a rather discrete economic and social unit. Trading centers were small with a few itinerant peddlers and craftsmen and their families trading in grain and livestock products (Stevenson 1985; Varisco 1984). Only within the past decade has a rudimentary road network begun to link village communities to each other and to Yemen's few urban areas.

Social Organization of Traditional Agriculture

The central institution in Yemeni agriculture has always been the independent producing household. Ideally, a farm household is an extended patrilocal family headed by a married male of the senior generation. Daughters are married into other households but usually retain the right to return to their father's home in the event of marital difficulties or widowhood. Brothers might share in a joint household after their parents' death or until their own sons have reached maturity. This composite kin group offers a labor pool and a level of sharing of productive and domestic tasks which is not possible in the nuclear family and has recognizable economic advantages (Tutwiler and Carapico 1981; Mundy 1979). It is also an insurance against death of a spouse or parent, major illness, or other personal disasters.

Joint ownership of the family land, housing, and instruments of production helps to prevent excessive land fragmentation. The recognition of private property and the inheritance rights of both male and female heirs has encouraged marriages among cousins to keep land intact. This results in a very high percentage of settlement clusters whose members share affiliation to localized descent groups. But fragmentation is inevitable, resulting in ownership of terraces scattered across a large area.

Kindred households of one village often pool labor resources so each household has its fields attended to in turn by a collective labor force. Men of different households usually cooperate when heavy labor is required while women often exchange labor in routine water, firewood, and fodder collection.

Traditional agriculture has been subsistence-oriented in that the value of produce was measured by its utility within the household and among similar households, rather than the price it might fetch at market (Tutwiler and Carapico 1981). Production for use is reflected in the intricate ecological relationships between people, land, and livestock in rainfed grain production (Figure 3). Human labor is used to work the land and tend livestock, producing food and fuel. These are inputs to the "kitchen", which represents food processing for human consumption. Most dietary staples come from this simple production system (Bornstein 1974).

There is a clear division of labor between the sexes, with women providing household consumption needs and nurturing young children, and men providing them with the material means to do so (Myntti 1979). Women collect and prepare direct use values (e.g., meals, fuel, fodder, and water) while men maintain the working capital (i.e., terraces, land, tools, and domestic

236

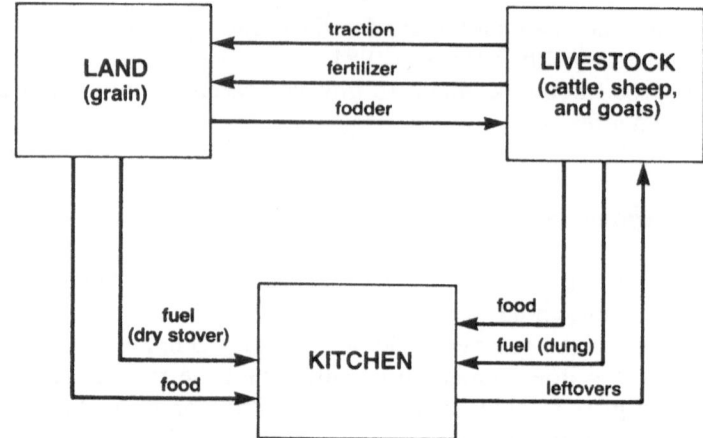

Fig. 3. Flow of use values in the subsistence household. *Source*: Tutwiler (1987).

structures). The main agricultural responsibilities of women are those which need daily activity, such as the care and milking of livestock, whereas men's activities are more seasonal. Both sexes may work together, for example during harvest and the post-harvest processing of grain. Women also participate in cultivation operations which provide animals with green fodder, especially thinning and pruning. Men may tend livestock or prepare meals when the women are unable to do so, and it is not uncommon to find women doing men's work when no man is available.

Land Tenure

There has been no land reform in Yemen, save for expropriation by the state of the former royal holdings, nor is any reform planned. There is no cadastre of landownership and there have been no major systematic studies of land tenure, which is highly variable across the country. The characteristics of one area may not hold in another.

Before wage employment became available abroad, 90% of rural households depended on access to arable land and even in 1985, 67–70% of rural households still derived most of their income from the land. FAO (1986a) estimated the average area of cultivated land per farming family to be 1.7 ha, but the agricultural census of 1979 gave a figure of 0.65 ha.

The land needed to adequately provision a household using traditional farming methods varies in the different regions depending on yield and labor availability. In 1976, average sorghum and millet yields in the Central Plateau were estimated at 600 kg/ha; in the Tihamah at 700 kg/ha; and in the Southern Uplands at 1,200–1,300 kg/ha (see MAF 1980; 1983b; Rees-Jones 1984; Rethwilm and Brandes 1979). In al-Mahwit province (Western Mountains), for example, 65% of farms are less than 1 ha, and 11% are less than

0.25 ha (MAF 1980). In the Southern Uplands, with much higher fertility and population density, the average farm is smaller (MAF 1983b), while in the Central Plateau, productivity per hectare is much lower and farms are larger (Rees-Jones 1984; Rethwilm and Brandes 1979).

Private ownership characterizes around 84–87% of arable land (Mujahid 1978). The owner may cultivate or make share tenancy agreements or fixed rent contracts; when land is jointly owned, one owner may cultivate with the permission of all. When the owner is the cultivator, production decisions are entirely his or hers. Female owners customarily allow close male relatives to manage land on their behalf (Mundy 1979).

Institutional ownership is held either by the Ministry of State Lands or by numerous *waqf* foundations (religious endowments), administered either by the Ministry of Waqfs or by individual supervisers. The land is cultivated by individual households possessing usufruct rights and making all or nearly all decisions themselves. Sharecropping is predominant on state and *waqf* land, but there are a few fixed rent contracts in which the tenant pays a lump sum for usufruct in perpetuity and a nominal annual fixed rent (Tutwiler 1987; Messick 1978). State and *waqf* lands together constitute 12–15% of the total arable area (Mujahid 1978).

Under usufructuary agreements, the degree of owner participation in production decisions varies according to contract, but minimally includes which crops will be grown. Sometimes the landowner will supply inputs such as seed, fertilizer, or a plow team. Landowners in share tenancy receive a proportion of the harvest rather than a fixed rent.

Communal ownership, in which land is periodically redistributed among households with communal rights, occurs almost exclusively in the Eastern Slopes and comprises less than 1% of total arable land. This system is found among semi-nomadic local descent groups or tribes that depend primarily on livestock which are the private property of individuals.

Production by owner-cultivators tends to be prevalent in the most productive areas. In many villages in Ibb and Taiz provinces, for example, all land is sharecropped (Messick 1978; Moczarski 1971) while most Southern Uplands farmers own none or only part of the land they cultivate (Omari 1979; FAO 1973; Hemosura 1973), one reason why many young men migrate (FAO 1986a). In al-Mahwit province, approximately half of the cultivators are owners, but significant areas in the most fertile regions are exclusively cultivated by share tenancy (Tutwiler 1987). In the less productive Central Plateau and in the Tihamah land is more likely to be owned by cultivators (Dresch 1984; Rethwilm and Brandes 1979; Mundy 1984) but where wadi irrigation is assured, landholdings are large and cultivated by non-owners. Because they manage the large-scale wadi irrigation systems these landlords have much more control over the production process in the lowlands (El-Basher 1986a) than in the rainfed highlands.

Additional information about land tenure may be found in Tutwiler (1987; 1984); El-Basher (1986a); UNECWA (1980); and Gerholm (1977).

Performance of the Agricultural Sector

The importance of agriculture in Yemen has steadily declined since 1962 and especially since 1975 due to international labor migration (Cohen and Lewis 1979). Development and growth of the economy are now directly related to the export of labor. The 1977–81 and 1982–86 5–year plans depended on expatriate remittances (CPO 1982a) which allowed unprecedented growth, with per capita GDP rising from 62 USD in 1964 to 528 USD in 1982. Annual remittances were estimated to be 1,200 USD for every man, woman, and child resident in 1983 (Lloyds Bank Group 1987; FAO 1986a), while total cash remittances rose from 272 million USD in 1975 to over 1 billion USD in 1980, where they stabilized despite falling oil prices (World Bank 1985). The personal savings of returning emigrants may be as much as two to three times the value of their yearly remittances (Shaw 1983; ASAJEC 1979).

The number of migrants abroad in any given month has been variously estimated at between about 366,000 (Birks and Sinclair 1980) and 1,394,000 (CPO 1982a). Migrants are generally transient guest workers in the oil states, subsidizing their families in Yemen and seeking to bring home as much money as possible. The usual stay is from 6 months to 2 or 3 years, although most migrants return annually for short visits, particularly during the fasting month of Ramadan or for the autumn harvest.

The government and underdeveloped banking sector cannot easily channel remittances and savings into productive investments (Kopp and Schweizer 1984; Saqqaf 1984) and despite.optimistic projections of agricultural growth, based on investment of remittances, agriculture (excluding qat) shrank from over 80% of aggregate GDP at the beginning of the 1970s to less than 29% in 1982. Annual agricultural growth averaged around 1% during the 1970s while the population grew at 2.7% and commerce, services, and construction grew at 16, 14 and 12%, respectively (Lloyds Bank Group 1987). Because agriculture was not an attractive investment, especially in rainfed areas, most returned migrants invested in consumption and small-scale business ventures or established independent households (Tutwiler 1987; Weir 1986).

Terms of trade for staple products have declined swiftly since the 1970s (Mundy 1984; Swanson 1979; Sinclair 1976). Domestic prices for traditional staple crops remained depressed due to import competition (World Bank 1979b) and retail prices of imported commodities were subsidized and prices regulated to prevent inflation. Imported cereals made up 40% of total cereal consumption in 1982. The market price of domestic cereals rose only 30% between 1975 and 1982 (Shaw 1983), while the consumer price index was rising at over 40% per year. As a result, cereals continued to be subsistence crops and no more than 10% of grain production reached the market. Since rural people had remittance income to spend on imported food, the need to produce cereals declined. This is reflected in the production figures for sorghum and millet, the principal rainfed crops (Table 2). However, over 70% of highland households not growing cash crops are committed to produc-

Table 2. Sorghum and millet production, 1964–82 (means of aggregate statistics).

Year	Area (×1,000 ha)	Yield (kg/ha)	Production (×1,000 t)	Economic events
1964–66	737	837	617	civil war, 1964–70
1967–70	930	720	669	drought, 1967–73
1971–74	909	740	679	migration boom 1974–onward
1975–78	792	838	653	migration stabilized, 1979
1979–82	683	884	603	recession, 1980 onward

Source: Lloyds Bank Group (1987); World Bank (1985); CPO (1982b); UNECWA (1973).

ing grains for family consumption (FAO 1986a; MAF 1985; 1983a; 1982). This would seem to be the case especially for poorer rural women.

Between 1964 and 1982 grain production was also affected by civil war, drought, and perhaps labor migration. After a brief upsurge in area cropped in 1974–75 there has been a steady decline which may be due to an increasing labor shortage (Enger 1986; Al-Kasir 1984). However, the traditional farming system might have disguised considerable underemployment of males, possibly as high as 20–40% (Shaw 1983), and migration may simply be absorbing this excess labor (Carapico 1985). Alternatively, women may be replacing the absent workforce (Adra 1983; MAF 1983a; 1982; World Bank 1979b).

The income structure of rural households has changed dramatically in less than 5 years, indirectly undermining agriculture (Carapico 1985). A limited survey in the Ibb/Taiz region revealed 32% of income is from qat, 24% from other crops, 30% from remittances, and 14% from other sources, mostly trade and construction (Shaw 1983). Studies in the extreme southern district of Hujuriyyah claim up to 70% of farm income is from off-farm employment although in 1982 the national average was probably around 32% (Shaw 1983). Male off-farm employment is becoming the main economic activity due to high urban incomes estimated at three times those in rural areas (World Bank 1979a).

Even rural wages have increased. In 1973 casual agricultural laborers accepted less than 1 USD per day but by 1980 demanded 12 USD, while unskilled construction workers commanded twice this figure. High wages discouraged farmers from hiring extra hands and even higher off-farm earnings drew marginal farmers into construction and related activities. Capital was used for speculative ventures rather than productive investments and labor-intensive enterprises, such as rainfed agriculture.

Several trends are clear. There has been an absolute decline in regularly cultivated land, as area cropped fell 32% in 1975–80 and a further 17% in 1980–85 (Enger 1986). This was largely due to abandonment of rainfed grain land, mostly marginally productive terraces in the highlands (Swanson and Hebert 1981). There has been a substantial increase in the production of higher value, income elastic crops like qat, fruits, and vegetables (see CPO 1982b) which are not subject to adverse competition from imports. Most of this increase has been on newly irrigated land and has offset only a small fraction of the rainfed land abandoned by grain cultivators. Land cultivated in potatoes, vegetables, and fruits increased from 35,600 ha to 48,500 ha in 1975–80 (CPO 1982b).

Abandonment of terraces can start an irreversible process of resource destruction (Alkamper et al. 1979). Most of the unattended terraces are on steeply sloping higher land, and if upper terraces are not maintained to divert and absorb runoff from monsoon downpours, more water, soil, and loose stones wash downhill and threaten the broader and more fertile lower terraces. Uncontrollable floods increasingly destroy crops and wash away terraces with fertile soils accumulated by generations of past labor. Groundwater regeneration is also affected in the middle elevations and lowlands, as more water flows wastefully into the sea. The long-term implications of abandoning terraced land are therefore much worse than short-term reductions in output (Kopp 1984; Aulaqi 1982).

Qat could help to preserve such areas as its root system holds tenuous highland soil and encourages water retention. It is a luxury crop with a bad reputation and its effects on the health of consumers and the household finances of habitual users are contentious issues (Weir 1985). But production has more than doubled in the past decade and in 1982 4.6% of total cultivated land or about 47,200 ha were under qat (MAF 1983b), slightly less than all other cash crops, except coffee, put together.

Qat requires fewer man-days of labor per hectare and has a higher monetary value than sorghum (Tutwiler 1987; Weir 1984; Swanson 1981; Gerholm 1977) and other cash crops. However, qat requires particular thermal and moisture conditions which exclude cultivation in the Tihamah and severely limit its range in the midlands and highlands. Initial costs are high, with land preparation requiring up to five times the labor required for sorghum and the first marketable harvest is 4–5 years after planting. Young shrubs must be hand watered if the first year's precipitation is less than 800 mm, although in a mature plantation the shrub becomes dormant in dry periods and grows quickly when moisture is available. Also, picked leaves must be sold within 48 hours, necessitating good transport and market connections, which are often lacking in remote villages. Finally, land tenure arrangements rule out qat cultivation for many sharecroppers and fixed-rent tenants (Tutwiler 1987).

New qat plantations were thought to be replacing mature coffee, thus undercutting agricultural export potential (World Bank 1979b), but village studies in the highlands indicate that qat is expanding onto former grain land and into newly irrigated areas (Weir 1985; Swanson 1984). Coffee may be being replaced with sorghum and in some areas unmaintained coffee trees are being kept as a source of firewood.

Animal production continues to be essentially traditional, but pasturage and veterinary services are inadequate and diseases prevalent. Cattle numbers have decreased from 1.26 million in 1961–65 to 0.87 million in 1977–81, while sheep and goats have declined from 11 million to below 4 million (CPO 1982b; 1975). Animals are rarely raised for market, except perhaps by some of the semi-nomadic populations in the Tihamah and Eastern Slopes which keep relatively large herds of sheep and goats.

For a typical farming household with three generations of women, FAO (1986a) calculates that livestock (typically a cow, several sheep and goats, and some chickens) require up to 13 hours per day of female labor. At market prices, the return to labor per day is at most 1 USD while an unskilled man's daily wage is 10–12 USD. But people continue to keep animals because they provide a regular supply of fresh milk and eggs, browse natural vegetation, stubble, and leftovers which otherwise would be of no use to humans, have low opportunity costs for the family labor of grain producers, larger animals can be used for traction and transport, and the female workforce is accustomed to raising livestock. Furthermore, livestock are often the personal property of the women who tend them and so represent the principal repository of women's disposable wealth.

Fodder produced on irrigated land is trucked to some rainfed areas and sold considerably more cheaply than local products. This enables many highland households to maintain animals that they would otherwise have to slaughter or sell in dry years, but requires male household members and their money, so women's economic independence with regard to livestock is lessened.

Commercial poultry operations and a few capital-intensive dairy farms selling processed milk products to Yemen's growing urban market have recently been established. However, rural women appear to be continuing to keep chickens and cattle.

Technological Innovations and their Sociological Impact

Diesel-powered pumps and drilled deep wells have dramatically improved production and many areas are under sustained cultivation for the first time. Pumps have been used since the late 1950s by large Tihamah landlords, and by the 1970s ownership extended to smallholders with migrant income (Mundy 1984). In the Eastern Slopes pump wells were augmented with bulldozed diversion structures that could withstand major spate floods (Brunner 1984). In the larger wadis of the eastern Tihamah, concrete diversion weirs and sluice gates were constructed.

Tractors first began to replace draft animals in the plains and broad wadis for land leveling, preparation, and plowing. Imported chemical fertilizers and pesticides became easily obtainable in the mid-1970s, and improved grain yields were attributed to these innovations (World Bank 1979b). It was predicted that migrants' savings would be absorbed in purchasing capital inputs to improve production and sustain rising rural incomes (World Bank 1979c).

Unfortunately, many mechanical and chemical inputs have limited applicability in Yemen. Irrigation and the use of tractors are feasible on less than a quarter of Yemen's farmland and are actually employed on less than one twentieth (Carapico 1985). In the highlands, terraces are too narrow, weak, or inaccessible to be plowed with a tractor or irrigated. Pump irrigation is feasible in the lowlands and, to a lesser extent, in plateau basins and broad wadi plains, but the groundwater is being depleted (Mundy 1984; Rethwilm and Brandes 1979).

Rainfed cultivation in particular faces constraints. Chemical fertilizers and pesticides require controlled and properly timed irrigation, while imported varieties usually show good results only if irrigated (McCuiston 1986). Research in the Central Plateau has shown that while irrigated hybrids respond to chemical fertilizers (Rees-Jones 1983), their efficiency is less certain for areas with less than 700 mm annual rainfall. It was also found that local dryland sorghum varieties performed better than imported varieties. Although innovations may be feasible (Walker et al. 1983), fragmented landholdings and the prevalence of sharecropping may discourage investment of money or labor in land improvement, irrigation, or new crops.

Much of the new technology has been applied initially to maintaining traditional cropping patterns (Mundy 1984; Rees-Jones 1985; 1984; Barrett and Morton 1982). Grains are estimated to account for half to two thirds of the irrigated area, not much less than sorghum in higher rainfall areas (Carapico 1985).

Prior to the revolution all dryland producers were differentiated according to the amount and quality of land owned. Now, producers are differentiated according to capital deployed in production. A small household on a small plot of irrigated land can increase production, produce high value crops regularly, and sell surplus water to neighbors. But the rainfed terrace owner must convert to qat to prevent his farm income from declining. Demographic studies, mainly in the Central Plateau, suggest that irrigated and cash crop farmers now migrate less and maintain larger households than rainfed grain farmers with traditional, large, extended households. With limited investment opportunities in farming, migrants returning to rainfed areas spend their savings on consumption, establishing an independent household, or starting an off-farm enterprise (Tutwiler 1987; Weir 1986), aggravating potential shortages of family labor.

Social relations of production are changing. In renegotiated contracts between sharecroppers and landlords, the latter receive a share for the pump in the harvest division (Carapico 1985). Although this may result in a lower return to the tenant on his labor and a higher return to the landlord on his capital, there is security in a share tenancy and the sharecropper will receive a larger absolute value from increased yields. Landlords favor share tenancy because of the low supervision costs and the guarantee of continued cultivation, important factors when wages are increasing and labor availability is uncertain.

In contrast, sharecroppers on unimproved rainfed land are successfully negotiating for higher shares, especially on more marginal terraces (Al-Kasir 1984; UNECWA 1980). In some cases landlords loan the land rent-free. Sharecroppers have considerable bargaining power because if a terrace is not cultivated, erosion will swiftly destroy its future value.

Wealthy entrepreneurs in the Central Plateau and Eastern Slopes now actively recruit sharecropping colonists from the densely populated mountains (Tutwiler and Carapico 1981). In return for the use of private plots watered by the landlord's pump, the settlers grow cash crops as directed by an overseer. In this way, landlords have a dependable workforce insulated from rising wage rates and uncertain labor markets while the sharecroppers have rent-free, irrigated private plots and additional income from working on the landlord's farm.

Integration into the world economy has not resulted in increased factor mobility in agriculture. Agricultural land and labor remain highly immobile, and there is continued reliance on unpaid family labor. The desire to translate money wealth into capital, coupled with a dearth of productive investment opportunities, has resulted in absurdly inflated land prices (Swanson 1979).

Labor Force Dynamics

With possibly one man absent for every farming household (Steffan 1978), emigration has resulted in a severe agricultural labor shortage. Wages for casual workers are rising with reduced worker availability and increased farmer demand for extra-household labor. In the late 1970s the urban workforce consisted primarily of rural migrants, many of whom had previously worked abroad (Meyer 1984). Others invested in small trading enterprises in rural marketplaces and along new roads (Schweizer 1984; Tutwiler 1984; Morton 1979). The result was a net labor drain on the agricultural sector, as migrants abandoned farming for non-agricultural work in towns and villages. Shortages of manpower have led to the importation of unskilled and semi-skilled foreign workers by large contractors (Kopp and Schweizer 1984). In local surveys, farmers complain of a labor shortage.

The labor shortage is often cited as a major factor in the poor performance of the agricultural sector over the past 15 years (Lloyds Bank Group 1987; World Bank 1979c; IMF 1975). Farmers are assumed to have cut back staple production and entrepreneurs to be hesitant to invest because of high costs and low availability of labor.

However, some recent studies challenge the existence of a labor shortage. One World Bank study (1985) concluded that there is considerable under-employment and an under-utilized rural labor reserve due to the declining area of cultivated land and increased use of labor-saving innovations. Serageldin et al. (1983) earlier predicted sharply falling demand for labor despite economic growth.

Table 3. Estimates of agricultural labor supply and demand, YAR 1985.

Variables	USAID	World Bank
Total population	7,271,029	7,700,000
less migrants[1]	(minus) 366,447	
Farm population[2]	(67%) 4,860,826	(70%) 5,400,000
Economically active farmers		
resident males	1,498,036	1,300,000
resident females	1,413,599	1,400,000
absent male migrants[3]		(minus) 600,000
total active farmers	2,911,635	2,100,000
Available man-years of labor		
male[4]	1,198,429	700,000
female[5]	141,360	700,000
total man-years available	1,339,789	1,400,000
Total man-days/year available[6]	335,000,000	350,000,000
Total man-years/month available	108,000	120,000
Estimated need, 1982 base-year[7]		
man-days/year, total demand	200,000,000	80,000,000
man-years/month, peak period	200,000	80,000

Source: Enger (1986); World Bank (1985).
[1]Based on figures in Steffan (1978) and Birks and Sinclair (1980) held constant for 1975 base year.
[2]Calculated as percentage of total population.
[3]500,000 working abroad, 100,000 in Yemeni towns and cities.
[4]USAID assumes 80%, World Bank assumes 100% male participation rate.
[5]USAID assumes 10%, World Bank assumes 50% female participation rate.
[6]Based on 250 working days per year.
[7]USAID figures include labor time for maintenance, marketing, and other functions.

The sufficiency of the agricultural labor force depends upon the size of the rural labor pool, male and female participation rates, labor requirements of the principal crops, and the social relations of production which mobilize and deploy rural labor. There is no agreement on any of these variables. For example, USAID and the World Bank use very different estimates of key variables to estimate labor supply (Table 3). USAID estimates for migration are probably too low, and the Bank's inclusion of internal migration to Yemen's urban areas is probably more correct. USAID estimates male participation at 80%, which is more reasonable than the Bank's 100%. However, USAID's estimate of 10% female participation disagrees with virtually all local farming system survey reports, which stress the importance of female labor in grain production (see, for example, FAO 1986a and 1986b). The Bank's figure of 50% is more realistic.

These contrasting assumptions largely cancel each other out, and the two estimates of total man-days/year available are remarkably close. However, by USAID's calculations this labor is 89% male, whereas according to the World Bank it is 50% female. This difference is highly relevant to the question of this labor's availability for certain kinds of tasks.

By estimating labor need in terms of man-days/year and man-years/month in the peak harvest months, a rough aggregate determination can be made of the sufficiency of the available labor pool, ignoring regional distribution of need and availability. USAID estimates of labor demand are 2.5 times those of the World Bank. The differences are explained by differing assumptions of man-day/ha requirements for the principal grain crops. For example, USAID estimates twice the World Bank labor figure for rainfed sorghum. As a result, USAID identifies a severe labor shortage during sorghum harvests in late autumn, a problem corroborated by other studies (Barrett and Morton 1982) while the World Bank concludes that there is no farm labor problem but rather substantial underemployment.

The USAID figures imply that with the current labor supply, expansion and increased production of labor-intensive crops is highly unlikely, and yield-increasing technologies will increase the need for scarce harvest labor. The World Bank considers the real problems to be pricing structures and traditional production of low value crops. Therefore, on one hand the World Bank analysis suggests production geared to market demand without considering labor availability while USAID's analysis suggests that stress be placed on substituting capital-intensive for labor-intensive methods of production.

Estimates of the labor requirements of rainfed sorghum vary widely (Table 4). Land preparation, terrace repair, and post-harvest processing are not consistently included, although they may constitute as much as half the labor traditionally invested in rainfed sorghum, especially in the Western Mountains and Southern Uplands. In these areas, there probably are labor shortages for traditionally cultivated sorghum.

In addition to the quantity of labor available, the way in which the existing labor force is organized and deployed through social relations of production is important. It is frequently assumed that in a fully monetized economy labor can be analyzed as a commodity, and that unskilled farm labor is an undifferentiated quantity which will seek the maximum wage. The assumption is that if commercial farmers are willing to pay workers an attractive wage, capital investments in agriculture will probably find the necessary labor. But the demographics in Table 3 are important. Women are far less likely to leave the household and seek wage employment than men for numerous cultural, social, and family reasons (cf. Myntti 1979). As women are increasingly important in the agricultural labor force, the factors which produced the labor mobility of Yemeni men in the 1970s may no longer have the same influence. In the Tihamah, for example, where it is more common for women to work for wages than in the highlands (Hebert 1981), hired

246

Table 4. Calculated labor requirements per hectare of sorghum.

Region	Source	Man-days/ ha	Yield (kg/ha)	% of dryland in sorghum
Yemen	World Bank 1985	45[1]	700	65
(entire)	Enger 1986	90[1]	800	
Tihamah	El-Basher 1986b	106[1]		
Western	Rees-Jones 1985	165[2]	1,300	90
Mountains	Tutwiler 1987	230	1,200	80
Central Plateau	Barret and Morton 1982	118[2]	700	82
Southern	Tutwiler and Carapico	240	1,300	
Uplands	1981; FAO 1986a	88[1]		

[1]Labor expended in plowing, post-harvest work, and terrace repairs not included.
[2]Terrace repairs not included.

labor is not geographically mobile, is primarily geared to harvest seasons, and is often paid in kind with grain and stover which the women use as feed (El-Basher 1986b). When women do take employment, the returns must be similar to those they would receive at home.

The Substitution of Capital for Labor

Traditional rainfed agriculture was an ecologically balanced system of labor-intensive grain and animal production in which productivity was proportional to the amount of labor invested. The objective was the self-provisioning of producer households, although by the 1960s as many as half the cultivators had to share their harvests with landlords.

All producer households have now become involved in the market economy. The degree of participation varies, but no households are entirely self-sufficient, even in food production. The effects on agriculture have included a smaller agricultural labor force, greater female participation (perhaps more than 50%), and less land and labor devoted to producing staple grains and animals. There is increased capital investment in wells, pumps, tractors, and chemicals. There have been modest increases in production of high value crops such as qat, fresh vegetables, and fruits, all of which require more capital and in some cases less labor than traditional grain and animals.

Households continue to be the fundamental units of production, the sexual division of labor has not changed, and the basic legal structure of land tenure is intact, although the balance of bargaining power has shifted to tenants on rainfed grain land and the suppliers of capital inputs on other types of land.

Jurisdiction over resource allocation still follows old patterns, but women have increased influence over self-provisioning and men are assuming firmer control over cash and market participation. Traditional rainfed grain and animal production are still the most prominent features of rural labor mobilization and land use, simply because they are the best economic adaptation to the highland environment.

The changes which have occurred are essentially indigenous responses to market incentives. They include the steady deterioration of the terrace system and the use of off-farm income to subsidize female subsistence production, which is a rational response to the present situation but an inefficient use of cash resources.

If the cost of casual labor rises or off-farm work supplies less disposable income, the household must either devote most of its resources to agriculture or follow a non-farming economic strategy. For agriculture to be attractive, it must be profitable and given current wage and price structures, with the exception of qat production, returns are much higher from off-farm employment.

Yemenis are not bound by tradition, either in their economic or their social behavior (Ghanem 1984). Rural communities have invested heavily in constructing a rudimentary infrastructure through a nationwide network of Local Development Associations, aiming to improve the quality of life and access to commodity markets (Carapico 1984). But farming has the security of expertise and past experience, and landowning families do not want to relinquish this asset, either to buyers or to the elements. Farm surveys report widespread experimentation with different production strategies and new crop varieties in response to labor constraints and market incentives (FAO 1986a; 1986b; Rees-Jones 1984; 1985; Barrett and Morton 1982; Tutwiler and Carapico 1981).

Given the changed labor force, improved farm incomes will depend on better market prices and on increasing capital inputs in the production process relative to labor. Farmers cannot influence local prices but the committment of capital to the production process is still the decision of individual households.

If capital is to play a significant part in household production, then the crops produced must provide a profitable return on investments. To date, only qat fulfils this requirement in rainfed farming and capital has been successfully integrated into the conversion of the production process to irrigation. Farmers have so far not applied modern technology to extensive rainfed grain production and ecological constraints would seem to rule out technological innovations for most of them.

Conclusions

Monetization has changed Yemen's agricultural geography. Traditional grain and livestock production continue in the moist highlands, but since the late

248

1960s agricultural development has taken place in more arid regions with gentler relief where modern, petroleum-fueled machinery can be used to exploit groundwater resources. What was once the most productive region in the Arabian peninsula is now in danger of becoming a permanent labor reserve for others, largely due to the changed composition of the local labor force and the constraints to irrigation in the mountains. The future of highland agriculture ultimately lies with the mountaineers themselves, for they continue to search for answers to their problems and to experiment with new technologies and crops which may yet allow them to participate in the new economy as successful commercial farmers.

References

Adra, N. 1983. The Impact of Male Migration on Women's Roles in Agriculture in the YAR. Sanaa: USAID.

Alkamper, J., Haffner, W., Matter, H.E. and Weise, O.R. 1979. Erosion Control and Afforestation in Haraz, Yemen Arab Republic. Giessen: Tropeninstitut Giessen.

Al-Kasir, A. 1984. The Impact of Emigration on the Social Structure in Yemen Arab Republic. Pages 122–133 in Economy, Society, and Culture in Contemporary Yemen (Pridham, B., ed.). London: Croom Helm.

ASAJEC (American-Saudi Arabian Joint Economic Commission). 1979. Survey on Employment and Wage Levels. Mimeo.

Aulaqi, N. 1982. Household Energy and Tree Seeding Demand Survey in the Southern Uplands of the Yemen Arab Republic. Sanaa: FAO.

Barrett, A. and Morton, J. 1982. Labour and Power Inputs in Crop Production in the Montane Plains. Surbiton: Overseas Development Administration.

Bartelink, Alexander. 1974. Yemen Agricultural Handbook. Eschborn: Federal Agency for Economic Cooperation.

Birks, J. and Sinclair, C. 1980. Arab Manpower. New York: St. Martin's Press.

Bornstein, A. 1974. Food and Society in the Yemen Arab Republic. Rome: FAO.

Brunner, U. 1984. Irrigation and Land Use in the Ma'rib Region. Pages 51–63 in Economy, Society, and Culture in Contemporary Yemen (Pridham, B., ed.). London: Croom Helm.

Carapico, S. 1984. The Political Economy of Self-Help; Development Cooperatives in the Yemen Arab Republic. PhD dissertation, State University of New York at Binghamton.

Carapico, S. 1985. Yemeni Agriculture in Transition. in Agricultural Development in the Middle East (Beaumont, P. and McLachlan, K., eds). London: John Wiley and Sons Ltd.

Cohen, John M. and Lewis, David B. 1979. Rural Development in the Yemen Arab Republic: Strategy Issues in a Capital Surplus, Labor Short Economy. Norfolk: AID/RD Report Distribution Center.

CPO (Central Planning Organization). 1975. Statistical Yearbook, 1974. Sanaa: Prime Minister's Office.

CPO (Central Planning Organization). 1982a. The Second Five-Year Plan, 1981–1986. Sanaa: Prime Minister's Office.

CPO (Central Planning Organization). 1982b. Statistical Yearbook, 1981. Sanaa: Prime Minister's Office.

Dresch, P. 1984. The Northern Tribes of Yemen and their Place in the Yemen Arab Republic. DPhil thesis, Oxford University.

El-Basher, A. 1986a. Crop-Sharing Practices in the Tihama Region. YAR-MAF/Tihama Development Authority. Sanaa: FAO-UNDP.

El-Basher, A. 1986b. Agricultural Labour in Tihama Region. YAR-MAF/Tihama Development Authority. Sanaa: FAO-UNDP.

Enger, W. 1986. YAR Agricultural Sector Assessment, Update 1985. Sanaa: USAID.

FAO (Food and Agriculture Organization). 1973. Draft Report of the Yemen Arab Republic Southern Uplands Rural Development Project Preparation Team. Rome: FAO.

FAO (Food and Agriculture Organization). 1986a. YAR Southern Uplands Rural Pioneer Women Development Project: Formulation Mission Report. Rome: FAO.

FAO (Food and Agriculture Organization). 1986b. YAR-MAF; Tihama Development Authority; FAO-UNDP. Project Findings and Recommendations (Agricultural Production and Marketing Economics). Mimeo.

Fergany, N. 1980. The Affluent Years are Over – Emigration and Development in the Yemen Arab Republic. World Employment Working Paper, WEP 2-26/WP50. Geneva: ILO.

Gerholm, Tomas. 1977. Market, Mosque and Mafraj. University of Stockholm.

Ghanem, I. 1984. Tradition and Change in Social Behaviour. Paper presented to the Symposium on Contemporary Yemen, Centre for Arab Gulf Studies, University of Exeter, England, 15–18 July 1983.

Harley, B. 1942. Dry Farming Methods in the Aden Protectorate. in Proceedings of the Conference on Middle Eastern Agriculture. Cairo: Middle East Supply Centre.

Hebert, Mary. 1981. Socio-Economic Conditions and Development Maghlaf, Hodeidah Governorate. Ithaca: Cornell University Rural Development Participation Project.

Hemosura, S. 1973. Highlands Farm Development and Training Centre, Draft Final Report. Taiz: FAO.

IMF (International Monetary Fund). 1975. Recent Economic Developments in the Yemen Arab Republic. Washington, D.C.: IMF.

Kopp, H. 1984. Land Usage and its Implications for Yemeni Agriculture. Pages 41–50 in Economy, Society, and Culture in Contemporary Yemen (Pridham, B. ed.). London: Croom Helm.

Kopp, H. and Schweizer, G. 1984. Entwicklungsprozesse in der Arabischen Republik Jemen. Weisbaden: Dr. Ludwig Reichert Verlag.

Lloyds Bank Group. 1987. Yemen Arab Republic Economic Report 1986. London: Lloyds Bank Group.

MAF (Ministry of Agriculture and Fisheries). 1980. (Results of the Agricultural Census in Hajjah and Al Mahwit Districts). In Arabic. Sanaa: MAF.

MAF (Ministry of Agriculture and Fisheries). 1982. Kingdom of the Netherlands-MFA Report on Farming Systems Research Survey in Wadi Tha. Rada' Integrated Rural Development Project Technical Note No. 9. Sanaa: MAF.

MAF (Ministry of Agriculture and Fisheries). 1983a. Kingdom of the Netherlands-MFA Report on Farming Systems Research Survey in Wadi Mansour. Rada' Integrated Rural Development Project Technical Note No. 13. Sanaa: MAF.

MAF (Ministry of Agriculture and Fisheries). 1983b. Summary of the Final Results of the Agricultural Census. Department of Planning and Statistics. Sanaa: MAF.

MAF (Ministry of Agriculture and Fisheries). 1985. Kingdom of the Netherlands Agricultural Sector Cooperation Program. Report of the Yemen-Netherlands Mission. Sanaa: MAF.

McCuiston, W. 1986. Realities and Possibilities of Wheat Production in the YAR. Agricultural Development Support Program. Sanaa: USAID.

Messick, B. 1978. Transactions in Ibb: Economy and Society in a Yemeni Highland Town. PhD dissertation, Princeton University.

Meyer, G. 1984. Labour Emigration and Internal Migration in the Yemen Arab Republic, The Urban Building Sector. Pages 147–164 in Economy, Society, and Culture in Contemporary Yemen (Pridham, B., ed.). London: Croom Helm.

Moczarski, S. 1971. Sample Socio-Economic Survey of Five Villages in Ibb Governorate. Taiz: FAO.

Morton, J. 1979. Agricultural Marketing in the Yemen Arab Republic with Special Reference to the Montane Plains and Wadi Rima. Surbiton: Overseas Development Administration.

Mujahid, A. 1978. (The Agricultural Cooperative, Doorway to Improvement in Yemen). In Arabic. Sanaa: Kitab al Ghad.

Mundy, M. 1979. Women's Inheritance of Land in Highland Yemen. Arabian Studies 5: 161–187.

Mundy, M. 1984. Agricultural Development in the Yemeni Tihama: The Past Ten Years. Pages 22–40 in Economy, Society and Culture in Contemporary Yemen (Pridham, B., ed.). London: Croom Helm.

Myntti, Cynthia. 1979. Women and Development in Yemen Arab Republic. Eschborn: German Agency for Technical Cooperation.

Omari, F. 1979. Statistical Bulletin for Taiz/Ibb Region. Rome: FAO.

Rees-Jones, H. 1983. An Economic Review of Fertilizer Trails at Dhamar Agricultural Improvement Centre. Surbiton: Overseas Development Administration.

Rees-Jones, H. 1984. Farm Systems Survey I: Qa Bakil. Surbiton: Overseas Development Administration.

Rees-Jones, H. 1985. Farm Systems Survey III: Hammam Ali. Surbiton: Overseas Development Administration.

Rethwilm, D. and Brandes, W. 1979. Proposals for Follow-On Measures for the Al Boun Project. Eschborn: Federal Agency for Economic Cooperation.

Saqqaf, A. 1984. Fiscal and Budgetary Policies in the Yemen Arab Republic. Pages 83–95 in Economy, Society, and Culture in Contemporary Yemen (Pridham, B., ed.). London: Croom Helm.

Schneider, W., Schroder, R., Speth, W. and Van Panhuys, H. 1981. Regional Development of the Province Al-Mahwit. Stuttgart: Institut Fur Projecktplanung.

Schweizer, G. 1984. Social and Economic Change in the Rural Distribution System: Weekly Markets in the Yemen Arab Republic. Pages 107–121 in Economy, Society and Culture in Contemporary Yemen (Pridham, B., ed.). London: Croom Helm.

Serageldin, I., Socknat, J., Birks, S., Li, B. and Sinclair, C. 1983. Manpower and International Labor Migration in the Middle East and North Africa. New York: Oxford University Press.

Serjeant, R. 1974. The Cultivation of Cereals in Medieval Yemen. Arabian Studies 1: 25–74.

Shaw, R. 1983. Mobilizing Human Resources in the Arab World. London: Routledge and Kegan Paul.

Sinclair, R. (ed.). 1976. Documents on the History of Southwest Arabia. Volumes I and II. Salisbury: Documentary Publications.

Steffan, H. 1978. Final Report of the Airphoto Interpretation Project. Berne: The Swiss Technical Cooperation Service.

Stevenson, T. 1985. Social Change in a Yemeni Highlands Town. Salt Lake City: University of Utah Press.

Swanson, J. 1979. Some Consequences of Migration for Rural Economic Development in the Yemen Arab Republic. Middle East Journal 33(1): 34–42.

Swanson, J. 1981. Socio-Economic Conditions and Development Bani 'Awwam, Hajja Governorate. Ithaca: Cornell University Rural Development Participation Project.

Swanson, J. 1984. Emigrant Remittances and Local Development Co-operatives in the Yemen Arab Republic. Pages 132–145 in Economy, Society and Culture in Contemporary Yemen (Pridham, B., ed.). London: Croom Helm.

Swanson, J. and Hebert, M. 1981. Rural Society and Participatory Development: Case Studies of Two Villages in the Yemen Arab Republic. Ithaca: Cornell University.

Tutwiler, R. 1984. Ta'awun Mahwit: A Case Study of a Local Development Association in Highland Yemen. Pages 166–192 in Local Politics and Development in the Middle East (Cantori, L.J. and Harik, I., eds). Boulder: Westview Press.

Tutwiler, R. 1987. Tribe, Tribute, and Trade: Social Class Formation in Highland Yemen. PhD dissertation, State University of New York at Binghamton.

Tutwiler, R. and Carapico, S. 1981. Yemeni Agriculture and Economic Change. Sanaa and Milwaukee: American Institute for Yemeni Studies.

Tutwiler, R., Murdock, M. and Horowitz, M. 1976. Problems and Prospects for Development in the Yemen Arab Republic: The Contribution of the Social Sciences. Binghamton: Institute for Development Anthropology.

UNECWA (United Nations Economic Commission for Western Asia). 1973. Tax Structure, Government Savings and Tax Problems: A Case Study of Yemen. Beirut: UNECWA.

UNECWA (United Nations Economic Commission for Western Asia). 1980. Crop Sharing and Land Tenancy Practices in the YAR. Beirut: UNECWA.

Varisco, D. 1984. The Adaptive Dynamics of Water Allocation in Al-Ahjur, Yemen. PhD dissertation, University of Pennsylvania.

Walker, S. Tjip, Carapico, S. and Cohen, J. 1983. Emerging Rural Patterns in the Yemen Arab Republic: Results of a 21–Community Cross-Sectional Study. Ithaca: Cornell University.

Weir, S. 1984. Economic Aspects of the Qat Industry in North-West Yemen. Pages 64–82 *in* Economy, Society and Culture in Contemporary Yemen (Pridham, B., ed.). London: Croom Helm.

Weir, S. 1985. Qat in Yemen: Consumption and Social Change. London: British Museum Publications Limited.

Weir, S. 1986. Labour Migration and Key Aspects of its Economic and Social Impact on a Yemeni Highland Community. *in* The Middle Eastern Village (Lawless, R., ed.). London: Assen Helm.

World Bank. 1979a. Yemen Arab Republic: Development of a Traditional Economy. Washington, D.C.: World Bank.

World Bank. 1979b. Agricultural Sector Report: Yemen Arab Republic. Washington, D.C.: World Bank.

World Bank. 1979c. Effects of Migration of Rural Labor on Agricultural Development in the Yemen Arab Republic. Washington, D.C.: World Bank.

World Bank. 1985. YAR Agricultural Strategy Paper. Washington, D.C.: World Bank.

12. Agricultural Labor and Technological Change in Tunisia

NICHOLAS S. HOPKINS

Department of Sociology, Anthropology and Psychology, American University in Cairo,
P.O. Box 2511, Cairo, Egypt

Introduction

Tunisia is the smallest of the North African Arab countries. It has always been primarily an agricultural country and one of its chief features is the considerable variety in agriculture from north to south, and within regions. The main crops include cereals, olives, fruits (citrus, grapes, dates), and livestock, especially sheep. Much has been accomplished in expanding and developing irrigated areas and tree crops, but there is a general sense of crisis in Tunisia with regard to dryland cultivation and animal husbandry.

For example, a Tunisian geographer recently pointed out that, "Tunisian agriculture is passing through one of the most difficult phases of its history. Its development does not cover the constantly growing food needs of a population growing at a rate exceeding 2.7% per year; our imports of grains, of meat, of milk and dairy products, of sugar, and so on, increase from year to year; we are importing 45% of our total consumption of grains, 50–60% of our needs in milk, 30% in sugar, 15–20% in meat, without counting exotic products such as tea, coffee, spices, etc., of all sorts. These imports drain away hard currency (more than 200 million dinars) which contributes in a spectacular way to our trade deficit.

"Although some subsectors, such as tree cultivation and market vegetable gardening have developed considerably, other subsectors, such as cereal cultivation, animal husbandry, fodder crops and industrial crops, remain atrophied and insufficiently developed" (Kassab 1986). He further argues that the state has concentrated investment in the irrigated sector, which now comprises 3.5% of farmland but yields 16% of the gross agricultural product and 20% of total agricultural production. Thus, a relatively small proportion of the rural population has benefited from this investment. Since attempts to improve rainfed farming through the cooperative/agrarian reform of the 1960s failed, nothing else has been done (Kassab 1986).

Kassab has offered a partial explanation. "Since 1960 and especially since 1970, the state has extracted an important surplus from agriculture, which has essentially benefited the secondary and tertiary sectors (trade, tourism, etc.), and as a result, the social classes which dominate these sectors: the industrial and mercantile bourgeoisie of the coastal cities" (Kassab 1983).

Dennis Tully (ed.), Labor and Rainfed Agriculture in West Asia and North Africa, 253–271.
© 1990 *ICARDA.*

254

Table 1. Characteristics of Tunisian agriculture.

Area	164,152 km^2
Agricultural land	approx. 5 million ha (30% of total)
Population (estimated 1986)	7 million (44/km^2)
Per capita income	1,280 USD
Labor force	approx. 1 million (1980) of whom half are employed full-time

Also, the price structure does not favor the producer and prices are too low to encourage production of wheat or beef (Kassab 1986). On the other hand, some problems are due to the social organization of farmers. "Isolated farmers, socially and politically neutralized, face up to the innumerable problems with broken ranks . . . The peasant world is disorganized, unstructured, scattered, and seems to find this situation agreeable, with each farmer full of suspicion of his neighbor and appearing to be convinced of the uselessness, even of the serious risks involved in any association" (Kassab 1983).

Aubry et al. (1986) also noted that, especially in rainfed areas, the standard of living is declining, particularly among the predominant small farmers, and the farm population is aging due to the migration of the younger generation away from agriculture. Off-farm activities are increasing, resulting in the feminization of agricultural work, while extensive production systems are being retained instead of undergoing intensification and diversification. Also, the national food deficit is becoming worse and there is a growing gap between urban and rural revenues.

A sound knowledge of dryland farming systems, in particular the structure and organization of labor within these systems, is important as a foundation for policy. This paper focuses on the social organization of agriculture, particularly at the level of the basic economic unit, rather than on natural conditions, seed, or soil. The underlying concerns are with the ways in which people organize themselves to produce crops and livestock and the ways in which the products are moved off the farm through marketing or consumption.

General Overview

The main characteristics of Tunisian agriculture are shown in Table 1. The chief source of variability is rainfall which is concentrated between September and May, with very little rain in summer. Annual rainfall varies from more than 1,500 mm in the Kroumir mountains in the northwest to less than 100 mm in the southern deserts (Figure 1). Terrain also is important; areas with most rainfall are generally mountainous and not suited for agriculture. Finally, access to communication routes which favors market-oriented agriculture is variable.

The best farmland, combining favorable soil and rainfall conditions, lies

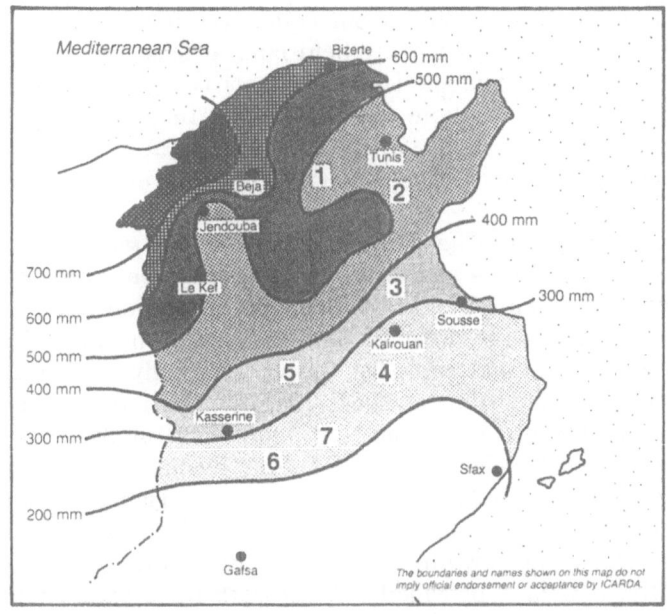

1. Testour
2. Bir Mcherga
3. Sbikha
4. El Menara

5. Hababsa
6. Ouled Sidi Ali ben Aoun
7. Sidi Bou Zid

Fig. 1. Rainfall distribution in Tunisia. *Source*: Johnson et al. (1983).

in a zone with average annual rainfall of 400–600 mm or sometimes 700 mm. A second important agricultural zone is the belt where average annual rainfall is 200–400 mm. These two zones run across the country from east to west and comprise roughly the northern half of the country. South of these areas there are oases and other pockets of irrigated agriculture, including some areas where farmers have adopted a terrace and water harvesting system to make best use of the scarce rainfall (El-Amami 1984).

Throughout the country, there are areas where irrigation is possible, which creates a very different pattern of agriculture. In some cases, irrigation is long-standing, and relies on springs, shallow wells, and rivers while in other areas, irrigation is the result of recent development projects and so reflects more modern conditions.

Land Tenure

The history of current land tenure patterns is complex, reflecting the changes which occurred during the colonial period (1881–1956) as well as an effort to collectivize land between 1964 and 1969. Many farmers do not have clear title to the land they farm, although they generally occupy it without challenge from their neighbors or the state. Since the beginning of the colonial period, there has been a shift from collective to individual landholding. Individuals took control of tribally held land at least two generations ago, and the state abolished religious endowments (*waqf* or *habus* lands) after independence. However, there is still a strong sense that land belongs to families more than to individuals, and so inherited land is often not divided. Thus collective farms persist and recur, even where management is individualized.

Important changes in Tunisian land tenure began with the institution of French land law in 1885, 4 years after the establishment of the protectorate. This allowed for freehold tenure with land registration and for European ownership of land. Large areas of the best land in northern Tunisia were transformed into colonial capitalist farms, especially after mechanization appeared in the 1920s (Poncet 1962; 1963). After independence in 1956, collective land was privatized and the *habus* land abolished. In 1964, all remaining colonial farmland was taken over by the state and an agrarian reform law was passed that promoted cooperative organization in agriculture. By 1968/69 there was a drive to collectivize all agriculture in Tunisia but this policy was reversed in 1969 due to resistance in various parts of the country. Subsequently, new laws provided a framework for agrarian reform. There have been sufficient changes in state policy towards land tenure to create an atmosphere of uncertainty (Kassab 1980) and the registration of all land and land transactions is still far from realized. Cooperativization ended in 1969 but it has been important in the history of Tunisian agriculture. The idea of a cooperative in any new venture remains unthinkable. Cherel (1964), Simmons (1970; 1971), and Stone (1971), and some of the Tunisian geographers and sociologists cited here give more details.

Some of the state land has retained a cooperative organization. In 1972 there were 216 Unites Cooperatives de Production employing 7,623 permanent workers, while in 1982 there were 212 with 6,786 workers, fewer than the strength of the Garde Nationale, and 26.4 ha of usable land per worker (Cherif 1986). The organization is bureaucratic and hierarchical, yet there is considerable slackness and poor financial management. In addition to the permanent members who have first call on available work, there are wage workers, in theory seasonal but in fact permanent, who do around 30% of the work. About 10% of the workers in 1977 have since left, mostly the younger ones moving to the cities. The average age of cooperative worker-members is higher than on private farms. Altogether about 47% of the workers live on or near the farms, 20% in nearby villages, and 33% in nearby towns, so one could say that they are semi-urbanized (Cherif 1986).

Farm Size

A high but regionally variable proportion of private farmland is owned by relatively few farmers (Table 2). The average farm size for the whole country is around 13 ha (Aubry et al. 1986), while 20–25 ha is generally cited as the

Table 2. Land distribution in Tunisia, 1980.

Farm size (ha)	No. of farmers		Area farmed	
	(×1,000)	(% of total)	(×1,000 ha)	(% of total)
0–20	296.0	83.3	1,832	39.4
20–100	54.6	15.3	1,947	42.0
>100	4.4	1.2	858	18.5
Total	355	100	4,637	100

Source: Ministry of Agriculture, cited Aubry et al. (1986).

minimum required for a viable farm under rainfed conditions. However, some irrigated farms are included in the figures. The best land is generally occupied by large capitalist farms, often the successors to the colonial farms, operated either by individual farmers or the state.

Approximately 8% of the farmland is owned by the state and cultivated as cooperatives (UCP or Unites Cooperatives de Production), state farms, etc. Medium-sized and small private farms are often found in marginal areas. Thus, where rainfall exceeds 400 mm there are more large, market-oriented farms than in the drier, more marginal southern zone.

Mechanization

Open field cultivation of cereals under dryland conditions has largely been mechanized since the 1920s, initially with tractors for plowing and seeding and subsequently with other machines such as combine harvesters and hay balers. At least 90% of farmers now use a tractor; the remaining farms are mostly either too small or the land is inaccessible to a tractor (Kassab 1983). There is an active machine rental market, so farmers who cannot afford their own tractor can rent from their larger neighbors and in fact about 90% of tractor users rent (Kassab 1983). The other machines are less common but their use follows the same pattern. Specialized cultivation, such as that of olives and other tree crops or under irrigated conditions, is less mechanized. Steep slopes and other limited areas are also often cultivated using animal power.

Cropping Patterns

Most Tunisian farmers of all sizes follow a strategy of mixed farming that includes cultivation of cereals under dryland conditions and investment of substantial resources (of labor if not of land) in animal husbandry. Wherever possible, trees are planted. The traditional combination was olive trees and cereal cultivation, but today one also often finds fruit trees (peaches, apricots, apples, pears, figs, quinces, pomegranates, etc., depending on natural conditions) or nut trees (almonds, especially in areas with marginal rainfall).

Finally, wherever physically possible, farmers attempt to develop irrigated areas for vegetables and other specialized crops.

Mixed farming can be construed as subsistence-oriented, in that the household attempts to supply most or all of its subsistence needs (olive oil for cooking, cereals, sheep for wool and meat, etc.). On the other hand, mixed farming can also reflect an effort to maximize returns in a money economy. The greatest returns are from irrigated farming and livestock, but households cannot concentrate on these alone because they may not have access to enough irrigated land and may not always be able to devote sufficient labor and pasture to animal husbandry. By producing a variety of different products, the farmer can make best use of his land and labor. This strategy is often combined with non-farm sources of income. For the small farmers, this may be paid labor either for larger farmers in the vicinity or outside agriculture, while the large farmers may have investments in industry, commerce, or urban real estate.

Farm Labor

Around 85% of farms rely entirely on household labor, particularly on small and medium-sized farms. The household is the basic decision-making unit and it controls land and other resources such as animals or machinery. Variations in family structure help to determine the choices of the family. For instance, a household without young children may not be able to maintain a flock of sheep since children usually herd the sheep. Hauling water home is also often a task for children when the distance is short. The ability of the husband to seek work off-farm, especially when that work is far enough away to require him to live away from the farm, is limited by the ability of the women to look after the farm, and thus seasonal or semi-permanent migration of the men can lead to the feminization of agricultural labor at the household level.

Wage labor has long been a feature of Tunisian agriculture, because of the large units formed during the colonial period when agriculture was mechanized. Before the colonial era, large holdings were not farmed as single units, but were distributed among sharecroppers who operated as independent farm households owing dues to the landowner. Large colonial farms had the same form of permanent workforce based on wage labor as contemporary state farms, production cooperatives, and large private farms. When this pattern first emerged, much of the labor was not recruited in the vicinity of these farms, because of people's reluctance to appear to be relinquishing their independence by working for another, but from the overpopulated mountainous areas in the north. Thus, historically, this kind of wage labor often represented an exogenous group in the local population. Recently, this pattern has been less pronounced, and wage labor now carries less of a stigma. The large farms now employ small farmers and others, predominantly adult males, as wage labor.

Medium-sized farmers may hire daily laborers at peak periods from what-ever labor pool may exist in the vicinity. This can lead to the use of female and child labor, where the work is appropriate and the labor available. Salaries tend to be higher for males. For example, in a northern Tunisian town in the early 1970s, female wages were 70% of male wages (Hopkins 1977). Local farmers spuriously justify this by arguing that women work less or accomplish less. However, the availability of female labor at lower rates can also exercise downward pressure on male wages so the wage differential serves the interests of those hiring the labor.

In a 1980 survey by the Institut National des Statistiques, 27% of the full-time labor force in agriculture was salaried, with the remainder being either farmers or family help. In the same year, the Ministry of Agriculture esti-mated that there were 41,360 permanent wage laborers and 93,060 temporary wage laborers (4 and 9% of the farm labor force, respectively). The remain-der were full- and part-time farmers (33%) and family members (54%). Farmers are more likely to be full-time than family members. Family labor is mainly female (between half and two thirds, depending on the type of labor and the rate of male emigration). Hired labor is overwhelmingly found on the large private farms, the cooperatives, and the state farms (agro-combinats) (Kassab 1983; Aubry et al. 1986).

Marketing

Marketing of crops is complex. Wheat is generally sold to the state wheat office, with tiny amounts exchanged among neighbors. At least in Testour, it was considered against Islamic law to buy grain in order to profit from its sale. Tree crops are typically sold on the tree to merchant-speculators (*khadhdhar*) who then organize the harvest and transport of the crop. Olive buyers are typically linked to the major olive presses, and can travel over the whole country, so in the early 1970s buyers from Sfax and the Sahel region appeared in Testour. Vegetables and other garden crops are often sold to itinerant merchants at the farm gate or in a nearby town at a *habbat*. There is a lively network of weekly markets throughout Tunisia where these crops may be bulked, but most of them are sent to the major wholesale markets in Tunis, Sfax, and other centers. Merchants with trucks purchase crops where they can find them, and farmers rarely transport their own crops to the main urban markets. Livestock sales typically take place at the weekly markets. There were an estimated 138 of these markets in the late 1960s (Stone 1976) and 275 10 years later (Michalak 1979).

Major Regional Divisions

Tunisia can be divided into regions although in such a varied landscape any division is an over-simplification. Sethom and Kassab (1980) and Achenbach (1983) give more details of the different geographical regions.

The Northern Mountains

This area receives 750 to over 1,500 mm of rain but is not considered to be favorable for crop production. Livestock and special forestry crops, such as cork oak from the forest around Ain Draham, are the main products. This area is the water reservoir for Tunisia, and there are an increasing number of dams and artificial lakes.

The Tell

This is the principal area of cereal cultivation in northern Tunisia. It includes most of the Medjerda Valley watershed, upstream from the newly irrigated areas near the mouth of the river, and certain adjacent areas to the south. It is bounded on the south by the 400 mm isohyet or geographically by the Dorsal, the long range of hills that extends from Jebel Chambi near Kasserine to the Cap Bon peninsula. Agriculture in this area is dominated by wheat cultivation combined with olives and fruit trees or with animal husbandry. Fields are cultivated every other year and used for pasturage in the intervening years. Herding adds flexibility to the economic activities of individual farmers. This is also the area where colonial farming was common and contemporary large-scale farms are found.

Early studies have been carried out in this region by Bardin (1965) and Cuisenier (n.d.; 1975). Kassab (1979) examined the wheat-growing areas of northern Tunisia in considerable detail. He distinguished three categories of farm size; less than 100 ha, 100–200 ha, and over 200 ha, and constructed a general model of the operation of each type. He concludes that, "The large and middle-sized farms are much more favored than the small, not only because they allow the rational use of agricultural machinery, but also because they lead to the production of a greater mass of marketable crops and easier access to all forms of bank credit".

Hamrouni (1981) analyzed the Amdoun, a mountainous area of tribally organized smallholders whose equilibrium was upset by the establishment of a few large capitalist farms during the colonial period. The people of the Mogods, near Sedjenane, have a history of mountain pastoralism combined with charcoal manufacture although recently they have begun some agriculture. The farms are organized in henchir of 300–500 ha (Jaafoura 1979). In the same area, a pilot project has experimented with common mountain pastures and with the institution of a local elected council to manage economic affairs, so far without much success (Venema and de Glas 1983).

Hopkins (1978a; 1983) studied the social organization of agriculture in one community in this zone. He stressed the contrast between a model of traditional agriculture based on shares and one of contemporary agriculture based on wage labor in the small town of Testour. Agriculture in Testour is a mixture of irrigated land (old and new), cereal cultivation, olive groves, and animal husbandry. The household plays a key role in the organization of agriculture. Dryland farming is based on a combination of wage labor and a high level of mechanization (tractors, combines, hay balers, and trucks). Machine owners seek to use machines to capacity, perhaps by renting land although some machine owners do not farm at all. The most prosperous households, and many not so prosperous, rely on a mixture of irrigated and rainfed cultivation. The modern farms are mechanized and have wage labor, bank credit, and increased security of land tenure (Hopkins 1978a). "The dominant mode of production is based on a mechanized agriculture in which large landowners farm their land directly with the assistance of wage labor, and produce crops on speculation directly for the market" (Hopkins 1977). Therefore, a class of wage laborers has emerged as the counterpart to the capitalist farming class. Thus land and capital (tools) determine agriculture, and labor is hired.

Other roles in the community include independent farmers, truck drivers, tradesmen, craftsmen, and merchants directly linked to agriculture, such as the *khadhdhar* who buy crops on the tree or in the field in order to harvest and market them, or the *habbat* who bulk small amounts of agricultural produce to attract traders from other towns or from urban markets.

In the 1960s, Ezzedine Makhlouf studied the extensive cereal cultivation patterns in three plains surrounding the city of El Kef, in the High Tell region (Makhlouf 1968; 1976). He traced the evolution of agriculture in the area from the precolonial through the colonial period, when this area had many large colonial farms, to the post-independence phase of transition towards a socialist agriculture. He focussed on the role of mechanization in the creation of large farms and on the effort to reorganize the relationship of the population to the land.

The High Plains

This area, which has predominantly a tribal form of social organization, extends south of the Dorsal and the 400 mm isohyet, on a plateau averaging around 600 m above sea level. Cereal cultivation is often attempted in this area, with success depending on rainfall, but herding and olives and other trees are more important. In this area there are small pockets of both old and new irrigation.

The area around Kasserine and Sbeitla was analyzed by Habib Attia (1977). This area combines substantial animal husbandry with opportunistic cultivation depending on rainfall. In this area, the cessation of the traditional migration to the north during the summer has led to more focus on local development. At the end of the colonial period, Bessis et al. (1956) carried out a study on the Ouled Sidi Ali ben Aoun at the southern limit of the cereal cultivation zone. They noted the sedentarization of the tribally organized population, and the growing importance of tree cultivation in addition to cereals. More recently, Ferchiou (1985) analyzed the increasing role of female labor in the Sidi Bouzid area, where small irrigated perimeters, both private and state-sponsored, are now dominant.

Salem-Murdock (1986; see also Schaefer-Davis 1985) carried out a study in Foussana and Sbeitla delegations of Kasserine governorate; she argues that most households participate in at least two of the three main ways of life: dryland farming along with herding, shallow-well irrigation, and participation in state irrigation projects. Dryland farming is the riskiest and the least rewarding of the three so that dryland farmers constantly aspire to transform themselves into irrigated farmers. In the meantime, the poorer farmers of all types are likely to supplement their farm income with wage labor within or outside the region. Success at irrigated farming of either type is also dependent on the ability to retain and mobilize household labor for this intensified cultivation. Thus when a household begins to rely on the sale of its labor, it becomes less able to take advantage of other development possibilities. Dryland farmers follow the strategy of diversifying production as much as possible, and minimizing the amount of labor spent on each crop, while selling labor and hoping for a chance to intensify. As the sale of labor power increases, women perform more of the actual farming tasks. In general, Salem-Murdock concludes that "While the affluent diversify to increase their wealth, the poor diversify to meet the daily requirements of maintaining and reproducing their households".

Hopkins (1978b) reported that in the high plains area there are many small holdings but few landless, reflecting the underlying tribal structure. In one area where titles were distributed (Bnanna II north of Kasserine), 80% of the recorded owners had 5 ha or less. Farmers are reluctant to sell land, to which they can retreat in case of difficulty. The strategy of these farmers is low risk and low gain and is aimed at survival, so they tend not to invest in agriculture. Rainfed farming involves a mixture of wheat and barley, supplemented by cactus, some trees, and animal husbandry. There is diversification, but with the goal of continuity or reproduction rather than profitability. Here, labor is more important than land or capital so the family cycle is of key importance as it determines how much labor is available to the household. Households with young children can be short of labor for a while, then the reverse before the next generation takes over. Households often adopt labor migration as part of their strategy. The family structure supports migration, which produces money that can reinforce the economic position of the family. The women and children may be left at home with an elder male such as a father-in-law, which ensures the eventual return of the migrant and also the flow of remittances (see also Salem-Murdock 1986).

Based on interviews in the same area, near Douleb, Zamiti (1982) showed how one generation succeeds another in the migration process, and how the elders use the control of marriage to retain the loyalty of their migrant sons. In addition to migration, activities such as charcoal preparation and beekeeping supplement wheat cultivation and animal husbandry. Zamiti suggests this is a combination of autarky and integration into the capitalist systems through migration.

The Low Plains

The flood plains around Kairouan and the flat hinterland of Sfax are included in this area. Animal husbandry and dry cultivation of orchards and olive groves predominate. The hinterland of Sfax in particular is known for its ingenious use of runoff to water olive trees. In the area around Kairouan, the spread of orchards is more recent, and partly reflects better control of the flood water. In the northern parts of the area, wheat or barley may be grown.

Martin (1983) analyzed the role of agriculture in the development of the delegation of Sbikha, in the semi-arid low steppes of the northern part of the Kairouan governorate. This area has an irrigated perimeter dating from the immediate post-independence period, an agro-combinat set up with Austrian technical assistance, a zone of private shallow-well farmers, and dryland farming combined with animal husbandry. These different types of land use represent different combinations of capital, machinery, and labor, ranging from capital-intensive agro-combinats to labor-intensive traditional dryland farming. Martin concludes that the private shallow-well farmers who are intermediate represent the most effective combination of factors but notes that this combination would be difficult to extend beyond the areas where there is an accessible water table.

The imada of El Menara is 55 km southwest of Kairouan. In 1975 it had a population of 6,603 and an area of 22,650 ha. Zamiti (1977) traced the history of this area from the time when tribal segments collectively owned grazing land while individuals owned animals, to the assignment of the best land to French colonial farmers in 1927. This was followed by the creation of cooperatives in the 1960s and a liberal period beginning in 1970. The result of this process is that most machinery (13 tractors, 2 combine harvesters, and 15 pickup trucks) is owned by 15 of the largest landowners, who each own over 20 ha, while 83% of landowners own between 5 and 20 ha. The family is still the main source of labor, but machinery owners control the situation through the rental of machinery. As a result, farmers with less than 20 ha cannot live from their farming, but must rely on the World Food Program. However, they can only have access to this source of food if they belong to the party, which in turn is controlled by the rich.

The Coastal Plains

The coastal plains extend from the Sahel area in the south through the Cap Bon and the Tunis plain to the Bizerte coast in the north. This area is quite variable but generally has mostly a village form of social organization (especially in the Sahel, parts of Cap Bon, and the Bizerte area), no large holdings (except around Enfidha and in parts of the Cap Bon), and farmers rely on production of both olives (as in Sfax) and vegetables. The presence of Tunis in this area has produced distinctive patterns of landholding, reflecting the urban penetration of the countryside, and considerable market gardening.

Sethom (1977a) divided the Cap Bon into a series of mini-regions consisting of a citrus belt on the eastern slopes of the Grombalia plain, a vineyard belt on the western slopes of the plain, the central plain, the Bled Takelsa on the western side of Cap Bon, an inland area of cereals and animal husbandry (the Dakhla des Maouine), and the gardens and orchards of the northern and eastern shores of the peninsula. He examined each mini-region historically and in terms of the major agricultural activities and problems. In another study (Sethom 1977b) he described the principal crops (citrus, grapes, cereals, vegetables, etc.) and constructed a general model of each one. This is an area of agricultural intensification, with considerable localized irrigation. In parts of the east coast, tourist development now competes with agriculture for space and water.

The special circumstances of the agricultural zone surrounding Sfax were analyzed by Fakhfakh (1976). The rainfall is generally insufficient for cereal cultivation, but olives are cultivated using ingenious methods of water harvesting. Indeed large olive farms predominate in this area. Fakhfakh's early study (1976) analyzes the period during which the Tunisian government was following a policy of cooperativization (1964–69) while the more recent study (1986) is concerned with the economic and geographic region around Sfax city, and analyzes olive production and agriculture in general through the pattern of urban dominance. Jedidi (1975) carried out a more specialized study of the town of Jebeniana, 36 km northwest of Sfax, and its coastal region. This area includes many small farms of less than 20 ha, generally given over to olives in the interior and to market vegetable gardening along the coast. It is intermediate between the large olive groves of the Sfax hinterland and the Sahel pattern of small subsistence-oriented farming.

The South

This region includes the area with less than 200 mm rainfall, roughly south of a line from Sfax to Gafsa and is therefore outside the area of rainfed agriculture. Cereal cultivation is only possible in isolated areas where the rainfall is slightly higher (the island of Jerba) or where the terrain allows for the concentration of scarce rainfall as in the Matmata mountains. This is also the area where the classic oases of Nefta, Tozeur, Gafsa, Gabes, and the smaller ones of the Nefzawa are found and agriculture is concentrated around these oases. Animal husbandry is possible here, but is generally not combined with agriculture, which is the speciality of the oasis farmers.

Contrasting Farming Systems: Bir Mcherga and Hababsa

The regional studies discussed above, mainly by Tunisian geographers, cover too broad an area to clarify the detailed operation of specific farming systems, though they are suggestive. This section contrasts two more intensive studies, one carried out in the Tell area of Bir Mcherga, the other on the mountainous northern border of the High Plains in Hababsa. The two areas are roughly comparable in size, but differ in many other respects. Bir Mcherga is a relatively prosperous grain-growing area with good transportation links to urban Tunisia, and is characterized by large state and private farms, with wage labor and off-farm employment. Hababsa is poor and isolated, and is characterized by small and medium-sized farms, no local paid labor, and labor migration. Rainfall is somewhat higher in Bir Mcherga than Hababsa, but other differences are probably more important.

Bir Mcherga

INRAT carried out a study in the imadas of Ghrifet, Bir Mcherga, and Sminja, north of al-Fahs in the governorate of Zaghouan (Aubry et al. 1986) where the annual rainfall is around 400–450 mm. Approximately 79% of the land is under cereals, 20% under trees, and only 1% is irrigated, but the irrigated land provides 13–15% of the value of agricultural production. Tree crops are more valuable than field crops.

The land of the three imadas is 35% state cooperatives and 65% private. Of the private land held by 500 farms, 8% is held in 250 farms of less than 20 ha, 31% is held in 224 farms of 21–100 ha, and 61% of the land is held in 40 farms (8%) of more than 100 ha. These three sizes correspond to the subsistence, peasant (partially market-oriented), and fully market-oriented types. Of the total number of farmers, 17% own a tractor and 79% of all tractors are owned by farmers with more than 35 ha. Only farms of more than 100 ha hire labor while the others rely on family labor, and thus their activities are dependent on the availability of such labor. For example, the presence of children to herd is a prerequisite for keeping a flock of sheep. The smaller farms cannot survive without off-farm income, but such income is also common in the larger farms, especially in the largest farms, whose owners often hold them as an investment while they work elsewhere.

The research team considered different paths of evolution – diversification through the introduction of irrigation and tree crops, intensification of cereal cultivation through mechanization and other inputs, and expansion of farm size. Of the three farm types, the peasant type seems to have the highest propensity to evolve, with the best combination of devotion to farming and resources. Overall, land is most likely to change hands between members of the same category, but labor exchanges occur between units of unlike size.

Aubry et al. (1986) conclude, "We have just presented several characteristics which, in our opinion, profoundly structure this local agriculture: the frequent need for non-agricultural resources, the variable place of agricultural activity in the hierarchy of activities in which the families of the zone partake, limited land transactions and then between units of equivalent size, exchange of work and products which strongly affect the manner in which many units produce . . . The existence of the unit of production is only really in danger in farms of 20–35 ha without outside resources and in farms of more than 100 ha whose owners, often absentee, distinctly prefer urban modes of life, consumption and work . . . Non-agricultural activities are going to play an even larger role in the reproduction of the production units in the satisfaction of family requirements."

Hababsa

Hababsa, in the governorate of Siliana, has less rainfall and poorer transportation than Bir Mcherga and has no state land or large farms. The rainfall is above 400 mm in the north and less than 400 mm in the south. In 1977/78, Hababsa had a population of around 5,000 divided almost exclusively into farming households (Zghidi 1978). The main agricultural activities are cereal cultivation (with wheat more important than barley) and animal husbandry, with a few small pockets of irrigation (on irrigation see Gallali et al. 1979).

The entire area was *habus* (religious endowment) until the Tunisian government abolished the *habus* in 1957 and the land was formally divided up among the inhabitants, who had probably already laid informal claim to their land. The occupation of the land reflects the tribal structure that dominates the area. Most of the inhabitants are the descendants of a saint (Nasr Ben Abderrahman Bel Habbes) and have been assimilated to the Ouled Ayar confederation, while others are immigrants from other branches of the Ouled Ayar and other tribes. Although habitation is scattered, close relatives live near one another. In Zghidi's study, 86% of the respondents said they had inherited their land, 14% had bought it, and 32% were co-owners.

The farms are largely independent; 85% of farmers did not hire labor and 96% farmed at least some of their own land themselves. Most farmers used their own inputs (only 40% had bought wheat seed and 36% barley seed in the previous year) and they consumed most or all of what they produced. The previous season was very poor, so comparing consumption and production was difficult. Only 18% of respondents were judged to have produced enough for their own consumption. The critical issue was survival rather than marketing of surplus.

The farm unit included land (divided into an average of 10 plots), simple tools such as the plow, animal and occasionally mechanical power, and family labor. The researchers estimated that on average 25 ha were necessary to provide a living for a household. Of a sample of 111 farms, the mean was 20 ha, with 31.5% owning over 25 ha, 46% with 10–25 ha, and 22.5% with less than 10 ha. The largest holders, with more than 40 ha, comprised 19%

of farms holding an average of 58 ha. Thus, by national standards, this area is one of small and medium-sized landowners.

Generally, the smaller the farm, the greater the proportion of the land under cereals which is a sign of a subsistence orientation. Only 33% of the farmers fallowed their land. The most common rotation was wheat followed by barley, although some farmers cultivated wheat continuously. Tractors came into the area in 1972, 5 years before the study, and in 1978 were still scarce. Most traction was therefore by animal power, although 15% of household heads said they had no animals or tractors of their own and so relied on others. In a typical sharecropping arrangement here, a farmer who has access to excess traction cultivates his neighbor's field in return for a share.

Another source of income and activity is animal husbandry. The data in Zghidi's study were affected by the bad year during which the survey was carried out. They showed that 50% of households were keeping cattle, most with 1–2 head, 62% were keeping sheep, most with 5–20 head, and 36% were keeping goats, most with 1–5 head. The herd or flock size reflects the household labor capacity, particularly the presence of children to herd the animals, take them to drink, etc. Cash income came from the sale of animals or tree crops such as olives or almonds, and from migration outside the area which was usually for relatively short periods (1–2 months), with some members of around one third of households migrating in any year. The small pockets of irrigation were isolated and did not produce much for sale. Thus Zghidi concluded, "Outside of almonds, subsistence predominates, and marketing is rare" (Zghidi 1978).

The dominant form of social organization in the area is kinship which is reflected in patterns of exchange between households. However, the nature of the wider society is reflected in growing monetization of the economy, undercutting the significance of kinship, as people now expect wages for their help. Thus a process of differentiation has begun. Further, the peasant in trouble now no longer looks to a kinsman for help, but to the state. Only the state can change the situation (Zghidi 1978). Thus the relative isolation of Hababsa is only apparent, and in fact the sector is progressively being integrated into the national structures. Since this study was completed, the road network into Hababsa and the water supply have been improved, both as a result of state investment.

Conclusion

The general picture of Tunisian agriculture is thus one of considerable variety. The relatively market-oriented north (the Tell) is polarized between large private farms and state cooperatives on the one hand and a mass of small farmers on the other. In the former, wage labor predominates, mechanization is advanced and farmers own their machines, most produce is sold on the market, and agricultural credit is available. In the latter, family labor predominates, machinery is rented, a large proportion of the produce is consumed at home, and agricultural credit is difficult to obtain. Researchers also identify a group of farms intermediate between these extremes; Aubry et al. (1986) show that this category is the most innovative and enterprising.

There is a contrast between the market-oriented north (roughly the zone with over 400 mm rainfall) and the more subsistence-oriented center (roughly the zone with 200–400 mm rainfall). In some ways, the small farmers of the center resemble the marginal farmers of the north, but probably rely more on animal husbandry because of the risky nature of grain cultivation. The small farmers of the center also live in a less polarized environment, and so are less constrained by the choices of neighbors with large farms.

In both areas, there is a tendency to supplement grain cultivation with tree crops (especially olives), animal husbandry, and irrigated gardening wherever possible. Off-farm employment is also common. In the north, some off-farm employment is available on large farms in the vicinity of the smaller farms. In the center, and sometimes in the north, men must migrate in search of paid work which appears to result in the feminization of farm work (see Ferchiou 1985) though more studies are needed. Thus it appears that, as part of a household strategy, off-farm income is added to farming income rather than becoming a substitute for it. Very large farmers are often reported to be absentees, preoccupied by their urban interests. Permanent and semi-permanent wage labor is concentrated on the large farms and state cooperatives and is rarely used by small farmers.

Certain kinds of machinery, particularly tractors, have been used since the 1920s in the Tell area of northern Tunisia. Ownership of tractors is highly concentrated, and there is an active rental market. Tractors are used outside agriculture, particularly for hauling water and other loads. The use of tractors for field preparation is practically universal and the use of combine harvesters nearly so, especially for wheat grown on relatively flat land in large fields. Current trends of mechanization badly need to be documented.

The coastal areas differ in some respects from the north and center (the Tell and Steppes). The northern mountains were historically an area of forest pastoralism, but are now being cultivated in some parts. The Sfax area and the Sahel are areas of olive groves and market vegetable gardening, with olives an industrial crop, and the vegetables essentially for local consumption. The Cap Bon, the Tunis ring, and the Bizerte area have a very mixed ecology with many citrus trees, vines, and similar crops, as well as vegetable gardening. Further studies are needed on the organization of labor in these areas.

In most areas, the household is the key decision-making unit. In the center, and sometimes elsewhere, the chief resource of the household is its labor, the subject of decisions on matters such as schooling, migration, wage labor, and home farming. In the north, land is sometimes more important. There is a conventional division of labor between men and women, which is changing, but we lack details on household dynamics. However, households are not always free to make the decisions which seem most rational because they are linked together as a community with specific values and a power structure (Aubry et al. 1986). The community leaders are often successful in imposing their solutions on their neighbors (Zamiti 1977; Bouzaiene 1985; Hopkins 1983). Thus local as well as national politics can be constraints to agricultural development.

Given the high level of monetization, marketing is a key activity. Prices, however, are not freely set but reflect government policy. Kassab (1986) argues that prices for grains and beef are too low to encourage much production. The animal population in particular reflects market conditions as well as rainfall and pasturage, and can fluctuate by a factor of four or five depending on these conditions.

The general features of Tunisian agriculture are: mechanization; market-orientation; credit for medium-sized and large farms; the dominance of the household for at least 85% of farms; the existence of an established agricultural proletariat; substantial off-farm income for all levels of independent farmers; combination of rainfed grain farming with tree crops, animal husbandry, and irrigation; and a stable but insecure land tenure system. However, from a farming systems perspective stressing labor organization, it is apparent that a number of questions require more research. These include household dynamics and especially decision-making; the role of female and child farm labor; current trends in mechanization; how households balance rainfed field crops with animal husbandry, tree crops, and irrigation as well as with off-farm activities and migration; how social, cultural, and political processes work at the community level to guide and give value to the different choices that farmers must make; the significance of differentiation among farmers and the implications of this for local social and political processes; the extent to which improved marketing conditions (both organization and pricing) can encourage more intensive agriculture; the extent to which co-operation among small farmers can improve the conditions for agriculture; an analysis of the role of the state, and the relationship between this role and the general social structure of Tunisia.

References

Achenbach, H. 1983. Agrargeographie-Nordafrika. Afrika-Kartenwerk, Beiheft N-11. Stuttgart: Gebrüder Bornträger.

El-Amami, S. 1984. Les Aménagements Hydrauliques Traditionnels de Tunisie. Tunis: Centre de Recherche du Génie Rural.

Attia, H. 1977. Les Hautes Steppes Tunisiennes de la Société Pastorale à la Société Paysanne. PhD thesis, Université de Paris.

Aubry, C., Elloumi, M. et al. 1986. Les Systèmes de Production dans le Sémi-aride: Première Approche de la Dynamique des Exploitations dans la Région de Zaghouan. Annales de l'Institut National de la Recherche Agronomique de Tunisie 59(1): 1–230.

Bardin, P. 1965. La Vie d'un Douar: Essai sur la vie Rurale dans les Grandes Plaines et la Haute Medjerda, Tunisie. Paris: Mouton.

Bessis, A., Marthelot, P., de Montety, H. and Pauphilet, D. 1956. Le Territoire des Ouled Sidi Ali Ben Aoun. Mémoires du Centre d'Etudes de Sciences Humaines. Volume I. Publications de l'Institut des Hautes Etudes de Tunis. Paris: Presses Universitaires de France.

Bouzaiene, B. 1985. Réaction des Communautés Rurales aux Modèles Proposés par le Programme de Développement Rural 'PDR'. Revue Tunisienne de Sciences Sociales No. 82/83: 29–74.

Cherel, J. 1964. Les Unités Coopératives de Production du Nord Tunisien. Review Tiers-Monde 5(18): 235.

Cherif, A. 1986. Secteur Organisé et Développement Agricole: UCP et Fermes Domaniales du Haute Tell Tunisien. Etude de Géographie Economique. PhD thesis, University of Tunis.

Cuisenier, J. n.d. L'Ansarine: Contribution à la Sociologie du Développement. Mémoires du Centre d'Etudes de Sciences Humaines, Volume VII. Tunis: Université de Tunis.

Cuisenier, J. 1975. Economie et Parenté: Leurs Affinités de Structure dans le Domaine Turc et dans le Domaine Arabe. Paris: Mouton.

Fakhfakh, M. 1976. La Grande Exploitation Agricole dans la Région. Tunis: Ecole Normale Supérieure.

Ferchiou, S. 1985. Les Femmes dans l'Agriculture Tunisienne. Aix-en-Provence: Edisud.

Gallali, T., Chabbi, A. and el Amami, S. 1979. Mode d'Irrigation dans le Fosse d'Effondrement d'Oued El-Hattab en Tunisie Centrale ou l'Exemple d'Adaptation des Techniques aux Conditions du Milieu. Pages 147–152 in Actes de IV Colloque de Géographie Maghrébine. Volume I. L'homme et la Montagne. Cahiers du CERES, Série Géographie 4. Tunis: CERES.

Hamrouni, T. 1981. La Paupérisation d'une Collective Montagnarde du Tell: les Amdoun. Revue Tunisienne de Géographie 8: 99–116.

Hopkins, N.S. 1977. The Emergence of Class in a Tunisian Town. International Journal of Middle East Studies 8: 453–491.

Hopkins, N.S. 1978a. Modern Agriculture and Political Centralization: A Case from Tunisia. Human Organization 37(1): 83–87.

Hopkins, N.S. 1978b. Elements for a Social Soundness Analysis of the Agricultural Interventions, Central Tunisia. Tunis: USAID.

Hopkins, N.S. 1983. Testour ou la Transformation des Campagnes Maghrébines. Tunis: Céres Productions.

Jaafoura, I. 1979. Le "henchir", Unité d'Organisation Agricole Traditionnelle des Mogods (Tunisie du Nord). Etudes Rurales 73: 145–146.

Jedidi, M. 1975. Jébeniana et sa Region: Etude Géographique. Geography-Sociology series, Volume III. Tunis: Faculté des Lettres et Sciences Humaines.

Johnson, W.F., Ferguson, C.E., and Fikry, M. 1983. Tunisia: The Wheat Development Program. Project Impact Evaluation No. 48, Document No. PN-AAL-022. Washington, D.C.: USAID.

Kassab, A. 1979. L'Evolution de la Vie Rurale dans les Régions de la Moyenne Medjerda et de Béja-Mateur. Geography series, Volume VIII. Tunis: Faculté des Lettres et Sciences Humaines.

Kassab, A. 1980. Etudes Rurales en Tunisie. Geography series, Volume XI. Tunis: Faculté des Lettres et Sciences Humaines.

Kassab, A. 1983. L'Agriculture Tunisienne. Revue Tunisienne de Géographie 10–11: 1–390.

Kassab, A. 1986. Les Secteurs en Développement et les Secteurs en Crise dans l'Agriculture Tunisienne. Paper presented at a conference on Le Developpement en Question: Dimensions, Bilan et Perspectives, Tunis, 24–29 November 1986.

Makhlouf, A.E. 1968. Structures Agraires et Modernisation de l'Agriculture dans les Plaines de Kef: Les Unités Coopératives de Production. Cahiers du CERES, série geographie I.

Makhlouf, E. 1976. Political and Technical Factors in Agricultural Collectivization in Tunisia.

Pages 381–411 *in* Popular Participation in Social Change (Nash, June, Dandler, J. and Hopkins, N.S., eds). The Hague: Mouton.

Martin, Catherine. 1983. Agriculture et Développement en Tunisie Centrale. Mémoire de Géographie. Lausanne: Institut de Géographie, Université de Lausanne.

Michalak, L.O. 1979. Vendor Strategies in Tunisian Weekly Markets. Paper presented at the meeting of the American Anthropological Association, Cincinnati.

Poncet, J. 1962. La Colonisation et l'Agriculture Européennes en Tunisie depuis 1881. The Hague: Mouton.

Poncet, J. 1963. Paysages et Problèmes Ruraux en Tunisie. Mémoires du Centre d'Etudes de Sciences Humaines, Volume VIII. Tunis: Université de Tunis.

Salem-Murdock, M. 1986. Household Dynamics and the Organization of Production in Central Tunisia. Working paper 28. Binghamton NY: Institute for Development Anthropology and Clark University.

Schaefer-Davis, S. 1985. Employment Generated by Projects of the Central Tunisia Development Authority. Working Paper 27. Binghamton, NY: Institute for Development Anthropology and Clark University.

Sethom, H. 1977a. Les Fellahs de la Presqu'île du Cap Bon (Tunisie): Etude de Géographie Sociale Régionale. Geography series, Volume IV. Tunis: Faculté des Lettres et Sciences Humaines.

Sethom, H. 1977b. L'Agriculture de la Presqu'île du Cap Bon (Tunisie): Etude de Géographie Sociale Régionale. Geography series, Volume V. Tunis: Faculté des Lettres et Sciences Humaines.

Sethom, H. and Kassab, A. 1980. Géographie de la Tunisie: Le Pays et Les Hommes. Geography series, Volume XII. Tunis: Faculté des Lettres et Sciences Humaines.

Simmons, J.L. 1970. Agricultural Cooperatives and Tunisian Development. Part I. Middle East Journal 24: 455–466.

Simmons, J.L. 1971. Agricultural Cooperatives and Tunisian Development. Part II. Middle East Journal 25: 45–57.

Stone, R. 1971. Tunisian Cooperatives: Failure of a Bold Experiment. Africa Report 16(6): 19–22.

Stone, R. 1976. Rural Markets: Organization and Social Structure. Pages 121–135 *in* Change in Tunisia: Studies in the Social Sciences (Stone, R.A. and Simmons, J., eds). Albany: State University of New York Press.

Venema, B. and de Glas, M. 1983. L'Application de la Réforme Agraire dans le Nord-ouest Tunisien: Priorités Nationales ou Locales? Maghreb-Machrek 100: 42–56.

Zamiti, K. 1977. Exploitation de Travail Paysan en Situation de Dépendance et Mutation d'un Parti de Masses en Parti de Cadres. Les Temps Modernes 375 bis: 312–333.

Zamiti, K. 1982. Dialectique de la Dissolution et du Maintien des Formes Communautaires en Tunisie. Peuples Méditerranéens 18: 195–217.

Zghidi, M. 1978. Monographie de Hababsa (Governorat de Siliana). Draft Report. Tunis: Centre National des Etudes Agricoles, Ministère du Plan.

13. Agricultural Labor and Technological Change in Morocco

AZZEDDINE M. AZZAM

Department of Agricultural Economics, University of Nebraska, Lincoln, NE 68583, U.S.A.

Introduction

In an attempt to increase agricultural output, and so relieve the population pressure on food supplies, less developed countries (LDCs) have become active seekers of new agricultural technology. This has been important in the transition of some LDCs to near or complete self-sufficiency (Darlymple 1977; Hayami and Ruttan 1985). Others have been less successful, despite increases in some outputs with new techniques. For example, the rate of self-sufficiency decreased in all but five African countries during 1971–80 (FAO 1984). The impact of labor-saving technologies on rural labor has become a major issue, in view of the vast surplus of agricultural labor in LDCs, the scarcity of non-farm employment opportunities, and the rate of rural-urban migration (see Richards, Commander, this volume). This paper presents an analytical review of available literature on agricultural labor and technological change in Morocco.

Agriculture in Pre-colonial Morocco

The pre-colonial (before 1912) Moroccan rural economy was largely self-sustaining (Stewart 1964). Nomads traded with farmers who in turn traded with the city dwellers. Foreign imports included tea, sugar, candles, cotton, cloth, and some hardware but the economic links with the outside world were limited to the vicinity of Atlantic ports (Hoover 1978; Godfrey 1985) and the impact of foreign ideas and practices on the economy was slight (Brace 1964). Land registration was almost nonexistent, and property was held in several ways. The most prevalent was *melk* land (private property), especially in the western plains where extensive cultivation was possible. The owners of private land did not necessarily cultivate their own land and absentee landlords were, and still are, common. Under the traditional *khammas* system, tenants supplied the labor for a fifth of the crops. Often, especially in years of drought, tenants could not survive on their share so they relied on landlords to provide credit, to be repaid from future harvests. "Frequently unable to get out of debt, the tenant was held in virtual slavery

Dennis Tully (ed.), Labor and Rainfed Agriculture in West Asia and North Africa, 273–295.
© 1990 *ICARDA.*

by the landlord, bound to the land unless released by his creditor" (Stewart 1964).

Domainial land, mostly expropriated from rebellious tribes, was owned by the Sultan. By 1912, state power was waning and there was less government-owned land, which meant less revenue, and so the Sultanate had to mortgage port customs to make up for lost agricultural taxes (Colburn 1927).

Habous lands, which belong to religious foundations, arose from charitable contributions. These lands are generally leased out with the income to be dispersed for religious purposes. The *jaysh* land system was introduced in the 12th century to reward those who offered military services against rebel tribes (Stewart 1964). The owners had permanent usufruct. Collective land was predominant in the mountainous and semi-arid areas, mainly in the High Atlas and Rif regions. It includes land less suitable for continuous cropping because, under Moslem law, land cannot be privately owned unless worked and cultivated (Nelson 1978).

According to one French account, "agricultural resources were very limited. Farming methods were archaic, the Biblical wooden plow and primitive hoes were the only implements used, and the extent of cultivated land was limited to the vicinity of towns and fortified villages because of the ever-present danger from warring tribes who burnt crops and carried off livestock" (Anonymous n.d.).

The *fellahs* (peasant farmers) were the main source of tax revenue. Muslims paid taxes on produce and livestock, while non-Muslims paid a head tax. Levies were also made for war and other special purposes. These taxes were often insufficient, necessitating foreign borrowing and other practices which eventually led to the collapse of the pre-colonial financial structure (Landau 1956).

In pre-colonial agriculture, therefore, production and distribution operated within a rigid, stratified institutional structure. The unstable central power merely extracted resources from the rural sector and so the stimulus for social and technological change was minimal. A distrust of central authority, which is often claimed to be among the major obstacles to agrarian progress, may have had its origin in this period, and been reinforced in the subsequent colonial era.

Moroccan Agriculture under the French

Technological change in the period 1912–56 can be studied through the chronology of legal, social, and economic programs designed to modernize the rural areas.

The registration of property was introduced in 1913 by the Torrens Act (Landau 1956) and peasants who were either unaware of the act or unable to pay the registration fees had their lands expropriated. Apart from the 1873–83 drought (Godfrey 1985), the 1913 land registration scheme was perhaps the most important factor in the emergence of a landless peasantry in Morocco.

In 1914, a royal decree was instituted to distinguish between alienable and inalienable land. *Habous*, *jaysh*, and collective lands were declared inalienable. *Melk* land was declared alienable and hence could be expropriated either by sale or seizure. State land was taken over by the administration and later sold or distributed to French settlers. According to Swearingen (1987), between 1917 and 1956 (the year of independence), 289,000 ha were distributed to settlers under official colonization policy. To encourage mechanization, the land was distributed in holdings larger than 150 ha. Private settlement amounted to 728,000 ha making the total area of land owned by Europeans 1,017,000 ha.

This represented 10% of the total crop land, 23% of the orchards and vineyards, and 4% of the fallow land (Stewart 1964), while Europeans accounted for only about 1% of the population. In addition, French settlers controlled land owned by Moroccans by supplying certain inputs under traditional types of contracts.

The driving force behind the French hunger for native land was irrigation. Moroccan and French speculators, attracted by the promise of government-supplied irrigation, acquired large holdings before the government could conduct land registration (Swearingen 1987). Speculators were most successful in acquiring unsurveyed collective lands. For small sums, Moroccans were "willing to sell land out from underneath their fellow tribal members or land of which they did not even share possession" (Swearingen 1987). In this process, landless peasants either worked for wages or migrated to major cities. The drought in 1936–37 and 1944 accelerated migration. Starvation drove the peasants to sell their land for a pittance. To counteract the rural exodus and protect farmers from dispossession, the authorities legislated that all transactions were prohibited for holdings of less than 7.5 ha for dry land, 1.5 ha for irrigated land, or 0.75 ha for orchard land (Swearingen 1987). This legislation, known as the *bien de famille*, was too little, too late (Stewart 1964).

Agencies to improve the technological level and productivity of Moroccan agriculture were first established in 1917, but these and later efforts were not effective due to the inadequate financial and personal resources provided and other factors (Van Wersch 1968). Credit agencies also did not reach the Moroccan small farmers until shortly before independence.

In 1945, *Secteurs de Modernization de Paysannat* (*SMP*) were formed to modernize collective land (L'Oeuvre de la France 1950). Membership of the SMP was compulsory. However, SMP mechanization schemes mainly benefited the rich in the rural areas (Anonymous n.d.), and under SMP modernization schemes, the natives lost control of their land with the promise that it would be returned to them when developed. In the words of one French official, "Faced with the inability of the Moroccan farmers to alter their ancestral habits, the administration of the protectorate decided in 1945 to take in hand the modernization of their means and methods of cultivation step by step" (Guillaume 1952). Ten years after the first SMP was formed there was no sign of land being returned to its owners (Stewart 1964).

While irrigation as introduced by the French has continued to be used, other new techniques were less suitable in independent Morocco. Settlers replicated their home-country farming systems, responded to demands outside the settled territory, and operated under European cultural endowments and institutions, with the result that the pattern of technological choice was inconsistent with the relative factor endowments of Morocco. Because the institutional structure was established in the context of an externally oriented agriculture, biases would persist today even if technologies correctly reflecting factor endowments were put in place. Swearingen (1987) confirms this and attributes the failure of Moroccan agricultural development strategy to practices that are not in harmony with existing environmental and institutional realities.

An equally important consequence of this uneven technological change was irreversible rural-urban migration. The reduction in arable land per capita caused many in the indigenous rural population to migrate to cities, establishing the shanty towns that still exist today.

Post-Independence Rural-Oriented Policies

Land Reform

On the eve of independence (1956), it was estimated that 5–10% of rural households owned 60% of the total land, 50% owned 40% of the land, and 40% were either landless or cultivated less than 0.5 ha (MEN 1960). According to Zartman (1964) large landowners were mostly either *colons* or "feudals" who had become rich during the protectorate. An agricultural commission appointed after independence found that 75–80% of the population subsisted on traditional agriculture, while 50% of rural families owned less than 3 ha, and the rate of rural unemployment (presumably including underemployment) was 50% (Demengeot 1975). In response to the pressure for land reform following independence, the government undertook several land distribution schemes.

Between 1956 and 1963 15,237 ha were distributed (Swearingen 1987). In 1963, the government announced that it would take over the land which had been owned by French settlers on a concessionary basis. The decree proclaimed the land state property and prohibited sale without prior government approval. Nevertheless, approximately 500,000 ha were transferred from French to private Moroccan farmers, mostly after 1963, of which as much as 100,000 ha may have been transferred illegally (Griffin 1975).

The government took over about 256,000 ha, but by the end of 1964 only

2,560 ha had been distributed to individual beneficiaries. The remaining area was put under newly created government management centers in an attempt to avoid further fragmentation and loss of productivity. However, the centers lacked the technical knowledge required by these modern farms. Subsequent land distribution programs failed for numerous reasons, including poor choice of beneficiaries, differences in cultivation practices, lack of cooperative spirit and organization, apathy to state recommendations, lack of understanding of contractual obligations, the surrounding traditional environment, interference by local authorities, constant turnover of extension workers, failure of farmers to pay their debts, and excessive mechanization (Demengeot 1975).

Increased production was the objective of the second (1965–67) plan, and little attention was paid to dryland areas or land reform. Resources were to be allocated to the sector with the highest potential rate of return. The traditional sector was considered to be resistant to change. The plan proposed a new program, based on "the success of an experimental project, in which farmers settled on distributed state land have been induced by economic incentives (credit, subsidies and production contracts) to grow sugar beets under modern conditions" (Demengeot 1975).

A land fund was created for the program in 1965 consisting of previously recovered land, state land, and land still owned by foreigners. However, by 1968 only 8,630 ha were distributed

The next 5–year plan (1968–72) suggested measures to improve the small traditional farms, but it devoted most financial resources to the irrigation of 1 million ha initiated during the colonial period.

In 1973, a new Moroccanization decree was announced to finish land decolonization and reduce foreign control (Nelson 1978). A total of 395,000 ha were expropriated under this decree, including 172,000 ha from foreign individuals and 94,000 ha from corporations. The 160,000 and 135,000 ha of rainfed and irrigated land, respectively, were to be distributed by 1977 (Griffin 1975).

All told, up to 1972 over 11,000 farmers had received 184,604 ha in redistribution schemes; of these 141,228 ha were distributed in 1970–72. The average farm size was 16.3 ha. By 1982, an additional 187,792 ha had been distributed bringing the cumulative total to 372,396 ha. Simultaneously, other processes led to much greater increases in the number of farms, especially larger ones (Table 1). Between 1961 and 1975 units over 20 ha quadrupled both in number and in land area controlled. The average farm size rose from 36 to 45 ha in less than 15 years. However, as Crawford and Purvis (1986) point out, there are two reasons why the data may not be representative of the actual distribution of farm sizes in Morocco. First, no distinction is made between landownership and land use. Second, the pre-1984 tax laws may have induced rural families to disburse landownership among several of their members to avoid tax liability, thus overestimating the number of farms below 5 ha.

Irrigation

Irrigation began in Morocco in the 1930s and by 1950, about 20,000 ha were irrigated by four major dams. In 1984, nearly 0.5 million ha were under irrigation, 20% being irrigated by sprinklers (Figure 1). In the future Morocco expects to have 825,000 ha under irrigation, leaving nearly 7.2 million ha rainfed. Of the total rainfed area, 4 million ha or 71% are under cereals

Table 1. Landownership in Morocco, 1961 and 1975.

Category (ha)	Number of farms (×1,000)		Percent of total		Land area (×1,000 ha)		Percent of total		Average farm size (ha)	
	1961	1975	1961	1975	1961	1975	1961	1975	1961	1975
Landless	543	452	32.9	23.4						
0–10	1,051	1,311	63.8	67.9	2,266	3,284	66.9	45.3	2.1	2.5
11–20	41	114	2.4	5.9	587	1,530	17.3	21.1	14.3	13.4
>20	15	54	0.9	3.1	538	2,436	15.8	33.6	36.2	45.1
Total	1,650	1,931	100	100	3,391	7,250	100	100		

Source: El-Ghorfi (1964); Piaison (1979).

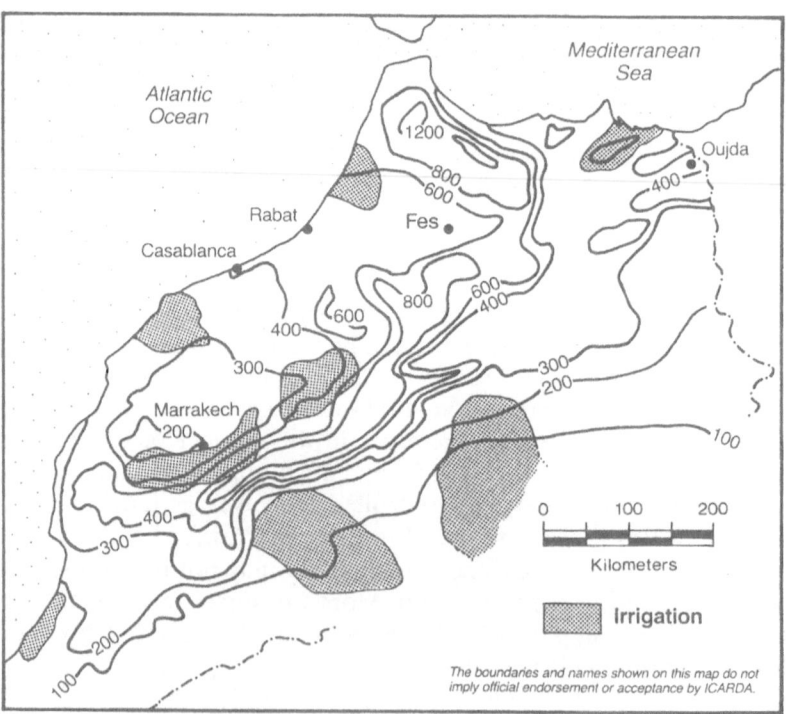

Fig. 1. Distribution of rainfall (mm) and irrigated areas in Morocco. *Source*: Gaussem et al. (1958); Nyrop et al. (1972).

Table 2. Land use, average 1973–78.

Item	Area (×1,000 ha)
Crop	
cereals	4,106
pulses	497
oilseeds	547
sugar cane, beet, cotton	84
vegetables	111
forage	82
sundry crops	14
tree shaded crops	−141
Total planted	5,343
Fallow	1,824
Orchards	400
Total cultivable land	7,565
Rangeland	7,100
Forests	5,300
Non-agricultural zones	51,034
Total area	71,000

Source: Piaison (1979).

(Table 2) and this sector poses the most complicated, costly, and urgent problem for planners. However, irrigated agriculture continues to receive the major share of public expenditure (Table 3). At the same time, the share of the agricultural sector in public expenditures has declined (Table 4).

Mechanization

In 1957, the government launched a revolutionary scheme, known as Operation Labor, to raise agricultural productivity. It arose from dissatisfaction with the French SMPs which, during 10 years, affected only 5% of the total cultivable area. The SMPs were renamed *Centres de Travaux* (CT) and increased from 62 to 110. Their reform function was given to the Ministry of Health which concentrated on economic reform. Operation Labor "featured a large-scale substitution of labor-saving machinery for peasant labor . . . [in spite of] the prevalence of rural underemployment" (Van Wersch 1968). This was justified by the restriction imposed on field operations by the physical environment and insufficient capital equipment.

    The specific objective of the operation was to increase grain production by making tractors available to traditional farmers. This involved the purchase of over 1,000 tractors to be used on consolidated lots of 400–1,000 ha (Ashford 1967). The CTs were responsible for implementing the program and most of the land involved was located in the major crop-producing areas. Although yields on CT farms increased by 45–60% during 1958–60 (Van Der Kloet 1975) the area cultivated under Operation Labor declined gradually from 751,000 ha in 1959 to 250,000 ha in 1961 (Ashford 1967).

    Ashford (1967) suggests several reasons for the failure of the scheme.

Table 3. Planned and actual public expenditures by agricultural subsector, 1968–85[1] (million MAD).

Subsector	1968–72		1973–77				1978–80				1981–85	
	Planned[2]	%	Planned	%	Actual	%	Planned	%	Actual	%	Planned[2]	%
Irrigation	1,760	70	2,599	57	2,021	63	2,421	60	1,885	68	4,753	40
Rural equipment	475	19	375	8	213	7	375	9	206	7	1,138	10
Dryland production[3]			685	15	350	11	492	12	294	11	1,982	11
Extension[4]											667	6
Livestock	105	4	235	5	139	4	235	6	116	4	1,275	11
Agroeconomic & cadastral studies			145	3	91	3	93	2	42	2	340	3
Forests and conservation	52	2	290	6	207	6	205	5	108	4	990	8
Teaching	47	2	120	3	69	2	87	2	61	2	177	2
Research	52	2	105	2	78	2	96	2	62	2	241	2
Legislation	39	2	43	1	31	1	26	1	6	0	233	2
Total	2,530	100	4,597	100	3,199	100	4,030	100	2,780	100	11,796	100

Source: MARA/AIRD (1986a).
[1]Excluding dams.
[2]Actual expenditures not available.
[3]Included in rural equipment in 1968–72.
[4]Included in dryland production until 1981–85.

Table 4. Public expenditures by sector as percentage of total, 1965–85.

Sector	Subsector	1965–67	1968–72	1978–80	1981–85
Agriculture	27.4	42.9	24.3	26.7	21.7
Infrastructure	21.8	21.5	19.6	19.6	17.7
Housing and social services			8.1	5.9	9.2
Regional development	22.5	17.0	9.9	11.5	8.3
General administration			10.8	6.8	8.9
Education			11.6	17.6	17.9
Other	28.2	18.6	15.7	11.8	16.2
Total expenditure (million current MAD)	2,132	5,478	17,923	15,960	61,440

Source: MARA/AIRD (1986a).

Table 5. Number of tractors and caterpillars, 1971–82.

Season	Irrigated		Rainfed		Total
	Tractors	Caterpillars	Tractors	Caterpillars	
1971/72	5,005	911	5,703	1,003	12,622
1972/73	5,602	931	6,498	1,022	14,053
1973/74	6,163	949	7,446	1,069	15,627
1974/75	6,985	964	8,446	1,159	17,554
1975/76	8,011	981	9,592	1,194	19,778
1976/77	7,913	763	10,116	1,264	20,056
1977/78	8,526	809	11,056	1,241	21,632
1978/79	9,544	764	12,278	1,319	23,905
1979/80	9,793	731	12,922	1,318	24,764
1980/81	11,449	770	16,174	1,324	29,717
1981/82	11,925	735	16,612	1,303	30,575

Source: MARA/AIRD (1986a).

Large landowners felt threatened by the new status of the small peasant farmers while sharecroppers felt threatened by the new labor-saving technology. More important was the general ignorance of credit and fertilizer use and how the new practices could fit into the whole farming system. In addition, fertilizer was initially distributed free but 2 years later farmers were asked to pay for it. As a result, many stopped using fertilizer.

Since then, the government has adopted a system of incentives to encourage the use of machinery. This resulted in a threefold increase in the number of tractors between 1968 and 1981/82 especially in rainfed areas (Table 5). These incentives, along with the attractiveness of custom charges relative to seasonal wages, have expanded the mechanization of farming practices. In some drier areas mechanical traction may account for 90–95% of land preparation, and combines may account for 80–85% of the cereal harvest (Crawford and Purvis 1986).

Fertilizers

In 1966/67, the government launched a fertilizer program (Operation En-grais), following the recommendations of FAO and USAID. This operation covered 200,000 ha in 1966, 359,000 ha in 1967, declining to 207,900 ha in 1972. Although yields were increased by 50% during the first three cropping seasons, the operation's effectiveness was reduced by the same complications experienced by Operation Labor. Yield increases were not consistent across all regions and farmers who combined mechanization and rotation obtained higher yields than those who used animal traction and practiced continuous cropping. Consequently, in 1969, the government demanded rotation con-tracts from farmers with holdings of 20 ha or more and those with less than 20 ha had to regroup into 80 ha units to qualify for a contract (Van Der Kloet 1975). To encourage farm consolidation, the authorities offered fertil-izer, seeds, and technical help, but as the contribution of the smaller farms was unsatisfactory, the area under the operation was further reduced and only those who were willing to practice rotation and market their surplus production qualified.

Since then, the supply of fertilizer has increased from 410,000 t in 1974/75 to 700,000 t in 1982/83, with imports providing 23–36% of the total. Utiliz-ation of available fertilizer has also increased from 41 to 65% in this period (MARA/AIRD 1986a). The value of output relative to the cost of fertilizer has also increased. In the case of wheat, for example, 1 kg of fertilizer in 1981 was equivalent to 1.1 kg of wheat which compares favorably with 4.9 kg in the USA in 1977, 3.3 kg in Turkey in 1978, 2.7 kg in Pakistan in 1978, and 1.9 kg in Jordan (MARA/AIRD 1986a). Since 1974, the fertilizer subsidy program has insulated farmers against fluctuating world prices, with an in-creasing burden on the treasury.

Although fertilizer use on dryland crops during 1975–84 was three times greater than that on irrigated crops in absolute terms, application per hectare for the latter was larger (MARA/AIRD 1986b). In the same period, nearly 50% of all fertilizer was applied to cash crops. Aggregate fertilizer consump-tion increased from 429,000 t in 1975/76 to 650,000 t in 1983/84. The share used by rainfed crops increased from 37 to 43% in that period, while the share for irrigated crops averaged 14%.

Modern Varieties

Interest in high-yielding varieties of semi-dwarf wheat began in 1968 with an area of 200 ha planted to Siete Cieros, Inia 66, Penjamo 62, and Tobari 66 (USDA 1971) which by 1982/83 was 462,000 ha (MARA/AIRD 1986a). Planned expenditure on the production and diffusion of modern varieties was 196 million MAD in the 1981–85 5–year plan. Modern varieties are marketed by the public monopoly SONACOS and individual farmers and cooperatives can currently purchase modern varieties with a 10 and 20%

subsidy, respectively. Nevertheless, adoption has been slow, possibly due to delays in delivery, the complex bureaucratic system of payments, and the production and marketing system.

Labor

The 1971 census (the most recent available) gives the economically active labor force as 53% of the total labor force aged between 16 and 64 (Lahbabi 1977), with 3.4 million males (85% of the total labor force and 44% of the total male population). Employed females numbered 605,000 (15% of the total labor force and 8% of the total female population) (ILO 1976).

In 1971, agriculture and related activities absorbed more than half of the economically active population (51.4%) or nearly 2 million people, 80% of whom were engaged in crop production, 17% in livestock production, and the rest in fishing and forestry. Since 1971, the number of workers engaged in crop production has remained fairly constant (1.6 million in 1971 and 1.7 million in 1982), while those engaged in livestock production have increased from 3.3 million to 4.5 million during the same period (Crawford and Purvis 1986). Any increase in employment of the rural workforce is likely to be in livestock production. Crawford and Purvis attribute the low level of employment generation in crop production to increased mechanization. Griffin estimated that 72% of rural labor is concentrated in the dryland areas.

The trend toward mechanization of crop production and agricultural diversification has also affected the configuration of seasonal labor. Labor demand is highest during May-June (cereal harvest) in the southern region, March-June (bean and potato harvest and corn planting) in the central region, October-November (apple harvest) in the Fes-Meknes area, and November-January (citrus harvest) in the Agadir region (Crawford and Purvis 1986). While there are no official statistics on the extent of underemployment caused by the seasonal demand for labor, Azzam (1980) estimated a national underemployment rate of 42% in rural areas.

Farming Systems in Morocco

The physical constraints on agriculture are climate, water, and land. The climate is characterized by moderate winters and dry summers and is affected by the country's latitude and proximity to the Atlantic and the Mediterranean, the desert, and the Canary currents. Rainfall is insufficient, irregular, and often delayed during the autumn sowing period. As a consequence, a shortage of animal power at sowing is a major constraint and has been an important factor behind the government's mechanization schemes.

There are essentially four agroclimatic zones in Morocco (Figure 1): the area north of Casablanca with average annual rainfall of 400 mm or more; the area south of Casablanca which has an average rainfall of 200–250 mm;

eastern Morocco, beyond the Atlas Mountains, where rainfall averages 150–350 mm; and southern Morocco, in the Souss Valley where rainfall is less than 200 mm. The zone with a rainfall of 400 mm and above covers 1.3 million ha or 34% of the cultivated land while the zones with rainfall less than 400 mm cover nearly 2 million ha (MARA/AIRD 1986a). The 3.3 million ha, which is half the total cultivated area, constitute what is referred to as *la surface agricole utile* (useful agricultural area).

It has been estimated that 80% of the rainfall is lost by evaporation due to the hot, dry climate and of the remainder only 5% is fully utilized. This represents a fraction of the 64% that can potentially be used for irrigation if properly caught and stored (Lahbabi 1977).

Of the total surface area of 71 million ha, about 7.5 million ha were cultivated during 1973–78 (Table 2). Of the total cultivated area, 70% is planted and 30% is fallow. Cereals are grown on 77% of all planted land, illustrating the importance of cereal cultivation as a source of staple food and agricultural employment. Pulses are grown on 497,000 ha or 10% of the total cultivated area.

Demengeot (1975) classifies Moroccan agriculture into three subsectors; traditional, transitional, and modern. The traditional subsector consists of the majority of farmers and is characterized by overpopulation, a high rate of illiteracy, a semi-closed economy, and very small farm units using traditional methods of production.

The distribution of the traditional rural population is uneven and poorly correlated with the location of natural resources. In 1960, population density for Morocco as a whole was 20/km^2. In farm areas, densities ranged from 26 in parts of the interior plateaux to 65 in the northwest, adjacent to the Atlantic, and in the Lower Rif (Van Der Kloet 1975). Populated areas such as the Anti Atlas and Rif have few resources, while the rich Gharb and Tadla are sparsely populated. This imbalance makes it extremely difficult to provide full employment for a rapidly expanding population (IBRD 1966). Space distribution also has an effect on the extent and intensity of cultivation in the rural areas. In the plains, 60–80% of the land is extensively cultivated, while in the interior plateaux of Rhamna and Ahmar only 25% of the land is cultivated.

To the poor Moroccan peasant, animals are an investment as well as a source of food, due to the lack of institutional credit and consumer goods. They may be sold to pay for harvest, settle a debt or a lawyer's fee, or cover the expenses of a daughter's wedding. Livestock are also a source of milk, draft power, meat, and even prestige. On the other hand, rich farmers tend to keep livestock for profit and prestige.

A government survey in 1961–63 identified the predominant occupations in the traditional sector as farming, livestock raising, fruit growing, and combinations of these activities (HNPC 1964). Those engaged in all three occupations constituted 36% of the total households, and those engaged in livestock raising and farming alone were 40% of the total. Households with

Table 6. Rural household occupations in the plains and hill regions, 1961/63.

Occupation	Plains		Hills	
	No. of households	Percent of regional total	No. of households	Percent of regional total
Farming	18,820	5.6	3,330	3.0
Husbandry	51,719	15.3	14,920	13.6
Orchards	589	0.2	1,085	1.0
Farming and husbandry	118,043	35.0	58,956	53.7
Farming and orchards	8,213	2.5	1,704	1.6
Husbandry and orchards	3,198	1.0	3,615	3.3
Farming, husbandry, and orchards	136,164	40.4	26,218	23.9
Total	336,746	100.0	109,828	100.0

Source: Cited in Van Der Kloet (1975).

farming as the sole occupation accounted for less than 5%. There appears to be a relation between the available space and the nature of activity (Table 6). In the plains, where density is high, a combination of farming and livestock raising is practiced by only 35% of farmers, whereas in the hills, where density is lower, 54% are in this category.

The land tenure system is one of the major institutional constraints as the farms are mostly small and fragmented, making adoption of some techniques more difficult. In some regions, fragmentation is the peasant's response to the risks imposed by variable soils and microclimates (Benatya et al. 1983). A local axiom is: "gather your sons but disperse your land".

The transitional subsector is composed of medium-sized farm units, increasingly using machinery and other conventional inputs. Its growth can be attributed in part to the sale of farms to Moroccans by foreign settlers, as well as government distribution of agricultural land (see above). The modernization of large and medium-sized farms in the traditional sector is another factor. All plots of 7 ha and more were considered by the government to have the potential for modernization. Finally, government investment in irrigated zones has led to modernization of formerly traditional farms (Demengeot 1975).

The modern subsector is composed of fairly large farms, formerly owned by Europeans. The present owners (all Moroccans) use capital-intensive techniques for production. On the eve of independence, this sector accounted for roughly 1.3 million ha or 16% of all cultivable land (Swearingen 1987), then in the early 1970s it occupied 1.4 million ha, 300,000 of which were managed by the government (Villeneuve 1971). This subsector mainly produces exportable goods.

Contractual Arrangements

According to Van Der Kloet (1975), 33–37% of private holders work their land directly, which accounts for 30–38% of the total area privately held. Most of those who own less than 3 ha work their land themselves since their assets are too limited to share under any type of contract. Of the 3.9 million ha cultivated in 1961/62, 700,000 were under the *khammas* type of sharecropping contract, in which the tenants are usually landless, they provide labor only, and their remuneration is one fifth of the harvest (El-Ghorfi 1964). The *bel-khobza* contract differs from the *khammas* contract in that the shares are proportional to the inputs invested in the partnership. Most of the farmers with this type of contract are large landlords, living in the city, or are elderly owners unable to cultivate their own land.

However, large owners' behavior appears to be changing. Van der Kloet (1975) reported that 30% of farmers owning more than 20 ha are actually renting more land, while in the past they would have contracted out their farms to sharecroppers. This change is attributed to mechanization. Large owners with tractors and expensive harvesters are gradually giving up their social status to make full use of their expensive machinery. Although this practice is not yet universal, it may be indicative of the role of technology in changing the rigid rural hierarchy.

With increasing population, land has become increasingly scarce, commanding a larger share of income in partnership arrangements. In the past, when the land more than supported the population, rent varied with yield but now rent is fixed regardless of yield. So with the prospect of increasing population pressure, the shares of landless renters will continue to diminish and land will receive a larger share of the output. This merits further study since it touches on the critical issue of employment and migration.

Livestock partnerships are agreed on the basis of two sharing methods. One provides for reimbursement of the original investment while the other involves sharing of income from butter, milk, and offspring. The contractual arrangement can either be oral or written for a period of 5–10 years and is terminated at the discretion of either partner. The sharing method also depends on the size of the animals, the region, and local customs (HCPN 1964).

Farm Practices

For climatic reasons, winter grains are more common than summer grains in dryland areas. In Morocco, harvest starts in mid-April in the south and mid-June in the north. A typical agricultural calendar for the four major crops is shown in Table 7.

The main crops planted in autumn are soft wheat, hard wheat, and barley. In the winter planting time, farmers plant areas missed in the fall. The main

Table 7. Agricultural calendar for four major crops in the Abda region, 1985/86[1].

Month	Crop			
	Corn	Soft wheat	Hard wheat	Barley
Aug.		T	T	
Sept.		T	F	T,F
Oct.		T	F	T,F
Nov.		F,P	P	T,F,P
Dec.				
Jan.	T			
Feb.	P	M	M	M
Mar.	M	M	M	M
Apr.	C			
May.	C			
June.	H,S	H,S		
July.			H,S	H,S

Source: Compiled from Rafsnider and Hammida (1987).
[1]The sequence of operations is tillage (T), fertilization (F), planting (P), maintenance (M), cultivation (C), harvest (H), and storage (S).

crops grown during this season are corn in the Chaouia and Doukkala regions, rice in the Gharb, and sorghum in the north.

The sequences of field operations in dryland areas differ by crop, farm size, machinery ownership, and climate. Benatya et al. (1983) identified as many as 29 possible combinations of field operations in wheat production in the Chaouia region (Table 8). In larger farms, a high degree of mechanization leads to more common field practices. For example, only 26% of the area in farms with less than 5 ha underwent the two most common sequences of field practices compared to 73% of the area for machine owners with more than 40 ha.

Chemicals are not extensively used in Chaouia. Benatya et al. (1983) attribute the lower fertilizer use to the location of the area surveyed and to the 1980/81 drought. Lands with failed crops in the drought year were considered as fallow, and therefore were not considered to need fertilization. Soil and moisture conditions are such that fertilizers may actually decrease yields in the south.

Mechanization and Labor

The most mechanized field operations are land preparation and harvesting. Of the three crops, hard wheat, soft wheat, and barley, the last is the least mechanized, as it is generally sown in areas inaccessible to modern machinery (Benatya et al. 1983).

Manual labor is relatively important in harvesting compared to land preparation, probably because crop yields are often too low to cover the combine rental. Also, the wheat plant may be too short to permit combine harvesting,

Table 8. Sequence of field operations by farm size for wheat.

Farm size (ha)	Number of farms	Most common sequences of field operations[1]	Hectares per sequence (%)
Machinery renters			
		BS-CC-HL-MT	15.4
0 to ≤5	18	BS-WP-MW-HL-AT	10.3
>5 and ≤10	24	BS-CC-HL-AT	13.6
		BS-WP-MW-HL-AT	12.3
		BS-WP-HL-AT	10.6
>10 and ≥20	29	BS-CC-HL-AT	14.2
		BS-WP-HL-AT	13.1
>20 and ≤ 30	24	BS-WP-HL-AT	19.3
		BS-CC-HL-AT	10.7
>40	20	BS-CC-CH-AT	21.8
		BS-CC-HL-AT	19.9
		BS-CC-CH-MB-MT	18.4
Machinery owners			
		BS-CC-HL-AT	22.6
		BS-CC-(MH + HL)-MT	14.0
≥20 and <30	13	BS-CC-(MH + HL)-AT	10.8
		BS-CC-CH-MT	10.8
>40	11	CC-BS-CC-MH-MB-MT	56.8
		CC-CF-BS-CC-MH-MB-MT	15.8

Source: Benatya et al. (1983).

[1] AT = Transport by animal   MB = Mechanical baling
BS = Broadcast sowing   MH = Manual harvesting
CC = Cover-crop   MT = Mechanical transport
CF = Chemical fertilizer   MW = Manual weeding
CH = Combine harvester   WP = Traditional plow
HL = Harvest by hand-lifting

especially in rocky areas where lowering the cutter bar increases the risk of damage to the harvester. The use of animal power to transport grain and hay from the field predominates when the distance from the homestead to the cultivated land is small.

The study by Benatya et al. (1983) suggests the degree of substitution between labor and machinery depends upon factors which are largely local. Among these are the terrain, cropping patterns on different terrains, availability and cost of machinery relative to the value of output, and the degree of fragmentation of landholdings and their size. In the area surveyed, the mean number of land parcels was 6.3 with a coefficient of variation of 70% and a maximum of 31 parcels. The average area per parcel was 1.3 ha with a coefficient of variation of 127% and a maximum of 27.5 ha.

One hypothesis is that in dryland areas like the Chaouia, the principal benefit of mechanical power, especially in land preparation, is increased yield. Examination of the yield effects requires more detailed data and appropriate analytical methods (see Jaeger 1986) and the data provided by Benatya et al. are not sufficient to test the hypothesis.

An attempt was made to test the hypothesis using aggregate regional data recently published by MARA/AIRD (1986b). Wheat yield (YD) was regressed on mechanical use (MH) in tractor hours required for land preparation, a binary variable representing the use of modern varieties (MV), and an interaction effect of MV and irrigation (IR). The results were:

$$YD = 724 + 62 \times MH + 48 \times MV + 693 \times MVIR$$

$$(10.4) \ (4.1) \ (0.43) \ (3.0)$$

where the numbers in parentheses are t-ratios. The r^2 of the model was 0.72. This indicates that a major portion of the variation in yield is explained by variation in mechanical power, represented by tractor use in land preparation, after adjusting for the use of modern varieties and irrigation. Another interpretation is that mechanical power was used on the best land.

Rafsnider and Hammida (1987) carried out a survey in the Abda region, Safi Province, to establish comparative enterprise budgets for barley, corn, and hard and soft wheat (Table 9). It involved 20 farmers who owned tractors. Fifteen of the farmers operated 60 ha or less, three had farms of 60–100 ha, and two farms were larger than 100 ha. The average farm size was 52.5 ha.

Total labor requirements per hectare of corn were about five times the labor requirements for wheat and six times those for barley. About 45% of total labor per hectare of corn was absorbed by harvesting compared with 17% for soft wheat, 19% for hard wheat, and 19% for barley. Of the total hours worked per hectare (human and mechanical), only 77% were provided by human labor for corn and 86, 83, and 87% for soft wheat, hard wheat, and barley, respectively.

It would be interesting to know to what extent labor-saving technology affects labor use, and if labor is displaced, what increase in outputs due to mechanization would be required if displaced labor is to be compensated by new employment.

The Abda survey did not include unmechanized farms so an area-specific comparison is not possible. However, data are available on unmechanized farms in the Settat region (MARA/AIRD 1986b). The labor requirements per hectare for four crops in Settat and for the sampled tractorized farms in the Abda region, and the number of tractor hours and combine hours per hectare are shown in Table 10. The combined tractor and combine hours displace labor by 77% for hard and soft wheat, and 74% for barley. Tractor hours used for corn production are approximately eight times those used in the production of the other grains, but only 10% of the corn labor is displaced. Therefore, it is likely that combines displace more labor than tractors.

Table 9. Hours per ha of family labor, hired labor, and machine and animal use for cereal crop operations by tractor owners in the Abda region, Safi Province, 1985/86.

Activity	Corn				Barley			
	Month	Family labor	Hired labor	Tractor use (animal)	Month	Family labor	Hired labor	Tractor use (animal)
Tillage	Jan.	3.8	0.6	1.9	Oct./Nov.	0.9	1.0	1.6
Fertilization	–	–	–	–	Oct./Nov.	0.7	1.1	0 (0.1)
Planting	Feb.	6.1	15.7	0.7 (19.4)	Nov.	1.2	3.6	1.0
Maintenance	Mar.	0.8	1.1	0	Feb./Mar.	10.3	0.2	0
Cultivation	Apr./May.	1.5	10.1	0 (25.8)	–	–	–	–
Harvest	July	6.9	13.5	0	June	0.1	0.5	0
Storage	July	19.1	29.4	2.8 (0.9)	June	1.3	3.1	0.9 (0.1)
Total		38.1	70.3	5.4 (46.2)		14.5	9.4	3.5 (0.2)

Activity	Soft Wheat				Hard Wheat			
	Month	Family labor	Hired labor	Tractor use (animal)	Month	Family labor	Hired labor	Tractor use (animal)
Tillage	Aug./Oct.	2.0	1.1	2.0	Aug.	2.3	1.0	2.0
Fertilization	Nov.	0.7	0.7	0.2	Oct.	0.7	0.7	0.3
Planting	Nov.	1.7	3.9	1.0	Nov.	1.6	3.4	1.0
Maintenance	Feb./Mar.	2.3	6.8	0	Feb./Mar.	4.3	1.1	0
Harvest	June	0	0.2	0	June	0.1	0.6	0
Storage	July	1.1	3.0	0.7	June	1.3	2.5	0.8
Total		7.8	15.8	3.9		10.3	9.1	4.1

Source: Rafsnider and Hammida (1987).

Table 10. Labor and mechanical inputs (hours) in mechanized and non-mechanized farms, Settat and Abda regions.

Crop	Type of farm			
	Non-mechanized farms (Settat)	Mechanized farms (Abda)		
	Labor	Labor	Tractor	Combine
Corn	120	108	5	0
Soft wheat	104	24	4	1.5
Hard wheat	88	19	4	1.5
Barley	88	24	4	1.5

Source: Rafsnider and Hammida (1987); MARA/AIRD (1986b).

Table 11. Proportion of labor displaced by mechanization and the required rate of increase in employment and rise in output to compensate for the displacement.

Crop	Proportion of displaced labor[1] (%)	Required rate of increase in employment to compensate for d[2] (%)	Required rise in output[3] (%)
Corn	10	11	15
Soft wheat	77	335	446
Hard wheat	77	335	446
Barley	74	284	378

[1]Displaced labor = d.
[2]The required rate of increase in employment to offset the displaced labor is R = d/(1−d).
[3]The required increase in employment is E = C/R where C is the percentage rise in employment associated with a 1% increase in output. We assumed that C = 0.75 (see Rao 1979).

Manual harvesting takes 4–12 days per hectare, depending on region, crop, yield, and terrain.

Using the labor displacement percentage for the various crops, Table 11 shows the rate of increase in employment required to compensate for displaced labor and the increase in output required to generate employment. For all crops, especially wheat and barley, output must rise faster than the required rate of increase in employment but it is unlikely that such increases in output can be obtained, despite the introduction of modern varieties. Moreover, the daily wage for manual harvesting is such that a hectare has to be harvested in 10 days for labor to be as efficient as combines.

Since the survey (Rafsnider and Hammida 1987) did not distinguish between the users of traditional and modern varieties, we do not know the extent to which they contribute to employment in that region. However, by using data across regions we can draw a general picture of the employment effect of this land-saving technology.

On average, the use of high-yielding varieties was associated with higher output per unit of land than traditional varieties for all crops. Farmers using modern varieties used more machinery hours per unit of land for all crops

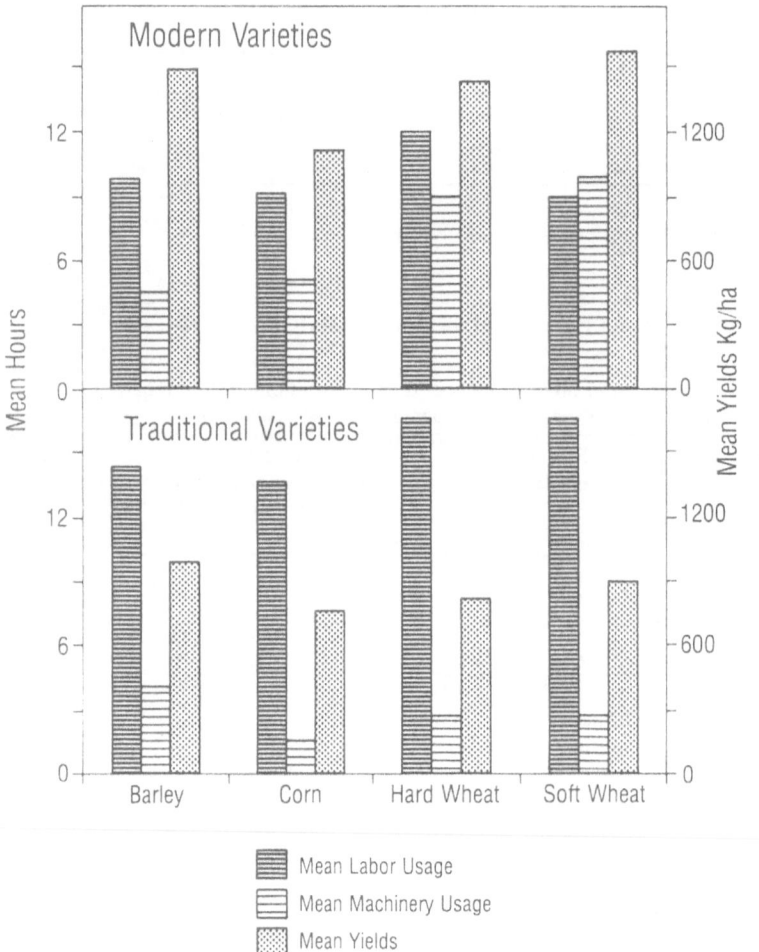

Fig. 2. Labor and machinery usage for modern and traditional varieties. *Source*: Compiled from MARA/AIRD (1986b).

except barley (Figure 2), and less labor per unit of land which suggests that in terms of employment, the output-raising effect of yield augmenting inputs may be outweighed by the substitution effect of other inputs.

To test this, labor use was regressed on a number of variables (Table 12). Farms using modern varieties required 8 days less labor than those using traditional varieties. The point estimate, however, may be more indicative of the substitution effect of other inputs, such as herbicides which replace labor for weeding, rather than direct labor displacement. Since all irrigated farms in the sample use modern varieties, tractor power, and combines, the net effect on labor use is also negative. The coefficient on the interaction term between tractors and combines is larger than that on tractors alone

Table 12. Regression on labor use in days.

Source	Parameter estimate	Standard error
Constant	24.56	3.9[c]
Daily wage/price of output	−0.63	0.26[b]
Use of irrigation[1]	6.96	2.81[c]
Use of modern varieties[1]	−8.82	1.98[c]
Use of tractors[1]	−0.72	1.98
Interactions		
Tractor × combine use	−4.81	2.17[b]
Tractor × draft animal use	2.97	1.91[a]
Variety × combine use	3.77	3.77
Regions		
Settat	−3.28	1.84[a]
Oulmanes Roummani	0.73	1.86
El Kalaa	4.52	2.09[b]

[1]Intercept shifters
[a]$P < 0.1$; [b]$P < 0.05$; [c]$P < 0.01$.
$R^2 = 0.61$; $F = 5.94$; $p < 0.01$; No. of observations = 48.

Table 13. Employment coefficient[1] by sector.

Crop	Employment coefficient	Crop	Employment coefficient
Hard wheat		Lentils	0.39–0.51
irrigated	0.50–0.52	Cotton	0.75
rainfed	0.36–0.56	Olives	0.19–0.28
Soft wheat		Oranges	0.41
irrigated	0.50–0.52	Clementines	0.45–0.39
rainfed	0.36–0.56	Tomatoes	0.59–0.69
Barley	0.33–0.49	Potatoes	0.59
Corn	0.29–0.40	Sugar beet	0.67–0.74
Beans	0.39–0.55	Sugar cane	0.61–0.85
Chickpeas	0.39–0.56	Soybeans	0.83

Source: MARA/AIRD (1986a).
[1]Defined as the ratio of labor costs to the costs of all factors of production.

which is insignificant at $P = 0.10$. The coefficient for the Settat region ($P < 0.10$) was negative, as expected, because of its proximity to Casablanca.

Some recent estimates of the employment coefficients of various sectors in Morocco are presented in Table 13. They indicate that the grain sector considered in this study is less labor-intensive than other sectors such as sugar cane, soybeans, sugar beet, tomatoes, and cotton.

Conclusions

Despite major trials and errors in modernizing Moroccan agriculture, the traditional rainfed sector shows remarkable resilience in preserving its norms and institutions. It is tempting to argue that it is precisely these institutional rigidities which hinder the adoption and diffusion of new techniques. However, there is every indication that, from the peasants' perspective, the new techniques are themselves rigid in that they limit the possibilities of maneuvering in a risky environment.

Most expenditure on technological reform in agriculture is heavily biased toward the irrigated sector. For those pockets in the rainfed sector that have adopted new techniques, our partial analysis showed that both mechanical and biological technologies displace labor, and output has to rise much faster than the rate of employment required to absorb the displaced labor. Of course, the results are tentative and tempered by the partial and static nature of the analysis performed. This suggests that a well designed analysis covering both the direct and indirect effects of technological change on rural employment in Morocco may be a fruitful direction for further research.

References

Anonymous n.d. Rural Modernization in Morocco. Rabat: Societés D'Etudes Economiques, Sociales et Statistiques.

Ashford, Douglas E. 1967. National Development and Local Reform: Political Participation in Morocco, Tunisia, and Pakistan. Princeton: Princeton University Press.

Azzam, Azzedine M. 1980. Moroccan Agriculture: Potential Contribution of the Moroccan Traditional Farmer. MS thesis, University of Wisconsin.

Benatya, D., Pascon, P., Zagdouni, L. and Magoul, O. 1983. L'Agriculture en Situation Aléatoire: Chaouia 1977–1982. Rabat: Institut Agronomique et Vétérinaire Hassan II, Direction de Développement Rural.

Brace, Richard M. 1964. Morocco, Algeria, Tunisia. New Jersey: Prentice-Hall.

Colburn, Dorothy B. 1927. France's Colonial Success in Morocco. Current History 26: 250–255.

Cranford, Paul R. and Purvis, Malcolm J. 1986. The Agricultural Sector of Morocco: A Description. Annexe C: Country Development Strategy. Rabat: USAID.

Darlymple, D.G. 1977. Evaluating the Impact of International Research on Wheat and Rice Production in Developing Nations. Pages 171–208 in Resource Allocation and Productivity in National and International Agricultural Research (Arndt, T.M., Dalrymple, D.G. and Ruttan, V.W., eds). Minneapolis: University of Minnesota Press.

Demengeot, Patrick D. 1975. Agricultural Development Policy Analysis and Planning in Morocco. PhD dissertation, University of Pittsburgh.

El Ghorfi, Nor. 1964. Contribution à l'Edification d'une Politique Agricole. Rabat: L'Institut Nationale de la Recherche Agronomique.

FAO (Food and Agriculture Organization). 1984. Socio-Economic Indicators Relating to the Agricultural Sector and Rural Development. Economic and Social Development Paper 40. Rome: FAO.

Gaussem, H., Debrach, J. and Joly, F. 1958. Planche No. 4a, Précipitations Annuelles. In Atlas du Maroc. Rabat: Comité de Géographie du Maroc, Institut Scientifique Chérifien.

Godfrey, John B. 1985. Overseas Trade and Rural Change in Nineteenth Century Morocco: The Social Region and Agrarian Order of the Shawiya. PhD dissertation, Johns Hopkins University.

Griffin, Keith B. 1975. Income Inequality and Land Distribution in Morocco. Bangladesh Development Studies 3: 319–348.

Guillaume, Augustin. 1952. The French Accomplishments in Morocco. Foreign Affairs 30: 625–636.

Hayami, Y. and Ruttan, V.W. 1985. Agricultural Development: An International Perspective. Baltimore: Johns Hopkins University Press.

HCPN (Haut Commissariat au Plan et la Promotion Nationale). 1964. Résultats de l'Enquête à Objectifs Multiples 1961–63. Rabat: Division du Plan et des Statistiques.

Hoover, Ellen T. 1978. Among Competing Worlds: The Rehamna of Morocco on the Eve of French conquest. PhD dissertation, Yale University.

IBRD (International Bank of Reconstruction and Development). 1966. The Economic Development of Morocco. Baltimore: Johns Hopkins University Press.

ILO (International Labor Office). 1976. Yearbook of Labor Statistics. Geneva: ILO.

Jaeger, W.K. 1986. Agricultural Mechanization: The Economics of Animal Draft Power in West Africa. Boulder: Westview Press.

Lahbabi, Mohammed. 1977. Les Fondements de l'Economie Marocaine. Casablanca: Les Editions Maghrébines.

Landau, Rom. 1956. Moroccan Drama, 1900–1955. San Francisco: The American Academy of Asian Studies.

MARA/AIRD (Ministère de l'Agriculture et de la Réforme Agraire/Associates for International Resources and Development Inc.). 1986a. La Politique de Prix et d'Incitations dans le Secteur Agricole: Rapport Finale. Somerville: AIRD.

MARA/AIRD (Ministère de l'Agriculture et de la Réforme Agraire/Associates for International Resources and Development Inc.). 1986b. La Politique de Prix et d'Incitations dans le Secteur Agricole: Annexe Statistique. Somerville: AIRD.

MEN (Ministère de L'Economie Nationale). Plan Quinquennal 1960–64. Division de la Coordination du Plan.

Nelson, Harold D. (ed.). 1978. Morocco, A Country Study. Washington, D.C.: Government Printing Office.

L'Oeuvre de la France au Moroc. 1950. Rabat: Editions Africaines Perceval.

Piaison, Frank J. 1979. Morocco: Agricultural Situation. Report No. 9–003. Rabat: Foreign Agricultural Service.

Rafsnider, G.T. and Hammida, Mustapha. 1987. Enterprise Budgets for Barley, Corn, Hard Wheat, and Soft Wheat Produced by Farmers Owning Tractors in the Abda Region of Safi Province During 1985–86 Cropping Year. MIAC Agricultural Economics. Bulletin No. 1. Settat: Aridoculture Center.

Rao, C.H.H. 1979. Farm mechanization. Pages 291–304 in Agricultural Development in India, Policy and Problems (Shah, C.H. and Vakil, C.N., eds). Bombay: Orient Longman Limited.

Stewart, Charles F. 1964. The Economy of Morocco, 1912–1962. Cambridge: Harvard University Press.

Swearingen, Will D. 1987. Moroccan Mirages: Agrarian Dreams and Deceptions 1912–1986. Princeton: Princeton University Press.

USDA (United States Department of Agriculture). 1971. High Yielding Varieties of Wheat in Developing Countries. Economic Research Service. ERS Foreign No. 322. Washington, D.C.: Government Printing Office.

Van Der Kloet, Hendrick. 1975. Inégalités dans les Milieux Ruraux: Possibilités et Problèmes de la Modernisation au Moroc. Genève: Institut de Recherche des Nations Unies Pour le Developpement Social.

Van Wersch, H.J. 1968. Rural Development in Morocco: Operation Labor. Economic Development and Cultural Change 17: 33–49.

Villeneuve, Michel. 1971. La Situation de l'Agriculture et son Avenir dans l'Economie Marocaine. Paris: Librarie Générale de Droit et de Jurisprudence.

Zartman, William. 1964. Morocco, Problems of a New Power. New York: Atherton Press.

List of Contributors

AKDER, HALIS

Department of Economics
Middle East Technical University
Inonu Bulvari
Ankara
Turkey

ASHRAM, MAHMOUD

Department of Agricultural Economics
Aleppo University
Aleppo
Syria

AYDIN, ZULKUF

Institute of Archaeology and
Anthropology
Yarmouk University
Irbid
Jordan

AZZAM, AZZEDINE M

Department of Agricultural Economics
University of Nebraska
Lincoln
NE 68583
USA

BURGESS, SIMON

University of Reading
London Road
Reading RG1 51Q
United Kingdom

COMMANDER, SIMON

Economic Development Institute
The World Bank
1818 H Street, N.W.
Washington, DC 20433
USA

GHOSH, JAYATI

Centre for Economic Studies and
Planning
Jawaharlal Nehru University
New Mehrauli Road
New Delhi 110 067
India

GURKAN, A. ARSLAN   Department of Economics
Middle East Technical University
Inonu Bulvari
Ankara
Turkey

HOPKINS, NICHOLAS   Department of Sociology
Anthropology and Psychology
American University in Cairo
P.O. Box 2511
Cairo
Egypt

KASNAKOGLU, HALUK   Department of Economics
Middle East Technical University
Inonu Bulvari
Ankara
Turkey

MAHDI, KAMEL A   Department of Economics
Yarmouk University
Irbid
Jordan

PANAYIOTOU, GEORGE S   Agricultural Research Institute
Nicosia
Cyprus

RAMEZANI, AHMED   Department of Agricultural and Resource
Economics
207 Gianni Hall
University of California
Berkeley, CA 94720
USA

RICHARDS, ALAN   Department of Economics
University of California
Santa Cruz
CA 95064
USA

TULLY, DENNIS   The International Center for Agricultural
Research in the Dry Areas (ICARDA)
P.O. Box 5466
Aleppo
Syria

TULLY, DORENE R

The International Center for Agricultural
Research in the Dry Areas (ICARDA)
P.O. Box 5466
Aleppo
Syria

TUTWILER, RICHARD

The International Center for Agricultural
Research in the Dry Areas (ICARDA)
P.O. Box 5466
Aleppo
Syria

Erratum

to *Labor and Rainfed Agriculture in West Asia and North Africa* (Dennis Tully, ed.).

In chapter 8 (pp. 163–183) the captions to Figures 1 and 2 were inadvertently interchanged. They should read:

Fig. 1. Geographical features of Syria. *Sources*: Syrian Central Bureau of Statistics (CBS 1986), Arab Center for the Studies of the Arid Zones and Dry Lands.

Fig. 2. Agricultural areas in Syria. *Source:* Syrian Central Bureau of Statistics (CBS 1986).

KLUWER ACADEMIC PUBLISHERS